EXS 89

New Approaches to Drug Development

Edited by P. Jollès

Birkhäuser Verlag
Basel · Boston · Berlin

Editor:

Prof. Dr. P. Jollès
Laboratoire de Chimie
des Substances Naturelles
URA C.N.R.S. No. 401
Muséum National d'Histoire Naturelle
63, rue Buffon
F-75005 Paris
France

Library of Congress Cataloging-in-Publication Data
New approaches to drug development / edited by P. Jollès.
p. cm. -- (EXS ; 89)
Includes bibliographical references and index.
ISBN 3764361298 (alk. paper). -- ISBN 0-8176-6129-8 (alk. paper)
1. Drug development. I. Jollès, Pierre, 1927-II. Series
RM301.25 .N49 2000
615'.19–dc21

Deutsche Bibliothek Cataloging-in-Publication Data
New approaches to drug development / ed. by P. Jollès. - Basel ; Boston ; Berlin : Birkhäuser, 2000
(EXS : 89)
ISBN 3-7643-6129-8

ISBN 3-7643-6129-8 Birkhäuser Verlag, Basel – Boston – Berlin

Printed on acid-free paper produced from chlorine-free pulp TCF ∞
Cover illustration: Ras signalling pathways (page 154)
Printed in Germany
ISBN 3-7643-6129-8

9 8 7 6 5 4 3 2 1

Contents

List of contributors

Gary W. Ashley, Department of Chemistry, Kosan Biosciences, Inc., 3832 Bay Center Place, Hayward, CA 94545, USA

Detmar Beyersmann, Department of Biology and Chemistry, University of Bremen, D-28334 Bremen, Germany; e-mail: beyers@chemie.uni-bremen.de

Michael Buerke, Department of Medicine, Johannes Gutenberg-University, Langenbeckstr. 1, D-55101 Mainz, Germany; e-mail: buerke@mail.uni-mainz.de

Christopher W. Carreras, Department of New Technologies, Kosan Biosciences, Inc., 3832 Bay Center Place, Hayward, CA 94545, USA; e-mail: carreras@kosan.com

Dean B. Evans, Oncology Research, Novartis Pharma AG, CH-4002 Basel, Switzerland; e-mail: dean.evans@pharma.novartis.com

Bernd L. Fiebich, Department of Psychiatry, University of Freiburg, Hauptstr. 5, D-79104 Freiburg, Germany

Mark A. Fletcher, Aventis Pasteur SA, 2, avenue Pont Pasteur, F-69367 Lyon Cedex 07, France; e-mail: mark.fletcher@aventis.com

Carlos García-Echeverría, Oncology Research, Novartis Pharma AG, CH-4002 Basel, Switzerland; e-mail: carlos.garcia-echeverria@pharma.novartis.com

Thomas C. Hardin, The University of Texas Health Science Center, 7703 Floyd Curl Drive, San Antonio, TX 78284-6620, USA; e-mail: hardint@uthscsa.edu

Bridget T. Hill, Division de Cancérologie Expérimentale, Centre de Recherche Pierre Fabre, 17, Av. Jean Moulin, F-81106 Castres Cedex, France; e-mail: bridget.hill@pierre-fabre.com

Mengfei Ho, Department of Biochemistry and Molecular Biology, Wright State University School of Medicine, Dayton, OH 45435, USA

Michael H. Hüll, Department of Psychiatry, University of Freiburg, Hauptstr. 5, D-79104 Freiburg, Germany; e-mail: michael_huell@psyallg.ukl.uni-freiburg.de

Giulia C. Kennedy, Affymetrix, 3380 Central Expressway, Santa Clara, CA 95051, USA; e-mail: giulia-kennedy@affymetrix.com

Sander Kersten, Institut de Biologie Animale, Bâtiment de Biologie, Université de Lausanne, CH-1015 Lausanne, Switzerland

Jean-Pierre Kraehenbuhl, Swiss Institute for Experimental Cancer Research, University of Lausanne, CH-1066 Epalinges, Switzerland; e-mail: jean-pierre.kraehenbuhl@isrec.unil.ch

Klaus Lieb, Department of Psychiatry, University of Freiburg, Hauptstr. 5, D-79104 Freiburg, Germany

Vorapann Mahaguna, College of Pharmacy, The University of Texas at Austin, Austin, TX, USA

Ronald G. Micetich, Synphar Laboratories, Inc., #2 4290-91 A St., Edmonton, Alberta T6E 5V2, Canada; e-mail: rmicetich@synphar.com

Dominique Perrin, Division de Cancérologie Expérimentale, Centre de Recherche Pierre Fabre, 17, Av. Jean Moulin, F-81106 Castres Cedex, France; (has since moved to Serono Pharmaceutical Research Institute, ch. des Aulx 14, CH-1228 Plan-les-Oates, Switzerland; e-mail: dominique.perrin@serono.com)

H.J. Rupprecht, Department of Medicine, Johannes Gutenberg-University, Langenbeckstr. 1, D-55101 Mainz, Germany; e-mail: rupprecht@2-med.klinik.uni-mainz.de

Pierre Saliou, Aventis Pasteur SA, 2, avenue Pont Pasteur, F-69367 Lyon, Cedex 07, France; e-mail: pierre.saliou@aventis.com

Eduardo L. Setti, Axys Pharmaceuticals, Inc., 385 Oyster Point Boulevard, Suite 3, South San Francisco, CA 94080, USA; e-mail: eduardo_setti@axyspharm.com

Peter Traxler, Oncology Research, Novartis Pharma AG, CH-4002 Basel, Switzerland; e-mail: peter.traxler@pharma.novartis.com

John J. Turchi, Department of Biochemistry and Molecular Biology, Wright State University School of Medicine, Dayton, OH 45435, USA; e-mail: john.turchi@wright.edu

Dominique Velin, Centre d'Immunologie Pierre Fabre (CIPF), 5 avenue Napoleon III, B.P. 497, F-74164 St. Julien en Genevois, France; e-mail: dominique.velin@pierre-fabre.com

Walter Wahli, Institut de Biologie Animale, Bâtiment de Biologie, Université de Lausanne, CH-1015 Lausanne, Switzerland; e-mail: walter.wahli@iba.unil.ch

John Wilding, University Hospital Aintree, Longmoor Lane, Liverpool L9 7AL, United Kingdom; e-mail: j.p.h.wilding@liv.ac.uk

Robert O. Williams III, College of Pharmacy, The University of Texas at Austin, Austin, TX, USA; e-mail: williro@mail.utexas.edu

Preface

The book contains a far-ranging collection of chapters devoted to a current analysis of new approaches to drug development for important diseases: Some have reared up with renewed virulence, others have eluded effective treatment despite the best efforts of academic and industrial researchers.

Opportunities came from new technologies, increasing understanding of the molecular basis of the disease from recent data of human genome research. Going through these papers reveals how rapidly the large field of drug development is moving thanks to the techniques recently developed in the areas of genomics, cellular and molecular biology.

Pierre Jollès
Paris, Summer 2000

New Approaches to Drug Development
ed. by P. Jollès

The impact of genomics on therapeutic drug development

Giulia C. Kennedy

Affymetrix, Santa Clara, CA 95051, USA

Summary. Genomics can be defined as a set of related technologies that are focused on the discovery of genes implicated in human disease. Although many of the estimated 100 000 genes in the human genome have been at least partially identified by nucleotide sequence, elucidation of biological function has been achieved for only a small percentage of these. An even smaller percentage of genes discovered by these methodologies have become valid drug targets. This review discusses the various genomics technologies and their likelihood of yielding therapeutic drugs. Emerging advances in microarray "chip" technology have allowed the parallel analysis of gene expression patterns for thousands of genes simultaneously. Sequence information derived from the genomes of many individuals is leading to the rapid discovery of single nucleotide polymorphisms or SNPs. Detection of these human polymorphisms will fuel the discipline of pharmacogenomics, resulting in an increase in the success of clinical trials, the rescue of drugs that have previously failed in clinical trials because of adverse reactions from patient subpopulations, and ultimately, in the development of more personalized drug therapies. The impending identification of all human genes will signal the end of the structural genomics phase and usher in the function genomics phase. Technologies have already begun to move toward high-throughput elucidation of gene relationships, interactions and, it is hoped, toward their functions.

Introduction

Genomics was a term originally coined to describe the discipline of science concerned with the mapping, sequencing and analysis of genomes, i.e., the complete set of genes for an organism [1]. While this definition is still accurate in general, the commercialization of genomics has resulted in a more narrow focus on the discovery of medically relevant genes as potential drug targets. This review will begin by discussing various genomics technologies and approaches, highlight their achievements and discuss their potential impact on the process of therapeutic drug development.

Genomics has been divided into roughly two phases: structural and functional. While structural genomics (i.e., gene discovery) is rapidly approaching its conclusion, functional genomics (i.e., gene function) is still in its infancy. Structural genomics has largely been driven by high-throughput sequencing technology, while functional genomics will probably be driven initially by high-throughput expression profiling, i.e., DNA microarrays. We may see a shift away from expression profiling as more traditional methods for elucidating gene function, for example, mouse model systems, are converted into high-throughput platform technologies capable of screening the functions of thousands of genes. There are also a number of newer technologies for elucidating gene function, namely antisense, ribozyme, aptamer and perturbagen technologies (see below). Pharmacogenomics is another field which has been fueled by the intensive sequencing efforts of the structural genomics era. The ultimate identification of the estimated 3 million single nucleotide polymorphisms (SNPs) in the human genome will allow more definitive genotype: phenotype correlations and, it is hoped, lead to a better understanding of human disease [2].

Structural genomics: the rate of *de novo* human gene discovery is slowing down

Structural genomics is concerned with elucidation of the complete nucleotide sequence of a genome, identification of all genes and elucidation of their physical organization. The Human Genome Project, officially launched in 1990, originally planned to sequence all three billion nucleotides in the human genome, with the subsequent identification of the structure of each of the approximately 100 000 genes, by the year 2005. This project has spawned an entire segment of biotechnology centered around genomics and gene research, whose commercial purpose is to identify more rapidly the thousands of therapeutic drug targets expected to be uncovered by this ambitious project [3]. Efforts to sequence the human genome are not limited to the public sector; the biotechnology company Celera has been formed with the goal of sequencing the human genome at an accelerated pace. Celera plans to use the shotgun-sequencing approach to complete the human genome by the end of 2001. Spurred on by competition from Celera and by additional funding from the National Human Genome Research Institute, the public effort has vowed to sequence 90% of the human genome by March 2000 [4].

Gene discovery in pathogenic microorganisms

The first complete genome of a free-living organism to be sequenced was *Haemophilus influenza* Rd, in 1995 [5]. The sequencing of this 1.8 million base pair genome successfully utilized shotgun sequencing and sophisticated bioinformatics. While traditional genome sequencing begins with the creation of physical maps of a particular genome, the undisputed successes of shotgun sequencing, as pioneered by the Institute for Genome Research (TIGR), have proven repeatedly that 90–98% of a genome can be elucidated without physical maps. However, whole-genome shotgun sequencing poses significant challenges in assembly and in final closure of the genome. Since 1995, the complete genome sequences of many more microorganisms have been reported, including the complete 12 million base pair genome sequence of the first eukaryote, the yeast *Saccharomyces cerevisiae* [6] and the first multicellular eukaryote *Caenorhabditis elegans* [7].

Whole-genome sequencing has ushered in the modern era of microbial genomics, and promises to identify targets for novel antimicrobials, and in some cases, vaccines. Sequence information from bacterial genomes faces many of the same issues demonstrated by human sequence information, namely the necessity for turning that information into knowledge by elucidating gene function. Bacterial genomics is often focused on the differences between the genes or genomes of virulent strains and their non-virulent counterparts, as well as on discovering orphan bacterial genes, both of which are accelerated by whole-genome sequencing. As has been accomplished in mammalian systems (see below), microarrays have been used to study expression patterns of bacterial genes in pathogenic conditions. Furthermore, genes involved in host interactions with microorganisms can also be addressed. Although several fluorescence-based technologies already exist to discover bacterial genes preferentially expressed when bacteria make contact with host cells [8], microarray technologies offer much higher throughput and can examine the expression of thousands of host and/or bacterial genes simultaneously. For instance, host epithelial genes regulated by contact with *Pseudomonas aeruginosa* have been identified using microarray analysis (CHI Conference on Gene Functional Analysis, San Francisco 1999). Thus, what is likely to emerge from bacterial genomics is the identification of genes involved in pathogenesis and host response. This information will probably be useful for the development of diagnostics, as well as in the development of vaccines against selected pathogenic organisms. Sequence information about antibiotic-resistant strains of bacteria could theoretically be useful for the development of novel antimicrobials.

Gene discovery using the EST approach

Fortunately, gene discovery has not had to rely solely on human genome sequencing efforts. The random sequencing of clones from cDNA libraries, by both public and private groups, has resulted in the identification of most, if not all, transcribed human genes. The EST (expressed sequence tag) approach is among the first of the genomics technologies to result in the identification of good drug targets and the first genomics technology to deliver molecules to the clinic for testing [9]. Utilization of nucleotide and protein-based bioinformatics tools enable researchers to discover genes whose homologies strongly suggest a biochemical activity. Scientists sift through sequence analysis information on thousands of novel genes and handpick those which have structural domains common to known therapeutics, such as seven-transmembrane receptors, ion channels, receptors, growth factors, etc. Researchers at Human Genome Sciences (HGS) used this approach to identify two molecules which are now being tested in clinical trials: a novel keratinocyte growth factor (KGF-2) for wound healing [10], [11] and a novel chemokine called myeloid progenitor inhibitory factor (MPIF-1) as a myeloprotectant [12].

Amgen, Inc. used a combination of EST/secreted protein discovery research to discover the novel protein, osteoprotegerin, or OPG [13]. This molecule was sequenced as a random cDNA from a foetal rat intestinal library, and found by sequence analysis to be homologous to members of the tumour necrosis factor receptor (TNFR) superfamily. OPG was found to be a naturally secreted glycoprotein involved in regulating bone density. Recombinant osteoprotegerin increases bone density *in vivo* and also protects rats from bone loss, in a model similar to the bone loss observed in postmenopausal women [13]. OPG will be tested in Phase-I clinical trials for osteoporosis and cancer-related bone metastasis.

In addition to their identification from EST databases by searching for homologies to known factors, secreted proteins may also be found by functional assays, such as the signal sequence trap. Investigators at Genetics Institute have identified thousands of putative novel secreted proteins by using an adaptation of this method [14].

Gene discovery by positional cloning

Positional cloning represents a highly targeted approach for identifying disease genes in that it uses genetics to study inheritance of disease in families. It has been used successfully to eludicate diseases caused by single genes, but has not had as much impact on the more common multigenic diseases, such as diabetes, obesity, asthma, cancer, Alzheimer's disease and heart disease [2]. In positional cloning, the culprit gene is localized to a specific region of a chromosome by genetic mapping, and mutations are ultimately verified by nucleotide sequencing. By definition, the genes discovered by this approach are causative of disease, and therefore of high biological relevance. Despite this clear advantage, positional cloning is a labor- and time-intensive technology and by necessity, low throughput (from a commercial prospective). With the exception of a small number of positional cloning projects underway at various genomics companies, much of commercial disease gene discovery has shifted away from this approach. As with any gene discovery methodology, gene function must follow positional cloning, and therefore is subject to the same vagaries.

SNP discovery and its potential application to pharmacogenomics

The discipline of pharmacogenomics is concerned with individual variation, caused by inherited differences, in response to effector molecules such as drugs, proteins, environmental toxins, endogenous ligands, etc. The concept of pharmacogenomics predicts that genetic polymorphisms determine whether or not, or to what degree, an individual responds to a given effector molecule. Historical work in pharmacogenomics has centered around polymorphisms in drug-metabolizing enzymes, and while much of the pioneering scientific work was laid down in the pre-molecular biology era [15], the imminent discovery of all human genes and the ability to re-sequence genes and study naturally occurring polymorphisms is likely to result in an explosion of pharmacogenomic data. Analogous to the situation in which scientists were inundated with vast amounts of sequence data emanating from the Human Genome Project, we will face a similar situation with SNP data. It is predicted that we will once again have the information before we have the tools and understanding to derive knowledge from that information. Currently no high-throughput technology exists for determining the functional consequences of these SNPs. Technology such as the re-sequencing DNA chip, pioneered by Affymetrix, has not only been used to test genetic variations in mitochondrial DNA, but can theoretically be used to test polymorphisms in any gene of known sequence [16]. Thus, information about genetic polymorphism immediately leads to diagnostic products, which can then be used to identify patients for whom a therapeutic regimen is likely to be successful.

There are various methodologies for discovering SNPs. The "gold standard" for both SNP identification and confirmation is actual DNA sequencing, although higher throughput methods exist for "candidate SNP" identification. For instance, single-stranded conformational polymorphism (SSCP), a gel-based assay, can be used to screen pools of possible SNPs, which are then confirmed by traditional sequencing. Chip-based approaches, as pioneered by Affymetrix, rely on previous SNP discovery, although they can be adapted to SNP discovery as well. Oligonucleotides corresponding to the locus containing the SNP are synthesized on chips by photolithography and hybridized to DNA from different individuals [17]. Genotype information at a particular SNP locus is assessed on the basis of the hybridization pattern. Thus, this methodology is useful for screening a patient population for previously identified SNPs. Mass spectrometry is now emerging as a useful and high throughput method for identifying and scoring genotypes, using both SNPs and other polymorphic markers. The resolution of this method is remarkable, in that it can detect mass changes corresponding to 0.1 nucleotide [18]. A related application that is nearing evaluation in the clinic is genotyping for the management of HIV infection. Patients infected with HIV often develop drug resistance due to mutations in the HIV genome. Visible Genetics, Inc. has developed a small "personal sequencer" for use in hospital labs [19] and will create a large-scale research database of HIV resistance profiles, along with clinical outcomes. The database will consist of resistance profiles for up to 30 000 enrolled patients under strict patient confidentiality, including drug history, drugs selected as a result of genotyping, and viral loads at eight weeks and sixteen weeks after genotype-directed drug selection. The database will be updated on a regular basis and will be made available to scientists, physicians and patients via the Internet.

Despite its relatively small genome size of 16 600 base pairs, mitochondrial variation has been implicated in dozens of human diseases. Precisely because of its small size and source of genetic variation, the human mitochondrial genome was chosen as the prototype for the DNA re-sequencing chip developed by Affymetrix to identify genetic polymorphisms [16]. Thus the DNA microchip has emerged as a multifunctional technology capable of studying genes whose expression is changed in disease (expression profiling), as well as identifying DNA polymorphisms in selected genes which have been implicated in genetic susceptibility to human disease.

Functional genomics

In the recent past, determination of nucleotide sequence was pivotal to many genomics technologies; however, there is now a shift away from sequencing, i.e., structural genomics, and towards functional elucidation, i.e., functional genomics. Functional genomics is in its infancy: following the impending identification of all 100 000 genes in the human genome, future technologies are being aimed toward understanding their functions and interrelationships. This will require a whole set of new high-throughput and parallel technologies to replace the traditional mode of elucidating the function of one or a few genes at a time. How does one even begin to elucidate the functions of thousands of genes? Fortunately, bioinformatics and microarray technologies have been developed to collapse and prioritize huge datasets into reasonably manageable subsets of interesting genes for further evaluation. Although it is outside the scope of this review to provide a detailed discussion of either bioinformatics or microarray technology, I will attempt to touch on a few of the key points. The reader is referred to many excellent sources of information in the literature [20], including a practical guide for setting up a microarray facility in-house (*Nature Genetics* supplement Jan 1999).

Bioinformatics can be used to predict the function of an unknown gene

Few genes are discovered in a complete biological vacuum. Even randomly sequenced genes are in fact cDNAs derived from a particular tissue, which was isolated from a particular stage of development, or from a particular disease state. Clearly this limited biological knowledge is insufficient to qualify these genes as drug targets. Fortunately, the genomics revolution has necessitated the development of sequence- and protein-based bioinformatics tools to extract knowledge from all this sequence information [20]. One can predict the function of many unknown genes using sequence analysis. The Pfam database (http://pfam.wustl.edu) contains over 1400 multiple sequence alignments for many common protein domains, for instance, protein kinases, ankyrin repeats, seven-transmembrane receptors, immunoglobulin domains, etc. In fact, the successes of the EST approach in identifying molecules which have advanced into the clinic have come about due to bioinformatics and homology searching.

Many newly-discovered genes have no discernible match to a protein profile and share no sequence homology or motifs with any other known gene in the database. In these cases, it may not be possible to predict function solely from sequence. It has been predicted that for the many thousands of unknown genes with no suggested function, microarray technology may serve as the next-generation "functionator" [21].

Microarray technologies

As most of the genes in the human genome are discovered, and microarray technology advances, it has become possible to look at *patterns* of expression of thousands of genes at a time, as opposed to the traditional approach of following expression of a few genes at a time. Thus, the true strength of expression profiling has shifted away from *de novo* gene discovery toward placing genes into biological pathways and thus helping to elucidate the function of a particular gene. Microarray or "DNA chip" technology, as commercialized by the two biotechnology companies, Affymetrix and Synteni, was, only a short time ago, too costly and inflexible to be used as an experimental tool. With the availability of hardware, microarray technology has been transferred successfully to multiple research organizations, both public and private [22]. Another promising application of microarray technology is for toxicological studies to identify potential human and environmental toxins [23]. A yeast total-genome array containing over 6000 genes, representing 97% of the yeast genome, was used to study the transcriptional program accompanying yeast sporulation [24] and similar approaches have been used to study transcriptional regulation in lymphocytes [25]. Recently, cDNA microarrays containing 8600 distinct human genes were constructed and used to study the transcriptional response of fibroblasts to serum [26]. Gene expression in quiescent fibroblasts was compared to that of fibroblasts stimulated with serum at 12 different time points (15 min to 24 h). This represents an experiment containing over 100 000 data points and would have been unthinkable before the advent of microarray technology. How are all these data analyzed and interpreted? How are complex relationships identified in this huge dataset? Once again, a new technology has been the driver for the building of better analytical tools. Several groups have developed useful informatics tools [27]. One method uses cluster analysis to group genes together which display similar patterns [28]. Another method uses a pattern-recognition algorithm to identify complex relationships in gene expression data [29]. A relational database, called ArrayDB, has been established which captures information beginning with the construction of the microarrays all the way to analysis of the data, through a web interface [30]. The URL for this resource is: http://www.nhgri.nih.gov/DIR/LCG/15K/HTML/dbase.html.

Additional functional genomics technologies

While microarray technology certainly possesses the scale and throughput for identifying a myriad of gene expression profiles, additional methodologies will be necessary to begin to "connect the pieces of the puzzle". Biologists have been doing this for decades, and we can expect to see many traditional biological tools being adapted to fill the need for functional elucidation. Many researchers will agree that projects can proceed rapidly in the gene discovery phase, but become quickly bogged down as these genes enter the function phase. While most genomics companies can often pursue several types of gene discovery approaches simultaneously, functional genomics companies are centered primarily around a single technology, such as the use of tractable genetic model systems to knock out expression in mice [31], yeast [32], flies [33] or worms (the new company DevGen n.v.), or the use of antisense [34] or ribozymes [35] to knock out mRNA expression. Newer technologies focus on perturbagens [36], aptamers [37], [38] and antibody display technologies [39]. Functional genomics companies often focus on a single methodology because of the large amounts of resources needed to scale up these traditional "benchtop methods" and convert them to high-throughput platform technologies. The complete genome sequence of *Saccharomyces cerevisiae* and *Caenorhabditis elegans* and identification of all their respective genes have served as prototypes to "test drive" some of these function-based technologies, as has the emergence of sequence data in other organisms.

Summary

Genomics has changed the way that genetic information is generated and processed, and will most certainly change the way it is used in the future to predict, prevent, diagnose, monitor and treat human disease. Structural genomics has forced the accelerated advancement of tools and technologies such as bioinformatics and DNA microarrays. In the near future, we will need to develop more tools for the analysis of data from complex massively parallel experiments, such as those performed with microarrays. The impending identification of all human genes will signal the end of the structural genomics phase and usher in the functional genomics phase. Technologies have already begun to move toward the high-throughput elucidation of gene relationships, interactions and hopefully, toward their functions.

References

1 McKusick VA (1997) Genomics: structural and functional studies of genomes. *Genomics* 45(2): 244–9
2 Chakravarti A (1999) Population genetics – making sense out of sequence. *Nat Genet* 21(1 Suppl): 56–60

3 Cohen J (1997) The genomics gamble [news]. *Science* 275(5301): 767–72
4 Pennisi E (1999) Academic sequencers challenge Celera in a sprint to the finish [news]. *Science* 283(5409): 1822–3
5 Fleischmann RD et al (1995) Whole-genome random sequencing and assembly of *Haemophilus influenzae* Rd [see comments]. *Science* 269(5223): 496–512
6 Mewes HW et al (1997) Overview of the yeast genome [published erratum appears in Nature (1997) Jun 12;387(6634): 737]. *Nature* 387(6632 Suppl): 7–65
7 Genome sequence of the nematode C elegans: a platform for investigating biology The C elegans Sequencing Consortium [published erratum appears in Science (1999) Jan 1;283(5398): 35]. *Science* 282(5396): 2012–8
8 Valdivia RH, Falkow S (1997) Fluorescence-based isolation of bacterial genes expressed within host cells. *Science* 277(5334): 2007–11
9 Marshall A (1998) HGS launches "first" genomics product in clinic [news]. *Nat Biotechnol* 16(2): 129
10 Beer HD et al (1997) Mouse fibroblast growth factor 10: cDNA cloning, protein characterization, and regulation of mRNA expression. *Oncogene* 15(18): 2211–8
11 Jimenez PA, Rampy MA (1999) Keratinocyte growth factor-2 accelerates wound healing in incisional wounds. *J Surg Res* 81(2): 238–42
12 Patel VP et al (1997) Molecular and functional characterization of two novel human C-C chemokines as inhibitors of two distinct classes of myeloid progenitors. *J Exp Med* 185(7): 1163–72
13 Simonet WS et al (1997) Osteoprotereger*In*: a novel secreted protein involved in the regulation of bone density [see comments]. *Cell* 89(2): 309–19
14 Tashiro K et al (1993) Signal sequence trap: a cloning strategy for secreted proteins and type I membrane proteins. *Science* 261(5121): 600–3
15 Nebert DW (1997) Polymorphisms in drug-metabolizing enzymes: what is their clinical relevance and why do they exist? [editorial; comment]. *Amer J Hum Genet* 60(2): 265–71
16 Chee M et al (1996) Accessing genetic information with high-density DNA arrays. *Science* 274(5287): 610–4
17 Lipshutz RJ et al (1999) High density synthetic oligonucleotide arrays. *Nat Genet* 21(1 Suppl): 20–4
18 Butler JM et al (1999) Reliable genotyping of short tandem repeat loci without an allelic ladder using time-of-flight mass spectrometry. *Int J Legal Med* 112(1): 45–9
19 Yager TD, Dunn JM, Stevens JK (1997) High-speed DNA sequencing in ultrathin slab gels. *Curr Opin Biotechnol* 8(1): 107–13
20 Ouellette BF, Boguski MS (1997) Database divisions and homology search files: a guide for the perplexed. *Genome Res* 7(10): 952–5
21 Schena M et al (1998) Microarrays: biotechnology's discovery platform for functional genomics. *Trends Biotech* 16(7): 301–6
22 Bowtell DD (1999) Options available – from start to finish – for obtaining expression data by microarray. *Nat Genet* 21(1 Suppl): 25–32
23 Nuwaysir EF et al (1999) Microarrays and toxicology: the advent of toxicogenomics. *Mol Carcinogen* 24(3): 153–9
24 Chu S et al (1998) The transcriptional program of sporulation in budding yeast [published erratum appears in Science (1998) Nov 20; 282(5393): 1421]. *Science* 282(5389): 699–705
25 Alizadeh A et al (1998) Probing lymphocyte biology by genomic-scale gene expression analysis. *J Clin Immunol* 18(6): 373–9
26 Iyer VR et al (1999) The transcriptional program in the response of human fibroblasts to serum [see comments]. *Science* 283(5398): 83–7
27 Bassett DE Jr, Eisen MB, Boguski MS (1999) Gene expression informatics – it's all in your mine. *Nat Genet* 21(1 Suppl): 51–5
28 Eisen MB et al (1998) Cluster analysis and display of genome-wide expression patterns. *Proc Natl Acad Sci USA* 95(25): 14863–8
29 Rigoutsos I, Floratos A (1998) Combinatorial pattern discovery in biological sequences: The TEIRESIAS algorithm [published erratum appears in Bioinformatics (1998) 14(2): 229]. *Bioinformatics* 14(1): 55–67
30 Ermolaeva O et al (1998) Data management and analysis for gene expression arrays. *Nat Genet* 20(1): 19–23

31 Zambrowicz BP, Friedrich GA (1998) Comprehensive mammalian genetics: history and future prospects of gene trapping in the mouse. *Int J Dev Biol* 42(7): 1025–36
32 Marton MJ et al (1998) Drug target validation and identification of secondary drug target effects using DNA microarrays [see comments]. *Nat Med* 4(11): 1293–301
33 Scangos G (1997) Drug discovery in the postgenomic era. *Nat Biotechnol* 15(12): 1220–1
34 Weiss B, Davidkova G, Zhou LW (1999) Antisense RNA gene therapy for studying and modulating biological processes. *Cell Mol Life Sci* 55(3): 334–58
35 Rossi JJ (1999) Ribozymes, genomics and therapeutics. *Chem Biol* 6(2): R33–7
36 Caponigro G et al (1998) Transdominant genetic analysis of a growth control pathway. *Proc Natl Acad Sci USA* 95(13): 7508–7513
37 Colas P, Brent R (1998) The impact of two-hybrid and related methods on biotechnology. *Trends Biotech* 16(8): 355–363
38 Osborne SE, Matsumura I, Ellington AD (1997) Aptamers as therapeutic and diagnostic reagents: problems and prospects. *Curr Opin Chem Biol* 1(1): 5–9
39 Vaughan TJ, Osbourn JK, Tempest PR (1998) Human antibodies by design. *Nat Biotechnol* 16(6): 535–539

New Approaches to Drug Development
ed. by P. Jollès

Regulation of mammalian gene expression

Detmar Beyersmann

Department of Biology and Chemistry, University of Bremen, D-28334 Bremen, Germany

Summary. The expression of mammalian genes is regulated primarily at the level of initiation of transcription. The regulatory structure of the mammalian genes consists of the sequence coding for a protein, a proximal upstream promoter sequence which binds the general (basal) transcription factors and a distant enhancer sequence which binds the inducible transcription factors. The general transcription factors are proteins which combine with the RNA polymerase at the promoter to form the initiation complex. Binding of the inducible transcription factors at short DNA sequences, named response elements, mostly enhances or rarely represses the formation of the initiation complex. Transcription factors share common structural motifs; the most frequent are zinc finger, leucine zipper and helix-loop-helix structures. Inducible transcription factors are activated to bind their target response elements on DNA by protein kinases, by binding of activating or removal of inhibitory factors, or by *de novo* protein synthesis. Inducible transcription factors are activated by hormones or growth factors addressing a number of genes which share common response elements. Steroid and thyroid hormones combine with intracellular receptors to form active transcription factors. Other transcription factors are activated by protein kinases which are themselves activated by hormones through cell membrane receptors and further cellular signaling paths. Whereas the main level of transcriptional control is the initiation of RNA synthesis, in some instances genes are also regulated by alternative splicing of the primary transcript or control of translation into proteins. Large-scale silencing of genes is mediated by the packing of DNA in highly condensed heterochromatin structures and DNA methylation at cytosines in defined guanine-cytosine (GC)-sequences.

Introduction

Individual metabolic steps and reaction sequences are adjusted quickly and precisely to metabolic needs by modulating the activity of preexisting enzymes. The more complex changes in the life of an organism, such as embryonal development, differentiation and proliferation, are regulated at the level of the genes allowing the adjustment of protein composition and mass. Typically, hormones or growth factors are released from tissues or cells different from the target cells. The hormones bind to cellular receptors in specific tissues which either function as transcription factors directly or activate transcription factors via protein kinases. These will bind to response elements in the promoters of genes and mostly activate, in rarer cases inactivate, their target genes. As a rule, a transcription factor does not address a single gene, but numerous genes. For example, a growth factor usually activates transcription of a group of early genes coding for proteins required for the start of DNA synthesis and cell proliferation. Possible levels of regulation of gene expression are initiation, elongation, termination, posttranscriptional processing (splicing), RNA transport into cytoplasm, RNA stability and translation into proteins.

This chapter predominantly deals with transcription, since this is the main level of gene regulation. Furthermore, it mainly treats the initiation of gene transcription, because regulation of mRNA elongation is of less significance and termination is not relevant to the control of mammalian gene expression. Regulation of gene expression by control of translation is observed in embryonic development where the initiation of translation of mRNAs is modulated by specific proteins.

The general aims of this chapter are to treat the basic facts about mammalian gene expression, especially the regulatory structure of the gene and the structure and properties of the transcribing apparatus. The structures, the modes of activation of transcription factors and the signaling pathways from hormones to individual genes are discussed in greater detail. The large-scale silencing of genes by chromatin and DNA structures are treated briefly. Furthermore, two modes of posttranscriptional control are discussed, i.e., alternative splicing of the primary transcript and the control of protein biosynthesis.

The regulatory structure of genes coding for proteins

Transcribed genes are composed of a DNA sequence coding for a protein and a regulatory sequence usually located upstream from the transcription start point which codes for the 5'-end of the nascent mRNA. Each gene is regulated separately, each at its own promoter. But usually, several genes are regulated as a group, because their promotors are addressed by the same transcription factors. Promoters are best defined by their function, i.e., by the binding of transcription factors and by being sites of initiation of RNA synthesis. The precise localization and orientation of the regulatory sequence elements vary considerably between individual genes, but a general scheme can be drawn as depicted in Figure 1. The regulatory sequence of many genes consists of the promoter itself, of the size of about 100 base pairs (bp) located upstream from the start point and a distinct enhancer sequence of similar size, often located

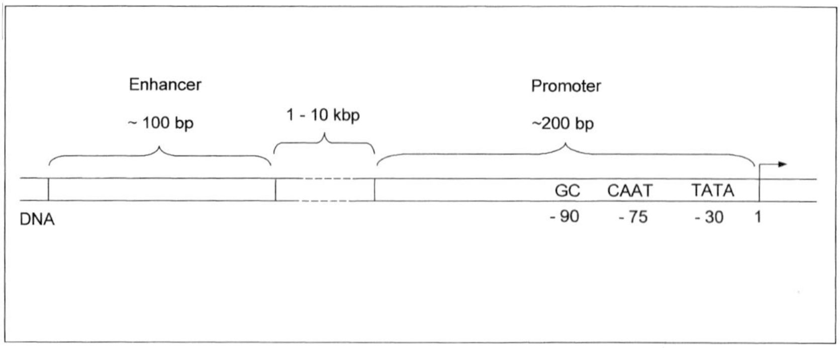

Figure 1. Regulatory structure of mammalian genes coding for proteins. bp: base pairs; 1: first transcribed base.

further upstream and separated by several thousand base pairs (kbp) from the promoter. Whereas the promoter is always located upstream in the vicinity of the start point, the enhancer is positioned at a distance and may be located up- or downstream from the start point.

Promoter and enhancer also differ with respect to the types of transcription factors they bind. Most, but not all promoters possess a sequence of about 20 bp located at about –30 bp. This sequence contains a common sequence motif called the TATA box. This box binds the general factors that are required for transcription of all genes to form the basal initiation complex with the RNA polymerase. Further upstream and separated by sequences without known function, promoters bear varying regulatory elements, also named promoter-near elements. These elements bind upstream factors that modulate tissue-specific transcription but are generally not inducible. At –75 bp a CAAT-containing box is often found which binds a specific transcription factor family, and at –90 bp a GC-containing sequence is frequently located binding the ubiquitous transcription factor SP1. Further upstream elements are found at varying locations in a cell type-specific mode. Whereas the upstream elements usually increase the efficiency of transcription initiation without being specifically induced, the enhancer is densely packed with response elements that bind the inducible factors responding to hormones or other signals received by their corresponding cellular receptors. The enhancer binding factors may interact with the general factors in the promoter sequence but not with the polymerase itself. It is assumed that these interactions are accomplished by bending the DNA to form a loop to locate the enhancer and the promoter binding proteins in proximity. A more detailed discussion of the gene structure and the transcribing apparatus can be found in the textbook of Lewin [1].

The transcribing apparatus

The initiation of transcription depends on stimulation by a signal. To a limited extent the elongation is modulated, too, whereas with mammalia the termination of transcription seems not to be regulated. After completion, the primary transcript is processed by attachment of a 5'-cap and a 3'-polyadenylyl tail, and splicing to remove the intron sequences. The processed mature mRNA is transported from the nucleus into the cytoplasm. Finally, the extent of translation into protein will depend on the stability of the mRNA in the cytoplasm and the rate of translation itself. All these levels may be levels of regulation of gene expression.

The transcribing enzyme: RNA polymerase II

There are three types of RNA polymerases found in eukaryotic cells. The enzyme catalyzing the mRNA synthesis is RNA Polymerase II. This enzyme is large (about 500 kD), and it consists of 8–14 subunits. The largest subunit (220 kD) is involved in the catalytic activity and is phosphorylated at its C-terminal domain during transcriptional initiation. The polymerase alone is not able to recognize the promoter; for initiation it depends on association with the general transcription factors.

The initiation complex

Initiation is the main level of transcriptional control and the initiation tools are very elaborate. The initiation complex consists of the polymerase II holoenzyme and a sequentially assembled complex of basal transcription factors named TFIIA, TFIIB, etc., which are generally required for gene activation [2, 3]. As depicted in Figure 2, the assembly starts with the binding of TFIID to a short attachment sequence of the promoter, the TATA box. TFIID is a large multisubunit protein of about 750 kD. It consists of the TATA-binding protein (TBP) and a complex of about 8 TBP-associated factors [4]. Subsequently, factors TFIIA and TFIIB are bound. Now the complex is ready to bind Polymerase II associated with TFIIF. Thereafter, TFIIE, TFIIH and TFIIJ join the complex. Finally, the functional aggregate consists of about 40 polypeptides and has a size of $>2 \cdot 10^6$ kD. TFIIH has a ATP-dependent helicase activity and probably functions as a DNA-unwinding factor for transcription [5]. Furthermore, TFIIH functions as a protein kinase which phosphorylates the 220 kD-subunit of the polymerase in the initiation process. The sequential assembly described above may be interrupted by an inhibitor protein present in the cytoplasm, which prevents the binding of the RNA polymerase.

Elongation

Although the major regulated process in transcription is the initiation, the elongation may also be modulated. Elongation has been observed to pause, or even to be blocked by certain DNA sequences or DNA-binding proteins [6]. Pausing occurs upon DNA damage and repair or heat shock. However, pausing or blocking does not cause dissociation of the ternary complex of DNA, polymerase and nascent mRNA, which is extraordinary stable. Hence, transcription may be resumed rapidly after repair of DNA damage.

Transcription factors

General properties

Transcription factors are defined as molecules that are either needed for transcription at a promoter or are able to stimulate transcription at enhancer ele-

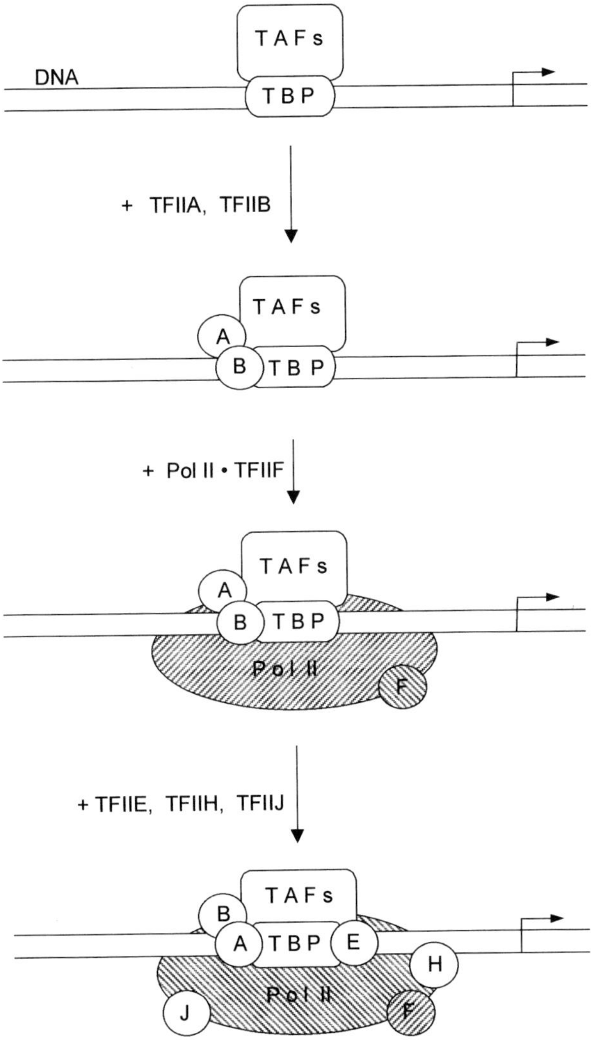

Figure 2. Sequential formation of the transcription initiation complex. A, B, E, F, H, J: transcription factors TFIIA, TFIIB, TFIIE, TFIIF, TFIIH, TFIIJ; TBP: TATA box-binding protein; TAFs: TBP-associated factors; Pol II: RNA polymerase II.

ments. They usually bind DNA and enhance the formation and/or stability of the initiation complex. Some of the general factors do so by interaction with other transcription factors or the RNA polymerase. The inducible factors bind to the response elements located usually in the enhancers but may also be interspersed in the promoter sequence. Inducible factors are activated by signaling paths that commence with the binding of a hormone or other first messenger to its receptor. With the exception of the steroid and thyroid hormones, activating signals are further processed through intracellular signaling cascades, including intracellular small second messengers and protein phosphorylation steps, and ultimately induce the phosphorylation of one or several transcription factors. One transcription factor generally controls several target genes which share common response elements specifically addressed by this factor. For example, the glucocorticoid receptor binds not only the glucocorticoid response element (GRE) but also the serum response element (SRE) and the TPA response element (TRE). *Vice-versa*, a specific response element is often addressed by several cellular signaling paths (see below).

The typical transcriptional regulators of bacteria are repressor proteins, but with mammalia activating transcription factors are the rule and repression seems to be exceptional. Recently, some transcriptional repressors have been identified in mammalia. For example, the REST protein binds to the promoters of many neuronal genes and represses neuronal gene expression in non-neuronal tissues [7].

Common structural motifs of transcription factors

The structure of the inducible transcription factors is modular: it comprises DNA binding, activating and variable domains. Most transcription factors bind to the major groove of the DNA as dimers. This property enhances the affinity for the specific DNA sequences and, with heterodimer formation, the regulatory potency is enhanced considerably. Transcription factors recognize the nucleotide sequence of responsive elements by interacting with bps without disrupting the hydrogen bonds within the double helix. The DNA-binding domains of these proteins have special characteristic folding motifs [8]. Structures frequently encountered with mammalian transcription factors are the zinc finger, the leucine zipper and the helix-loop-helix structures (Fig. 3).

Zinc finger proteins

Zinc finger proteins comprise two classes, the Cys_2 His_2 Zn and the Cys_4 Zn class. Zinc fingers of the Cys_2 His_2 Zn type are found with numerous transcription factors, e.g., in the general transcription factor Sp1. These Cys_2 His_2-zinc fingers have the following consensus sequence

$$- \mathrm{Cys} - X_{2\text{–}4} - \mathrm{Cys} - X_3 - \mathrm{Phe} - X_5 - \mathrm{Leu} - X_2 - \mathrm{His} - X_3 - \mathrm{His} -.$$

They usually are found in a repetitive array with two to ten repeats connected by linker sequences of 7–8 amino acids. The zinc ion stabilizes the protein folding within this sequence by the formation of a tetrahedral complex with the conserved cysteine and histidine residues. These zinc fingers contain about 20 amino acids which form a short α-helical and a short β-pleated structure (Fig. 3A). By X-ray crystallography of a trimeric fragment of the zinc finger

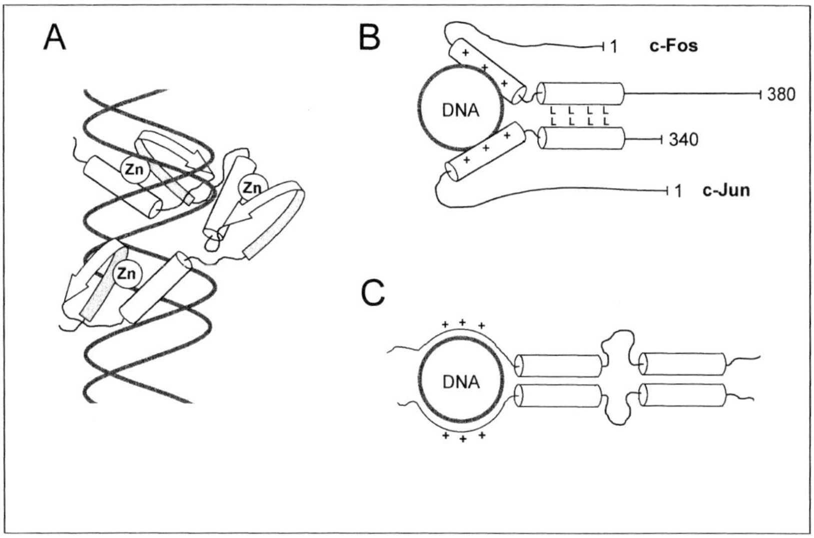

Figure 3. Common structures of transcription factors binding DNA. (A) threefold repeat fragment of a Cys_2His_2 zinc finger protein bound to DNA; (B) proteins c-Fos and c-Jun forming the heterodimeric leucine zipper protein AP-1 bound to DNA; L: leucine residue.(C) model of a dimeric helix-loop-helix protein bound to DNA.

protein Zif268 with an oligonucleotide, it could be demonstrated that a trimeric Cys_2 His_2-zinc finger winds along the major groove of DNA-forming interactions with the DNA bases through the α-helical parts [9].

The second class of zinc fingers contains zinc bound to 4 cysteines (Cys_4 Zn) and is typical for the DNA-binding domains of the steroid and thyroid family of receptors. This family comprises the receptors for estrogens, glucocorticoids, progesteron, mineralcorticoids, androgens, thyroids, vitamin D and retinoic acid [10].

Leucine zipper structures

Leucine zipper structures are also frequently found in DNA-binding domains of transcription factors, e.g., with the early gene products c-Jun and c-Fos which form a heterodimer, the transcription factor AP-1. Leucine-zipper proteins are dimeric by definition and contain two helical structures, one of each dimer, which mutually interact by hydrophobic interactions between leucine residues found in every seventh position (Fig. 3B). The interaction with DNA is achieved by the adjacent basic helices, one of each subunit, binding to the major groove of the DNA from two sides like forceps [11]

Helix-loop-helix domains

Helix-loop-helix domains are found with a class of dimeric transcription factors (HLH proteins) which control genes functioning in embryonic development and differentiation [12]. For example, if premyogenic fibroblasts are stimulated to differentiate to myoblasts by serum deprivation, a myogenic factor, MyoD, is activated. MyoD binds to DNA as a homodimer or a heterodimer with another HLH protein (E12 or E47). The common sequence motif of HLH proteins consists of a stretch of 40–50 amino acids containing a basic region and two α-helices of 15–16 residues, each separated by a linker region (the loop) of 5–20 amino acids. Helix-loop-helix proteins form homo- or heterodimers by hydrophobic interaction between both helices of each monomer. The positively charged residues of the basic region interact with theDNA in a sequence-specific mode (Fig. 3C)

The occurrence of these structural motifs in a protein is usually taken as a hint for the possible function of the protein in gene regulation. A common feature of most types of transcription factors is the close interaction between an α-helical region and the major groove of the DNA of the corresponding responsive element. Amino acid side chains specifically interact with the phosphate backbone of DNA by charge interactions and with DNA bps by H-bonds. The variability of possible regulatory activities of transcription factors is considerably enhanced by the formation of heterodimers, usually of the same structural type. These combinatorial features greatly enhances the regulatory possibilities.

Modes of transcription factor activation

Inducible transcription factors are activated to bind their target DNA elements in at least four different ways. (i) The most frequent type of activation is by phosphorylation by specific protein kinases. Examples are the serum response factor and many others. (ii) Binding of specific ligands is the manner in which steroid and thyroid hormones activate their receptors to function as transcription factors. (iii) Removal of an inhibitor is observed in the case of the transcription factor NF-κB, which stimulates the transcription of Igκ genes in B-lymphocytes after removal of its specific inhibitor I-κB. (iv) *De novo* synthesis of transcription factors is found in several instances, like synthesis of the early gene product c-Fos which is part of a dimeric transcription factor itself, or the tissue-specific synthesis of transcription factors which is used to activate transcription in embryonic development.

Signaling paths from hormones to genes

With mammalia, hormones and growth factors activate cellular signaling reactions which ultimately render transcription factors active. In the simplest case, the hormone itself forms the active transcription factor by binding its intracellular receptor, whereas in complex cases a large number of steps in a signaling cascade is involved.

The steroid and thyroid receptors/nuclear receptor family (Fig. 5A)

The steroid and thyroid hormones typically address a large number of genes containing the same response elements [13]. These lipophilic hormones penetrate plasma membranes and interact with cytosolic or nuclear receptor proteins. If such a hormone binds its intracellular receptor protein, it usually displaces an inhibitor protein from the receptor and activates the receptor. The hormone-receptor complex interacts with its corresponding response element. The steroid and thyroid receptors share a common structure which is found with the receptors for estrogens, glucocorticoids, progesteron, mineralcorticoids, androgens, thyroids, vitamin D and retinoic acid. It consists of a variable transactivating N-terminal domain, a DNA-binding and transcription activating domain and a hormone-binding C-terminal domain (Fig. 4). The three domains are functioning independently of each other, as has been shown by domain-exchange experiments [14]. Thus, the DNA-binding domain from the glucocorticoid receptor coupled to the hormone-binding domain of the estrogen receptor will constitute an receptor responding to estrogens but binding to the GRE at a promoter usually responding to glucocorticoids.

Receptor to hormone	Amino acids	General structure
		Domains A/B C D E F
Estrogen	395	
Androgen	918	
Glucocorticoid	777	
Retinoic acid	462	
Thyroxin	456	

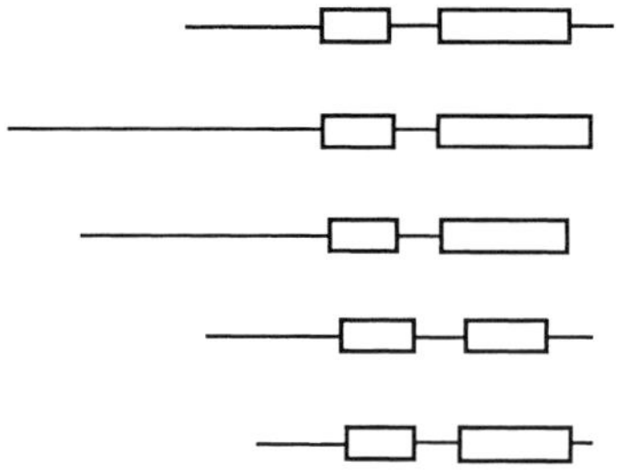

Figure 4. General structures of some members of the steroid receptor family. Domains A/B: activating; C: DNA binding; D: nuclear localization; E: hormone binding, F: carboxy-terminal.

Receptor tyrosine kinases (Fig. 5B)

Hormones and growth factors that control cell proliferation and differentiation mostly address receptor tyrosine kinases (RTK) of the plasma membrane [15]. These receptors consist of an N-terminal extracellular hormone-binding domain, a transmembrane helix and a C-terminal intracellular protein kinase domain. The extracellular domains of RTKs typically bind peptide growth fac-

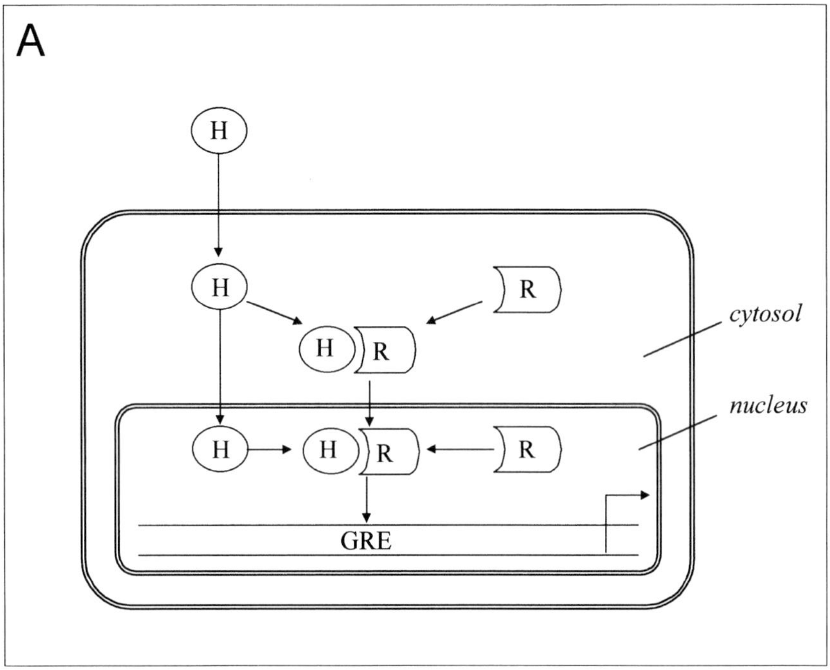

Figure 5A. Schematic representation of major signaling paths from hormones to genes. (A) Steroid-nuclear receptor path. H: hormone; R: receptor; GRE: glucocorticoid response element. (B) Receptor tyrosine kinase (RTK) path. P~Ras: phosphorylated Ras protein; MAPK: mitogen-activated protein kinase; MAPKK: MAPK-kinase; SRF: serum response factor; TCF: ternary complex factor; SRE: serum response element. (C) Cyclic AMP (cAMP)-dependent protein kinase path. G: G protein; AC: adenylate cyclase; PKA: cAMP-dependent protein kinase; CRE: cAMP response element; CREB: CRE binding protein. (D) Protein kinase C (PKC) path. PLC: phospholipase C; DAG: diacylglycerol; nucPKC: nuclear PKC; P~AP-1: phosphorylated transcription factor AP-1; TRE: TPA response element. (E) Calcium-calmodulin-dependent protein kinase (CaMK) path. IP3: inositol-1,4,5-trisphosphate; CaM: calcium-calmodulin complex.

tors like epidermis growth factor (EGF), platelet-derived growth factor (PDGF) or insulin-like growth factor (IGF). Upon binding of the hormone, a conformational shift is induced that causes a dimerization of the receptor protein. As a consequence, the catalytic domains of the two subunits mutually phosphorylate tyrosine residues at the C-termini. Thus, the tyrosine kinase domains are activated to phosphorylate target proteins, like the signaling protein Ras, which thus is induced to activate further protein kinases in a cascade manner, finally producing phosphorylated active mitogen-activated protein kinase (MAPK). This kinase phosphorylates the transcription factors SRF (serum response factor) and TCF (ternary complex factor), to form an active ternary complex binding the SRE, which is part of the promoters of a series of early genes whose transcription is required for the onset of the S phase of the cell cycle.

Mutation of the genes for RTKs may generate oncogenic variants of these proteins that no longer depend on hormones for activity but are constitutively

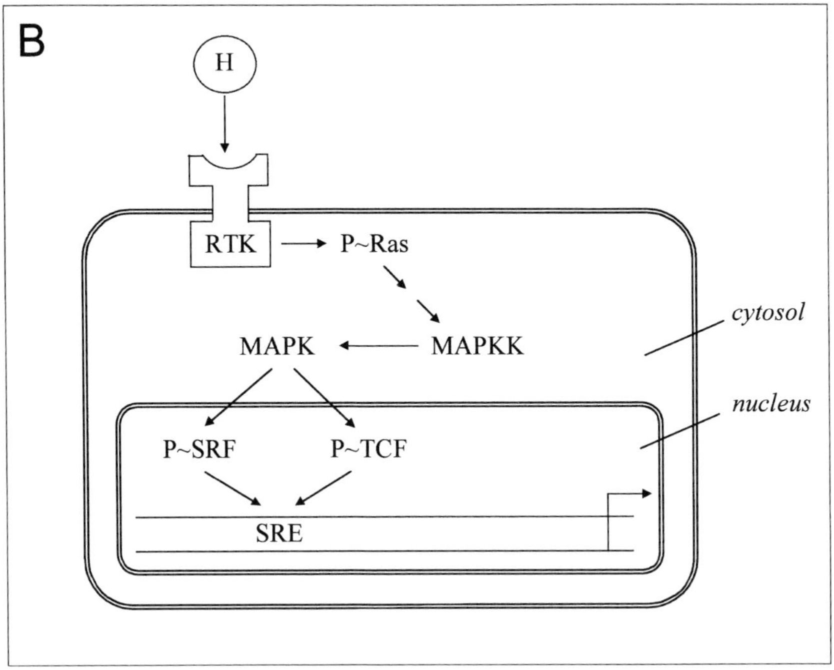

Figure 5B. (Legend see Fig. 5A)

active. For example, the avian erythroblastosis virus carries an oncogene named v-erb B which codes for a membrane tyrosine kinase lacking the N-terminal hormone-binding domain and is active as a tyrosine kinase without hormonal stimulation.

Cyclic AMP-dependent protein kinase (Fig. 5C)

Various hormones and growth stimulating factors bind to the extracellular domains of seven-helix transmembrane receptors linked to G-proteins at the cytosolic side of the plasma membrane. The binding of cAMP-linked hormones induces conformational transitions of the receptor proteins which in turn trigger a G-protein to activate the enzyme adenylate cyclase. This enzyme catalyzes the formation of cyclic adenosine monophosphate (cAMP) [16, 17]. This second messenger molecule induces the dissociation of the cAMP-dependent protein kinase (PKA) into catalytic and regulatory subunits. Through this reaction, the catalytic subunit is activated and translocated to the cell nucleus where it phosphorylates the transcription factor CREB (cAMP-response element binding protein). The binding of phosphorylated CREB to a CRE (cAMP response element) enhances the transcription of various genes,

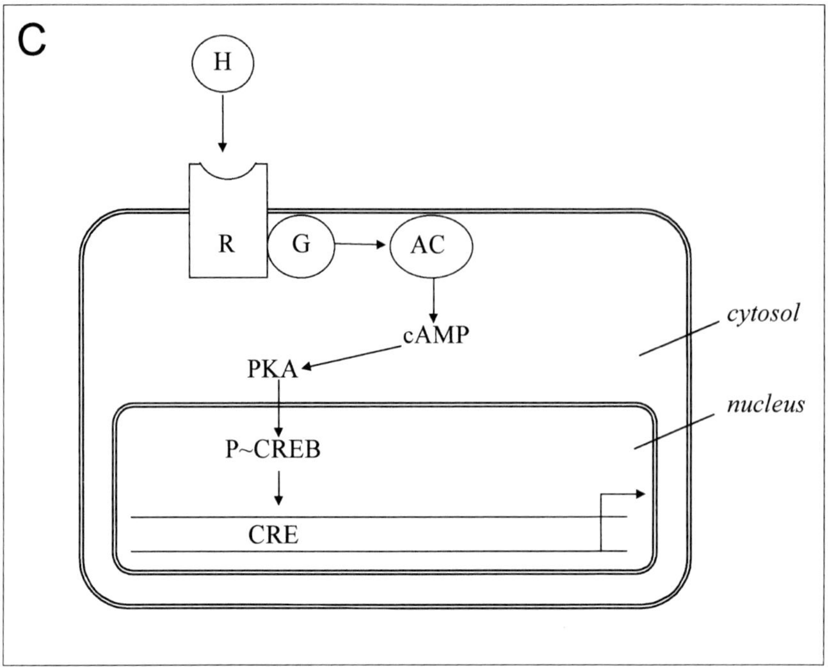

Figure 5C. (Legend see Fig. 5A)

including the somatostatin gene of endocrine cells, genes, coding for enzymes of gluconeogenesis in liver cells and the early gene c-fos in fibroblasts. By alternative splicing of the pre-mRNA coding for CREB, a negative regulator of CRE called CREM (CRE modulating protein) is created. CREM lacks the activating domain of CRE, hence it binds to CRE but cannot be activated by phosphorylation.

Protein kinase C (Fig. 5D)

The protein kinase C (PKC) path also is frequently involved in the regulation of genes controlling cell proliferation and differentiation [18]. Hormones binding to the extracellular domain of certain seven-helix-transmembrane receptors induce intracellular G-proteins to activate the enzyme phospholipase C via intracellular G-proteins. This enzyme catalyzes the hydrolysis of phosphatidyl-inositol-bisphosphate (PIP_2) to 1,4,5-inositol-trisphosphate (IP_3) and diacylglycerol (DAG). IP_3 is a second messenger which mobilizes intracellular free Ca^{2+} from storage vesicles, whereas DAG activates PKC. The DAG-PKC-complex is translocated into the nucleus where it causes the phosphorylation of the proteins c-Fos and c-Jun which form the transcription factor AP-1. AP-1 binds the TRE, named after the phorbol ester TPA which is an exogeneous

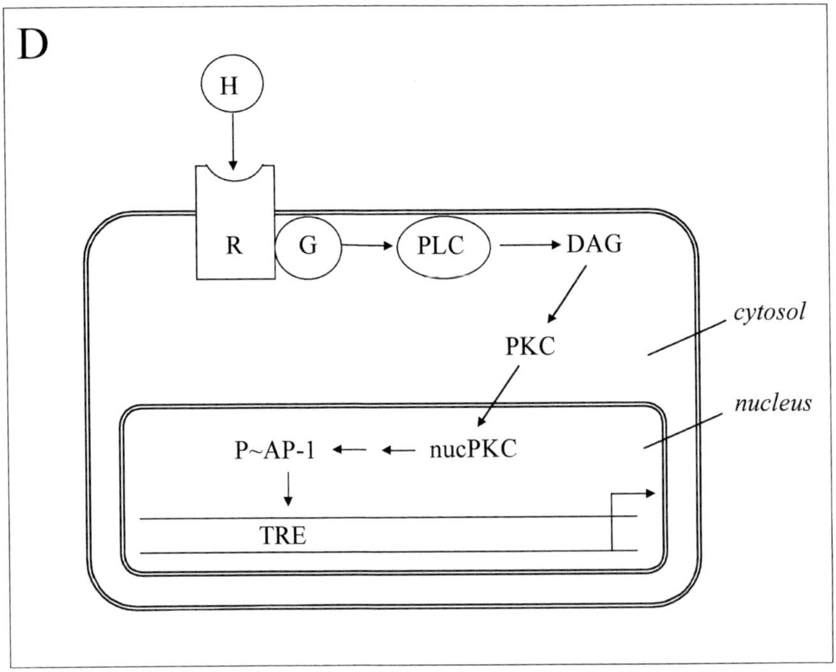

Figure 5D. (Legend see Fig. 5A)

activator of PKC. The TRE is located in various early genes controlling the onset of the S-phase of the cell cycle.

Calcium-calmodulin kinase (Fig. 5E)

Intracellular free calcium ions serve as activators of a large number of cytosolic enzymes and second messengers for the activation of numerous genes, e.g., genes controlling neural functions, hormone secretion by endocrine cells or cell proliferation. These various functions are differentially addressed by the spatial and temporal pattern of Ca^{2+} mobilization within cells [19]. In neural tissues, the main mechanisms to increase intracellular Ca^{2+} are by voltage-operated channels (e.g., L-type channels) or receptor-operated Ca^{2+} channels (e.g., of N-methyl-D-aspartate receptors) of plasma membranes [20]. In fibroblasts, hormones bind to G-protein linked receptors which trigger the formation of IP_3 by phospholipase C. The second messenger IP_3 triggers the release of Ca^{2+} ions from intracellular vesicles. Subsequently, the signal is mediated by binding of Ca^{2+} ions to the mediator protein calmodulin and activation by the calcium-calmodulin complex (CaM) of the CaM-dependent protein kinase family [21]. CaM-dependent protein kinases phosphorylate target proteins in the cytosol for the tuning of cytoplasmic functions or phosphory-

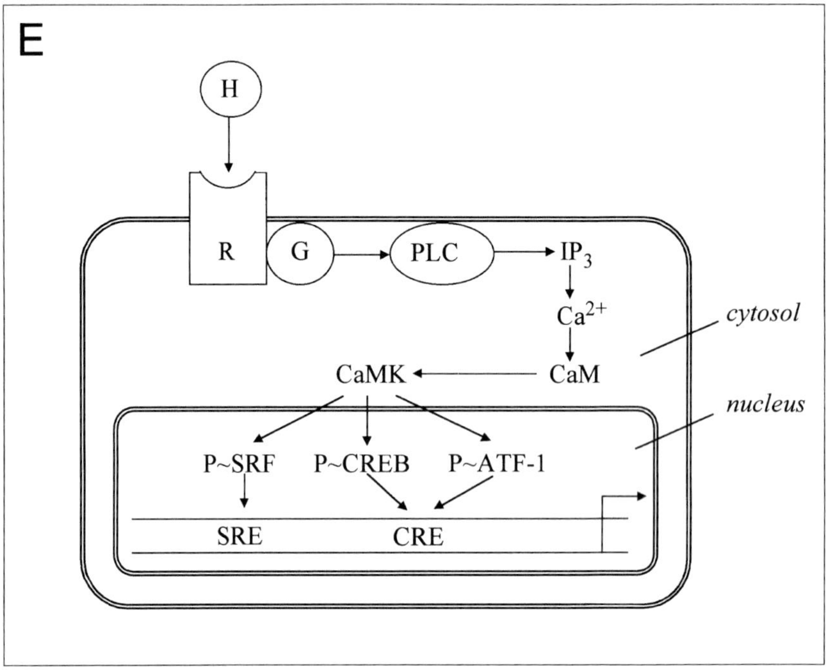

Figure 5E. (Legend see Fig. 5A)

late transcription factors in the nucleus to induce transcription of genes. CaM-dependent kinases activate one of the general transcription factors (CPT) binding the upstream promoter sequence CAAT as well as the specific transcription factors CREB, ATF-1 (activating transcription factor 1) and SRF. CREB and ATF-1 bind CRE, whereas SRF binds SRE. By a different mechanism, i.e., activation of PKC, Ca^{2+} also induces TRE. All three response elements are located in the promoter of the early gene c-fos, which controls the onset of cell proliferation [22].

Crosstalk between signaling paths

As already evident from the discussion of the individual signaling paths, signal transduction is an interlinked network; this feature is named "crosstalk" (Fig. 6). One type of signal may address several genes which contain the same response element, e.g., the CRE. And one and the same gene may be induced by multiple signals, if the gene contains response elements for different signals, as found in the promoter of the early gene c-fos which contains the SRE, CRE and TRE. Moreover, one type of signal may activate further signaling factors, as in the case of calcium, which activates three response elements, SRE, CRE and TRE. The crosstalk between signaling paths greatly enhances

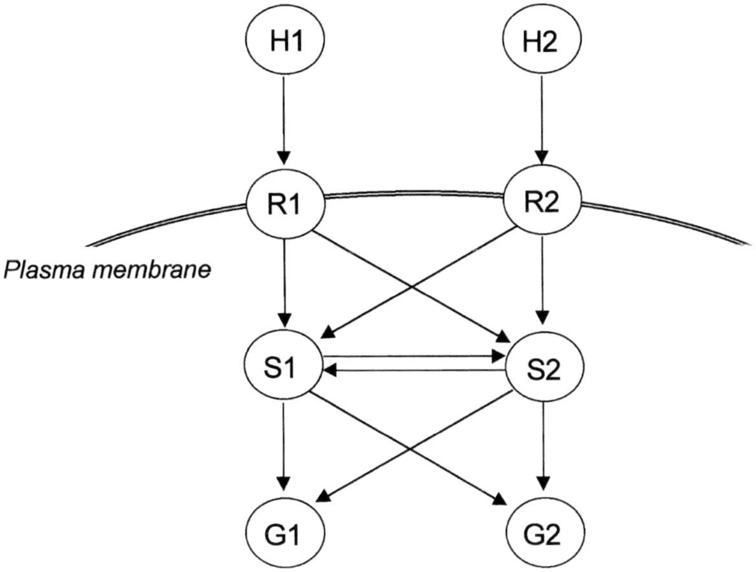

Figure 6. Scheme of crosstalk between two signaling paths. H: hormone; R: receptor; S: signaling factor; G: gene.

the potential for gene regulation. But for the same reason, the study of signaling to genes has to obey a caveat: It is rarely possible to study the effect of an agonist or inhibitor on the expression of one gene without affecting the expression of interlinked genes, too.

Transcriptional control by chromatin structure and DNA methylation

In a variation to the mechanisms of specific control of expression of defined genes, a large-scale silencing of genes is achieved by chromatin structure and DNA methylation. In chromatin, the DNA is complexed with histone proteins and organized in nucleosome structures. In heterochromatin, the DNA is packed in a highly condensed form where the DNA is relatively resistant to nucleases. In this form, genes are silenced on a large scale. An example of gene regulation by heterochromatization is the inactivation of one of two X chromosomes present in the cells of females. In an early phase of embryonic development, one of two X-chromosomes is inactivated by packaging in heterochromatin. During the further cell divisions, the daughter cells conserve this imprinted pattern. In contrast, euchromatin represents a more loosely organized type of chromatin which is transcribable and characterized by a higher susceptibility to degradation by DNase. In DNA regions characterized by their hypersensitivity to DNase, the euchromatin seems to be kept free of packaging in nucleosomes by a stable association of basal transcription factors. This form is found in active promoter and enhancer regions. On the other hand, genes present in heterochromatic regions may also be activated by transcription factors by an ATP-consuming process that disrupts nucleosomes. An example of silencing and activation by chromatin organization is the control of temporal pattern of expression of the foetal and adult globin genes during development of the erythroid system.

Gene activities may be further regulated by acetylation of histones. In several eukaryotic organisms, histone structure and gene expression is modified by acetylation of lysine residues in histones which reduces the positive charge and thus the interaction of the modified histones with DNA [23].

A further tool of large-scale repression is DNA methylation. A few percent of the cytosines in DNA are present as 5-methyl-cytosine, the majority of methylation located in so-called CG-doublets. Methylated genes are inactive and have to be demethylated for activation. In early embryonic development, soon after fertilization, a massive wave of demethylation erases almost all methyl groups in the genome, but the basic pattern is reestablished at about the time the embryo is implanted in the uterus [24].

DNA methylation represents a form of epigenetic inheritance, since it is conserved even during DNA replication. At the time of DNA replication, cytosines are methylated by a DNA methylase preferentially at those sites where the parent strand contains a methylated cytosine in a CG-doublet, whereas the non-methylated, transcribable DNA is protected from methylation. DNA methylation reduces the binding of many transcription factors to their response elements, e.g., of CREB, Myc, E2F, NFκB and others. On the other hand, demethylation is observed in tumour development where normally silent growth genes are activated. Recently, a mammalian DNA demethylase has been discovered that specifically demethylates mCpG islands in DNA [25].

Regulation of mRNA processing (alternative splicing)

The expression of genes may be further regulated after the termination of the synthesis of the primary transcript by alternative splicing of the pre-mRNA. It is estimated that only about 5% of all pre-mRNAs in eucaryotes are regulated by alternative processing [26]. A good example is the tissue-specific expression of the fibronectin gene. Fibronectin is a protein synthesized by fibroblasts and by hepatocytes in two different forms. Whereas the protein synthesized by fibroblasts contains protein sequences interacting with cell surface receptors, the protein synthezised by hepatocytes lacks these sequences and is secreted in a form soluble in serum. These differences are explained by alternative splicing, which excises two exons coding for the membrane-adhesive protein sequence during the processing in hepatocytes but not in fibroblasts. Alternative splicing in a similar mechanism occurs in the synthesis of immunoglobulin M which exists in a membrane-bound and in a secreted form. If in the process of splicing of the corresponding pre-mRNA, two exons coding for the membrane attachment are excised, the mRNA coding for the soluble form is generated. A further example for regulation by alternative splicing is the generation of both the positively acting factor CREB and the negatively acting factor CREM from one primary transcript. Whereas CREB is rendered active by phosphorylation at its activation domain, CREM lacks this domain and can only bind to DNA but not be activated [16].

Translational control of gene expression

Although the most important and of course most economic manner of gene control is the regulation of transcription, control of translation at the ribosome level is also observed. A classic example is the coordination of heme and globin synthesis for the formation of haemoglobin in erythroblasts. If the supply of heme is declining, an inhibiting enzyme is activated that inactivates the initiation of protein synthesis by phosphorylation of the initiation factor IF2. A more general example is the S6 kinase, a protein kinase that phosphorylates and activates the ribosomal protein S6 which functions in the translation of a group of mRNAs coding for proteins of the ribosomal translation apparatus. S6 kinase is activated by mitogenic signals. The corresponding signaling cascade involves a sequence from growth factors via receptor tyrosine kinases, phosphatidyl-inositol-3-kinase, further effector kinases and finally S6 kinase [27]. By this mechanism, the translation of mRNA into proteins is enhanced at the level of the ribosome.

Modulation of gene expression and its regulation by specific drugs

In the recent years there has been considerable progress in the development of substances that activate or inhibit signaling pathways from hormones to genes. The elucidation of gene regulation has been greatly assisted by the use of specific cell-permeable modulators for various receptors and protein kinases, often analogues of second messenger molecules. Other chapters of this volume address the more detailed description of the use of drugs.

Acknowledgement
The author is grateful to Wim Waetjen for the design of figures and for critically reading the manuscript.

References

1 Lewin B (1997) *Genes*. Oxford University Press, Oxford, New York, Toronto
2 Zawel L, Rheinberg D (1993) Initiation of transcription by RNA polymerase II: a multistep process. *Prog Nucl Acids Res Mol Biol* 44: 67–108
3 Burley SK, Roeder RG (1996) Biochemistry and structural biology of transcription factor IID

(TFIID). *Annu Rev Biochem* 65: 769–799
4 Hahn S (1998) The role of TAFs in RNA polymerase II transcription. *Cell* 95: 579–582
5 Goodrich JA, Tjian R (1994) Transcription factors IIE and IIH and ATP hydrolysis direct promoter clearance by RNA polymerase II. *Cell* 77: 145–156
6 Uptain SM, Kane CM, Chamberlin MJ (1997) Basic mechanisms of transcriptional elongation and its regulation. *Annu Rev Biochem* 66: 117–172
7 Thiel G, Lietz M, Leichter M (1999) Regulation of neuronal gene expression. *Naturwissenschaften* 86: 1–7
8 Pabo CO, Sauer RT (1992) Transcription factors: Structural families and principles of DNA recognition. *Annu Rev Biochem* 61: 1053–1095
9 Pavletich NP, Pabo CO (1991) Zinc finger DNA recognition: crystal structure of a Zif268-DNA complex at 2.1 A. *Science* 252: 804–817
10 Schwabe JNR, Rhodes D (1991) Beyond zinc fingers: steroid hormone receptors have a novel structural motif for DNA recognition. *Trends Biochem Sci* 16: 291–296
11 Vinson CR, Sigler PB, McKnight SL (1989) Scissors-grip model for DNA recognition by a family of leucine-zipper proteins. *Science* 246: 911–916
12 Davis RL, Cheng P-F, Lassar AB, Weintraub H (1990) The MyoD DNA binding domain contains a recognition code for muscle-specific gene activation. *Cell* 60: 733–746
13 Tsai M-J, O'Malley BW (1994) Molecular mechanisms of action of steroid/thyroid receptor superfamily members. *Annu Rev Biochem* 63: 451–486
14 Evans RM (1988) The steroid and thyroid receptor superfamily. *Science* 240: 889–895
15 Fantl WJ, Johnson DE, Williams LT (1993) Signaling by receptor tyrosine kinases. *Annu Rev Biochem* 62: 453–481
16 Sassone-Corsi P (1995) Transcription factors responsive to cAMP. *Annu Rev Cell Dev Biol* 11: 355–377
17 Montminy M (1997) Transcriptional regulation by cyclic AMP. *Annu Rev Biochem* 66: 807–822
18 Nishizuka Y (1992) Intracellular signaling by hydrolysis of phospholipids and activation of protein kinase C. *Science* 258: 607–614
19 Dolmetsch RE, Lewis RS, Goodnow CC, Healy JI (1997) Differential activation of transcription factors induced by Ca^{2+} response amplitude and duration. *Nature* 386: 855–858
20 Bading H, Ginty DD, Greenberg ME (1993) Regulation of gene expression in hippocampal neurons by distinct calcium signaling pathways. *Science* 260: 181–185
21 Heist E, Schulman H (1998) The role of Ca^{2+}/calmodulin-dependent protein kinases within the nucleus. *Cell Calcium* 23: 103–114
22 Tsuda T, Hamamori Y, Yamashita T, Fukumoto Y, Takai Y (1986) Involvement of three intracellular messenger systems, protein kinase C, calcium ion and cyclic AMP, in the regulation of *c-fos* gene expression in Swiss 3 T3 cells. *FEBS* 208: 39–42
23 Grunstein M (1997) Histone acetylation in chromatin structure and transcription. *Nature* 389: 349–352
24 Kafri T, Gao X, Razin A (1993) Mechanistic aspects of genome-wide demethylation in the preimplantation mouse embryo. *Proc Natl Acad Sci USA* 90: 10 558–10 562
25 Bhattacharya SK, Ramchandani S, Cervoni N, Szyf M (1999) A mammalian protein with specific demethylase activity for mCpG DNA. *Nature* 397: 579–583
26 McKeown M (1992) Alternative mRNA splicing. *Annu Rev Cell Biol* 8: 113–155
27 Pullen N, Thomas G (1997) The modular phosphorylation and activation of p70S6k. *FEBS* 410: 78–82

New Approaches to Drug Development
ed. by P. Jollès

New trends in antimicrobial development

Eduardo L. Setti[1] and Ronald G. Micetich[2]

[1] *Axys Pharmaceuticals, Inc., South San Francisco, California 94080, USA*
[2] *Synphar Laboratories, Inc., Edmonton, Alberta T6E 5V2, Canada*

Summary. So far, two strategies have been applied to develop new anti-infective agents: (a) the synthesis of analogs of classical antibiotics with enhanced activity against resistant pathogens and (b) the screening of naturally occurring substances and libraries of synthetic compounds for antimicrobial activity in whole-cell assays. Today, the same principles are being used; however, the search for antimicrobial compounds with novel modes of action is based on targeting specific resistance and virulence factors. Novel targets for anti-infective agents are currently being discovered as a consequence of a better understanding of cell biology, the molecular basis of bacterial resistance, the gene-pathogenicity relationship and the mechanism of the infection process.

Introduction

The development and use of antibiotics for the chemotherapy of bacterial infections was one of the most remarkable accomplishments in medicine of this century. The use of antibiotics produced an immediate important improvement of the public health by reducing dramatically the mortality and morbidity of people infected with microbial pathogens. However, antibiotic-resistant bacteria were found in clinical isolates soon after the introduction of the earliest antimicrobial agents into the market. At present, the problems associated with the continuous emergence and spread of bacterial resistance have been recognized as serious problems that threaten the effectiveness of many anti-infective agents currently in use [1].

Today, the emergence of methicillin-resistant *S. aureus* (MRSA), *S. pneumoniae*, vancomycin-resistant enterococci (VRE) and multiresistant *M. tuberculosis* is a major problem in hospitals and medical institutions worldwide. Consequently, there is a clear and urgent need to develop new anti-infective agents that will permit the effective treatment of the most resistant microorganisms.

Nowadays, pharmaceutical research scientists are using new tools that enable them to significantly accelerate the drug discovery process. For instance, automation applied in chemistry and biology is allowing the rapid synthesis and screening of thousands of compounds in a short period of time. Advances in structural biology and chemistry make it possible to design more specific and effective inhibitors with great potential in the therapy of infectious diseases.

Modern approaches to develop new antimicrobial agents

The modern tendency in antimicrobial design is based on a better understanding of the infection process, cell biology and the mechanism of bacterial resistance at a molecular level. The traditional approach was based on developing analogs of drugs that were already in clinical use. At present, most of the new antimicrobial agents are being developed by targeting specific bacterial-resistant factors and/or new/unexploited bacterial targets.

Antibiotics usually eradicate bacterial infections by killing the pathogen that has already been successful in invading, colonizing and damaging the host. One of the latest approaches to combat infectious diseases relies on the use of chemical entities that act at a very early stage of the infection process [2]. This can be achieved by inhibiting adherence or other virulence factors (VF), especially those that are essential for bacterial survival.

Virulence factors as novel targets for new antimicrobial agents

The ability of the bacteria to elicit adaptive responses to changes in environmental parameters such as temperature, osmolarity, pH, nutrients, etc., is an essential requirement for the bacteria to evolve into a pathogen. VFs (virulence factors) include the production of bacterial toxins, adherence factors, capsules, siderophores, outer membrane protein expression, chemotaxis and antigenic variation. The expression of VF requires a coordinated regulation of a series of genes that allow the pathogen to adapt to the hostile environment in the host [3]. For that purpose, the bacteria use a histidine protein kinase and a response regulator, a two-component system (TCS) that affects responses to environmental cues [4]. The prevention and control of the expression of virulence genes by interfering with the bacterial sensory system has become one of the novel approaches to identifying a new generation of antibacterials.

For example, quite recently, it has been shown that the proton-pump inhibitor, lansoprazole, interacts with several of *Helicobacter*'s VFs, including motility, urease and adhesins [5] and that the proper combination of proton-pump inhibitors/antibiotic is highly effective in eradicating the bacteria from gastric mucosa.

Recent studies of gene regulation in the *Pseudomonas* spp. have resulted in the identification of, at least, seven different TCSs that regulate the expression of well-characterized VFs. One of them is the TCS that controls the alginate biosynthetic pathway. Alginate is synthesized by *P. aeruginosa* as a component of the exopolysaccharide coat and is thought to play an important role in the pathogenesis of this microorganism in the lungs of patients with cystic fibrosis [6]. Alginate synthesis is controlled by a TCS, AlgR2/AlgR1, and its inhibition would render the bacterium more susceptible to conventional antibacterial agents. Quite recently, this goal has been achieved by discovering the compounds **1** and **2** which inhibit the autophosphorylation of AlgR2 and the

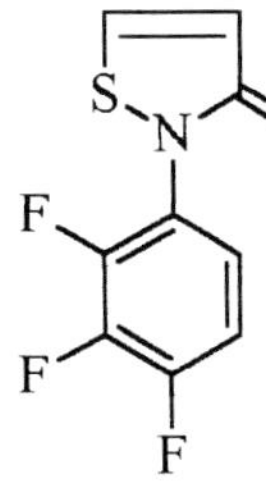

1

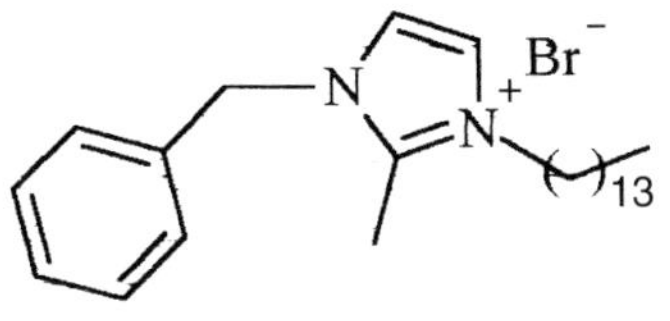

2

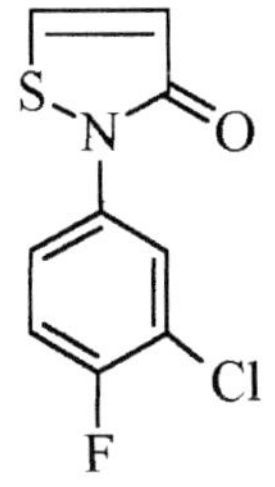

3

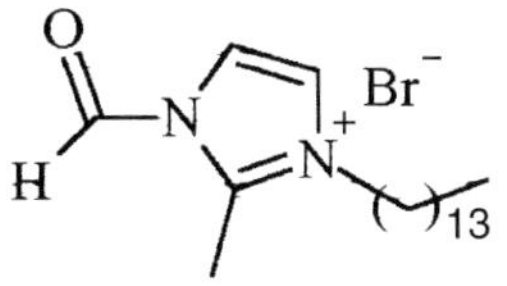

4

Figure 1. Different chemical entities that inhibit alginate synthesis controlled by TCS AlgR2/AlgR1.

phosphotransfer step to AlgR1, the DNA –binding response regulator. On the other hand, compounds **3** and **4** were found to interfere with this TCS by inhibiting the binding of AlgR1 to the algD promoter [7] (Fig. 1).

Recent reports have shown that many TCS inhibitors have potential in the treatment of different pathogens. Compounds **5** and **6** have been identified as TCS inhibitors of a series of gram-positive organisms. The minimum inhibitory concentration (MIC) of these compounds for *B. subtilis*, *S. aureus* and *S. pyogenes* was found to be between 1 and 4 μg/ml [8].

Efforts in this area have yielded a number of different chemotypes having *in vitro* gram-positive antibacterial activity. For example, structure RWJ-49445 **7** was found to have MICs of 1 to 4 μg/ml for gram-positive bacteria and to be rapidly bactericidal [8] (Fig. 2).

The ability of the microbes to adhere to the host cell surface is an important factor in the infection process of bacteria and viruses; therefore the search for compounds that prevent or reduce bacterial attachment is an attractive strategy to obtain a new generation of antimicrobial agents [9]. It has already been found that mice infected with a mannoside-specific strain of *E. coli* showed a reduction in the colonization of the urinary tract when they were treated with methyl-α-D-mannopyranoside, a compound that inhibits bacterial attachment by blocking the mannose-binding activity of the bacteria [10].

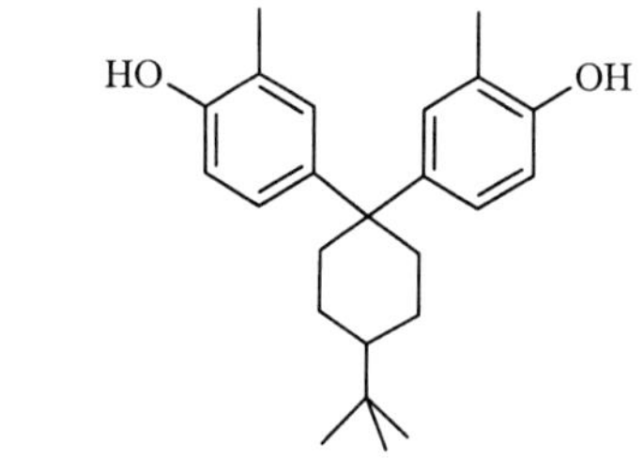

5

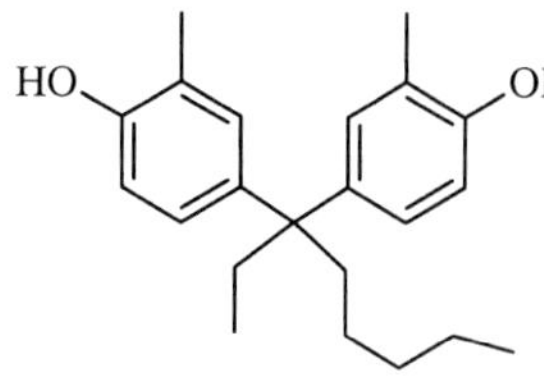

6

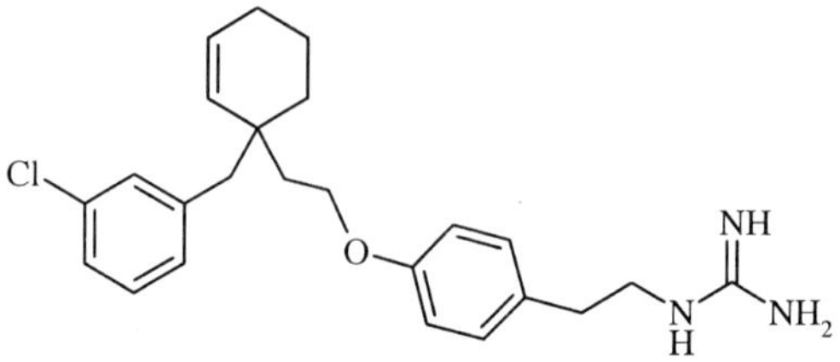

7

Figure 2. Structure of various TCS inhibitors.

It has also been reported that compounds **8** and **9** have a protective effect against *Helicobacter* infections by inhibiting specific molecular interactions between specific bacterial adhesins and cellular sialic acid-containing glycoprotein receptors [11] (Fig. 3).

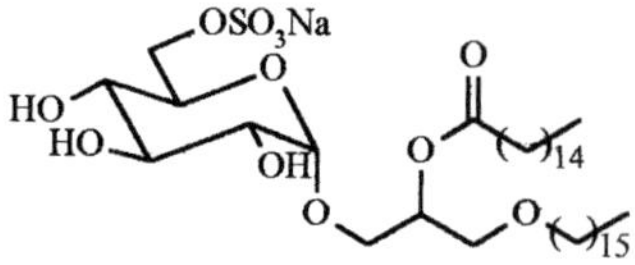

8

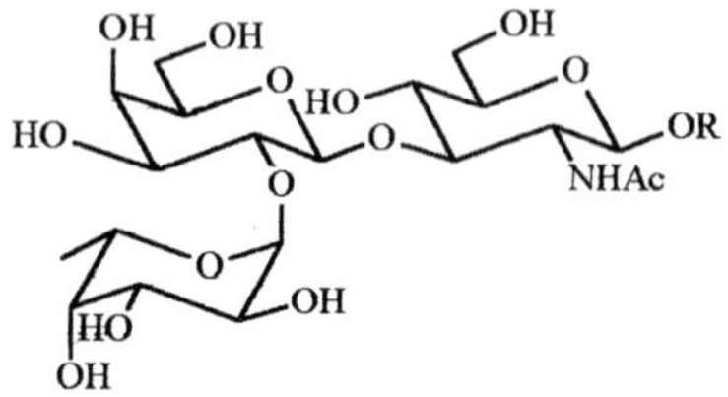

9

Figure 3. Two carbohydrate derivatives that interfere with *H. pylori* attachment to gastrointestinal mucosa.

Bacterial resistance factors as targets to develop new "classical" antimicrobial agents

Bacteria may become resistant to antibiotics as a result of chromosomal mutations, inductive expression of silent genes or acquisition of foreign DNA carried in plasmids, bacteriophages or transposable elements [12].

Bacteria possessing different resistance-determinants are known to render antimicrobial agents ineffective by deploying three major resistance mechanisms:

(a) Inactivation by bacterial enzymes (drug inactivation)
(b) Loss of antibiotic binding affinity through target modification or target overproduction (target modification) and
(c) Reduction of drug accumulation at the action site (decreased accumulation)

Drug inactivation

Drug inactivation by bacterial enzymes is perhaps the major mechanism of bacterial resistance to antimicrobial agents of natural origin. Bacteria have been using this mechanism to resist β-Lactam antibiotics, aminoglycosides, aminocyclitol, chloramphenicol, fusidic acid, fosfomycin and macrolides-lincosamides-streptogramins (MLS) [13].

So far, two distinct approaches have been applied to address this type of bacterial resistance:

(a) Development of enzyme inhibitors that act in synergy with the antibiotic and
(b) Development of analogs that are resistant to enzyme inactivation.

β-*Lactams*
β-Lactam antibiotics belong to a group of antibiotics that exert their bactericidal effect by inhibiting the last stage of the peptidoglycan synthesis [14]. They inhibit a family of cytoplasmic membrane-bound proteins – collectively termed penicillin-binding-proteins (PBPs) – which catalyze the peptidoglycan cross-linking, the final stage in the cell wall synthesis. This action affects the cell wall integrity and triggers mechanisms of bacterial autolysis. However, the continuous use of this kind of antibiotic has fostered the emergence of bacterial resistance attributed mainly to the production of β-lactamases, a bacterial enzyme that renders the antibiotic inactive by catalyzing the hydrolysis of the β-lactam ring. Clavulanic acid **10**, isolated from cultures of *S. clavuligerus* [15], and sulbactam **11** of synthetic origin, were the first two broad-spectrum β-lactamase inhibitors to reach the market. Tazobactam **12**, an analog of sulbactam, was the latest synthetic β-lactamase inhibitor to enter into the clinic

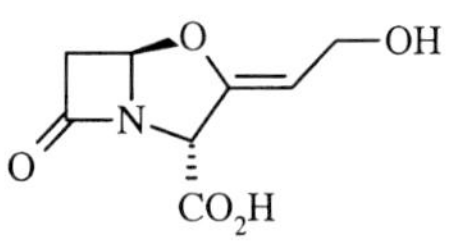

Clavulanic acid **10**

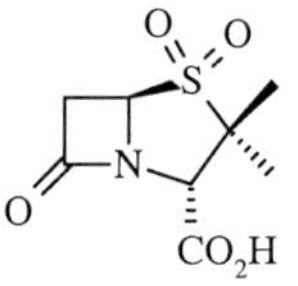

Sulbactam **11**

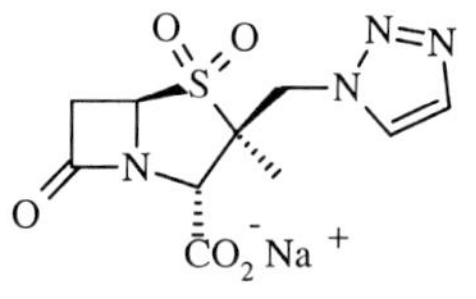

Tazobactam **12**

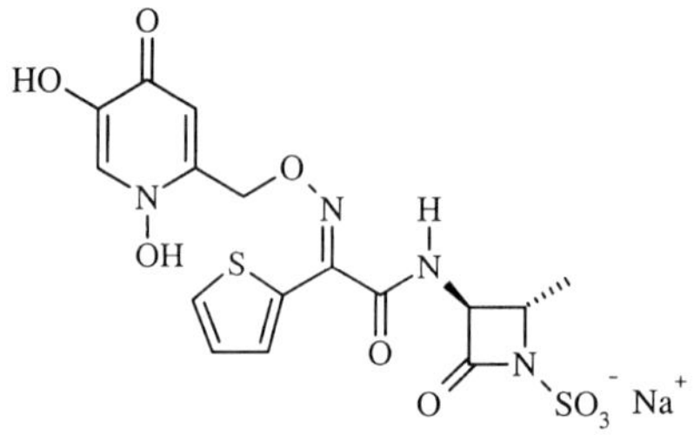

SYN-2190 **13**

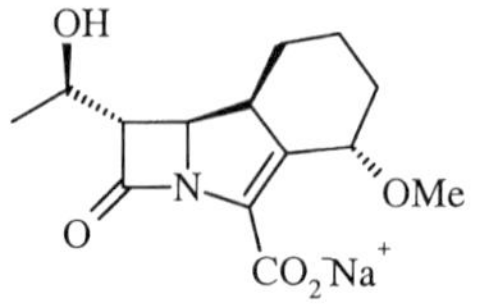

GV-104326 **14**

Figure 4. Clavulanic acid, sulbactam and tazobactam are beta-lactamase inhibitors in clinical use. SYN-2190 is a potent class C beta-lactamase inhibitor that shows synergy in combination with cleftazimide and cefpirome. GV-104326 is a new beta-lactam antibiotic with enhanced beta-lactamase stability.

with a broader spectrum of activity compared to their predecessor. These three inhibitors are effective potentiators of their partner β-Lactam antibiotic against class A β-lactamases-producing microorganisms [16] (Fig. 4).

Recent reports indicate that the synthesis of a potent inhibitor of class C β-lactamase was achieved. SYN-2190 **13** is a novel monobactam derivative that inactivates specifically the chromosomally encoded class C β-lactamase of *Enterobacter, Citrobacter, Pseudomonas* and *Morganella* species. This new inhibitor proved to have a synergistic effect in combination with ceftazidime and cefpirome against resistant *P. aeruginosa* strains [17].

Bridged monobactams are another novel class of potent inhibitors of Class C β-lactamases which was found to display synergy when combined with β-lactamase-labile antibiotics, especially against *Enterobacteriaceae* that produce large amounts of this enzyme [18].

The second approach, which is the synthesis of analogs resistant to enzyme inactivation, can also be found in β-Lactam antibiotics with the development of the 3rd and 4th generation of cephalosporins. These antibiotics proved to be useful in clinical settings due to their particular stability to penicillinases and to most common plasmid β-lactamases (TEM-1), which accounts for the resistance of *N. gonorrhoeae*, *E. coli*, *H. influenzae*, *Shigella* and *Salmonella* spp. to penicillins as well as to 1st and 2nd generation of cephalosporins.

Sanfetrinem (GV-104326) **14** belongs to a relatively new class of antibiotics – collectively called trinems – with enhanced β-lactamase stability, in particular towards β-lactamase-hydrolyzing cephalosporins. This compound proved to have high activity against resistant gram-positive pathogens and appears to be a promising compound suitable for further studies [19] (Fig. 4).

Aminoglycosides

Aminoglycosides (AGs) belong to an important group of broad-spectrum antibiotics that act by inhibiting the bacterial protein synthesis. The primary mechanism of AG resistance is due to enzymes that modify the antibiotic by acylation (aminoglycoside N-acetyltransferases, AAC enzymes), nucleotidylation (aminoglycoside O-nucleotidyltransferases, ATN enzymes), adenylation (aminoglycoside adenyltransferases, AAD enzymes) and phosphorylation (aminoglycoside O-phosphotransferases, APH enzymes) [20]. Since all clinically important AGs are susceptible to modification by more than two enzymes, to find a single inhibitor capable of inactivating most of the AG-modifying enzymes (AME) has been a difficult goal to achieve and little progress has been made to date. There are examples of approaches to inhibit each of the three more prevalent classes of AME. For example, 7-hydroxytropolone **15** was proven to potentiate the *in vitro* activity of several AGs against AG-resistant bacteria possessing a 2”-O-adenyltransferase [21].

It has also been described that several structural analogs of kanamycin act as mechanism-based inhibitors of the 3'-phophotransferases, an ubiquitous APH enzyme that inactivates a variety of AGs such as amikacin **16** [22].

The development of semisynthetic AGs insensitive to the most common AME has also been reported. For instance, arbekacin **17** (ABK) is a semisynthetic AG with very good activity against a wide variety of known AG-resistant bacteria, including MRSA with AME [23].

Another example of resistant AG can be found in isepamicin **18**, a relatively new broad-spectrum antibiotic that is active against endemic or epidemic strains of gram-negative bacilli and staphylococci. Its superior spectrum of action compared to amikacin was attributed to the fact that this antibiotic showed a higher stability against AAC(2'), a prevalent AG enzyme that confers resistance to tobramycin, netilmicin and amikacin [24] (Fig. 5).

Vancomycin

Vancomycin inhibits peptidoglycan synthesis –an essential bacterial cell wall component- by binding to the D-ala-D-ala terminus of the peptidoglycan intermediate of the lipid-linked N-acetylglucosamine-N-muramyl pentapeptide. The mechanism of resistance of enterococci to vancomycin involves an alteration of the peptidoglycan precursor in which the terminal D-alanine is replaced with D-lactate. This change reduces the affinity of the antibiotic to its target and renders the glycopeptide inactive. Lately, vancomycin resistance in enterococci has appeared and has been traced to a set of five genes: VanA, VanH, VanR, VanS and VanX [25].

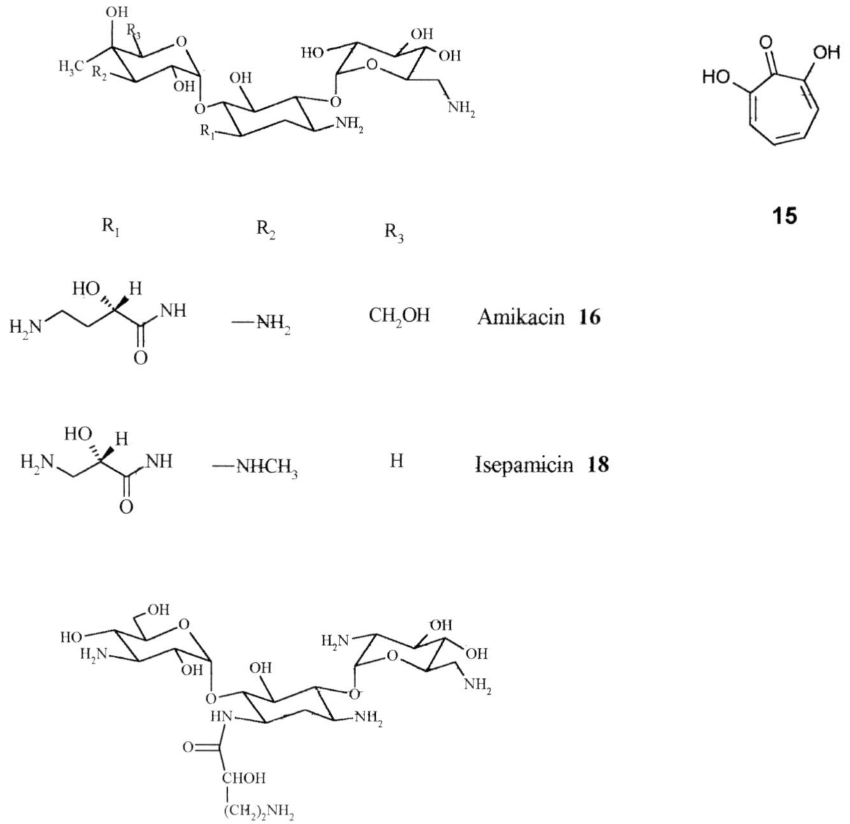

Arbekacin **17**

Figure 5. Aminoglycosides with activity against known aminoglycoside-resistant bacteria. Compound **15** has proven to be an effective potentiator of several aminoglycosides against aminoglycoside-resistant bacteria possessing a 2''-O-adenyltransferase.

The use of inhibitors of resistance factors as antibiotic potentiators is also being considered as a viable strategy to reverse the phenotypic gram-positive bacterial resistance to vancomycin. VanX is a Zn-dependent D,D-dipeptidase which is essential for vancomycin resistance in *E. faecium*. It has been reported that several phosphinates are slow-binding inhibitors of VanX [26] and that certain dithiol compounds are able to inhibit this enzyme in a time-dependent fashion, thus opening the prospect of finding more specific inhibitors with potentiating action on vancomycin [27].

Target alteration

The production of altered forms of normal biological targets that have low affinity for the antibiotic is a common source of resistance mechanisms used by different organisms against various anti-infective agents [28]. This mechanism of resistance has been shown to play a major role in the emergence of resistant pathogens against β-Lactam antibiotics, glycopeptides, macrolides, tetracycline, chloramphenicol, rifamycin, quinolones, β-lactamase inhibitors, aminocyclitol, novobiocin, trimethoprime, steroids and sulfonamides [13].

β-Lactam antibiotics

β-Lactam antibiotics kill the bacteria by targeting PBPs 1, 2, and 3 –called essential PBPs. However, certain gram-positive pathogens are able to resist the action of this class of antibiotics either by producing an extra PBP, which does not bind to the antibiotic, overproducing a low-affinity PBP or through a combination of both mechanisms.

Susceptible strains of *S. aureus* have five PBPs (1, 2, 3, 3' and 4) and the maximum antibacterial effect of β-Lactam antibiotics was found to be related to the simultaneous binding to PBPs 1, 2 and 3. However, MRSA can resist β-Lactam antibiotics by producing an extra PBP, PBP 2a (or PBP 2') which does not bind to the antibiotic and is able to perform all the PBP catalytic activities necessary for cell wall synthesis [29]. On the other hand, high-level ampicillin resistance in *Enterococcus faecium* has been shown to be associated with the overproduction of a modified PBP 5 with a reduced penicillin-binding affinity [30].

Extensive modification on the cephem and carbapenem nucleus allowed the identification of a number of β-lactam compounds possessing enhanced affinity for PBP 2a of MRSA strains, therefore showing potent activity against this opportunistic pathogen. For example, compound SM-17466 **19**, a novel β-methyl carbapenem, has been identified as a potent anti-MRSA antibiotic. This compound has been shown to have between 10 to 20 times more binding affinity to PBP 2a than meropenem and imipenem, respectively [31]. Similarly, the increased potency of the carbapenem L-695,256 **20** against MRSA strains compared to known β-Lactam antibiotics has been attributed to its high binding affinity to PBP 2a [32] (Fig. 6).

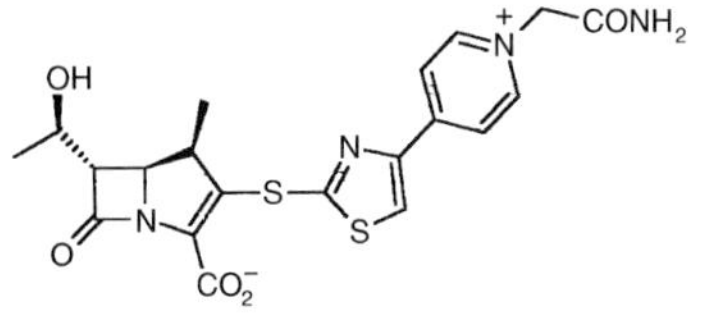

SM-17466 **19**

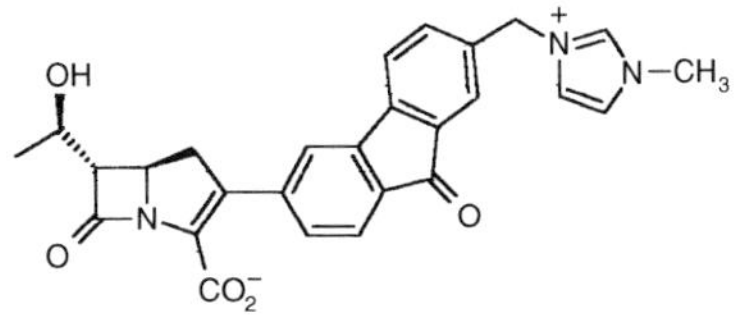

L-695,256 **20**

Figure 6. Two new carbapenem antibiotics with high potency against methicillin-resistant *S. aureus*.

Vancomycin

Efforts to overcome problems related to the emergence of VRE have yielded a number of semisynthetic glycopeptides (GP) with potent activity against VRE. For instance, LY191145 **21** is at least 500 times more potent than vancomycin against vancomycin-susceptible and VR enterococci. Mechanistic studies performed on this GP suggested that this compound inhibits transglycosylation in a different fashion to that of vancomycin [33]. Further modification on the epi-vancosamine sugar moiety of LY191145 produced GPs such as LY307599 **22** and LY333328 **23** with exceptional potency against pathogenic streptococci, including penicillin-resistant isolates. The enhanced inhibitory activity of LY333328 on the peptidoglycan synthesis in resistant microorganisms was attributed to its dimerization and enhanced membrane-anchoring properties rather than to its increased binding affinity to the peptidoglycan precursor [34] (Fig. 7).

Macrolides

Macrolides antibiotics inhibit the RNA-dependent bacterial protein synthesis by reversibly binding to the 50S subunit of the ribosome. Although resistance to macrolides can ocurr due to low-frequency mutational changes in ribosomal proteins, the most frequent resistance mechanism observed in clinical isolates is due to the presence of a specific methylase which is responsible for the dimethylation of the adenine 2058 in the 23S rRNA sequence [35]. This enzymatic reaction alters ribosome structure in a manner that prevents the binding of MLS.

The synthesis of novel macrolides structures that bind to the methylated ribosome subunit has already been achieved. For instance, it was found that certain 6-O-methyl-11,12-cyclic carbamate derivatives of erythromycin are able to bind to a dimethylated MLS-resistant 50S ribosomal subunit, thus showing enhanced activity against erythromycin-resistant strains. For example, it has been shown that the 11,12-carbamate analogs **24** of clarithromycin **25** [36], their 3-keto derivatives **26** and more recently the tricyclic ketolide **27** have increased activity against certain macrolide-susceptible and macrolide-resistant pathogens [37] (Fig. 8).

Fluoroquinolones

Fluoroquinolones are potent antibacterial agents that act by inhibiting DNA synthesis. They target DNA gyrase and DNA topoisomerase IV, two essential enzymes for bacterial growth. The action of quinolones on bacteria brings about a decreased introduction of negative supertwists into DNA, decreased decatenation of interlocked circles, damage to DNA and inhibition of DNA synthesis [38]. Single step-mutations in DNA gyrase have been shown to produce between 16- to 32-fold increases in the MICs of the newest fluoroquinolones such as norfloxacin, ciprofloxacin and ofloxacin [39]. Structure-activity relationship studies performed on quinolones allowed the identification of new fluoroquinolones having good binding affinity for mutated DNA gyrase from quinolone-resistant clinical isolates of *P. aeruginosa* [40].

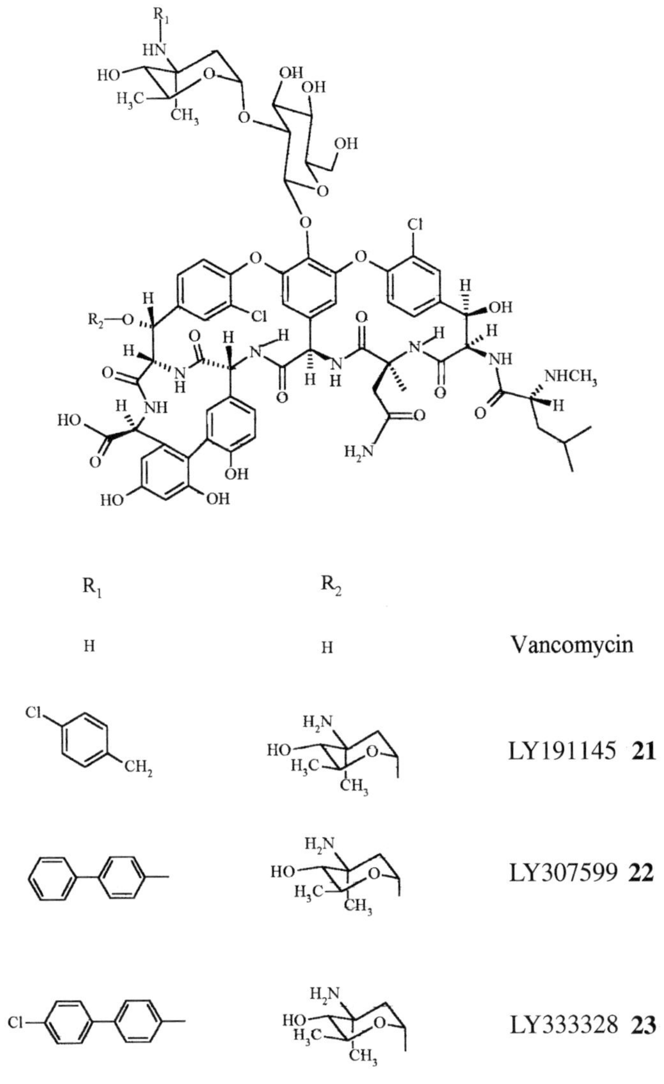

Figure 7. New semisynthetic glycopeptides with enhanced activity against vancomycin-susceptible and -resistant enterococci.

Tetracyclines

Tetracycline antibiotics prevent bacterial growth by inhibiting peptide synthesis. They act by binding to a single high-affinity site on a bacterial 30S ribosomal subunit, therefore precluding the attachment of the aminoacyl-tRNA to the ribosome A-site [41]. There are currently two major mechanisms responsible for tetracycline resistance: (a) the production of cytoplasmic proteins (tetM-tetO) that reduce the sensitivity of ribosomes to tetracyclines – called

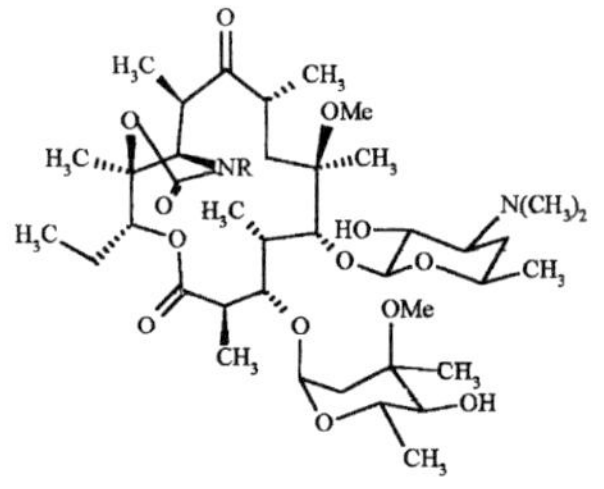

24

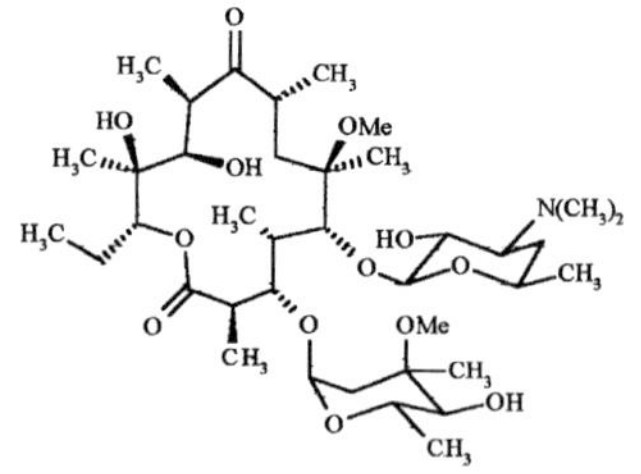

Clarithromycin **25**

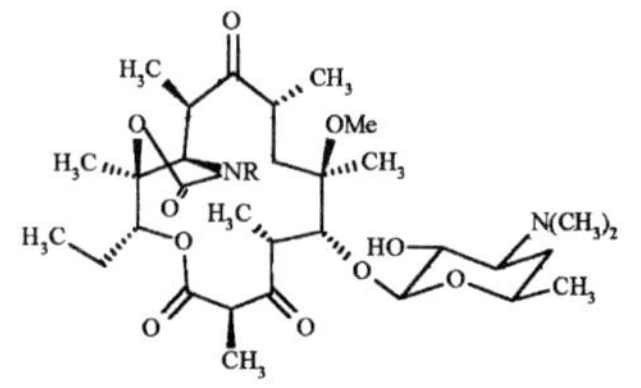

26

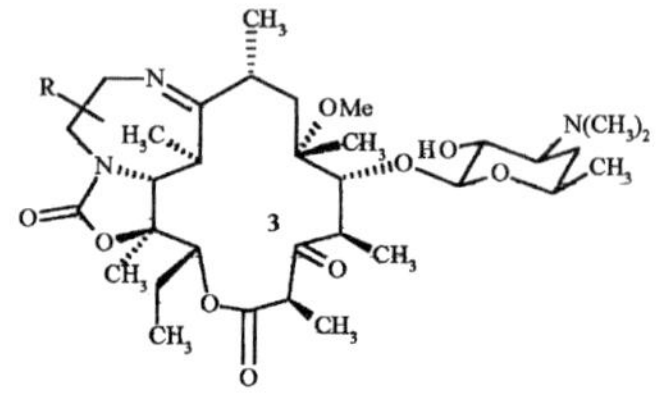

27

Figure 8. Structure of new macrolide antibiotics with enhanced activity against erythromycin-resistant strains.

ribosome protection – and (b) the reduction of the intracellular concentration of the antibiotic due to the action of membrane-associated protein pumps that efflux the antibiotic out of the bacteria in an energy-dependent fashion [42].

Efforts directed towards finding new tetracycline antibiotics effective against tetracycline-resistant isolates have already paid off with the discovery of a new class of tetracycline derivatives called glycylcyclines. For example, DMG-DMDOT **28** and DMG-MINO **29** are two novel tetracyclines that are unaffected by the ribosomal tetracycline factors Tet(M) and Tet(O), therefore showing high potency against a wide variety of gram-positive and gram-negative aerobic and anaerobic resistant strains [43] (Fig. 9).

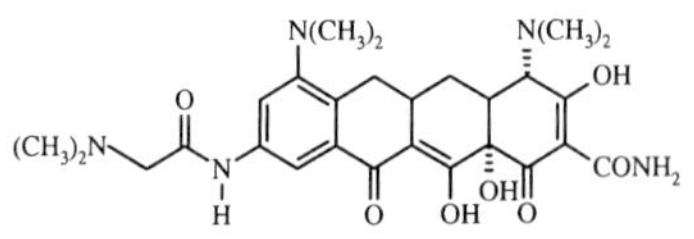

DMG-DMDOT **28**

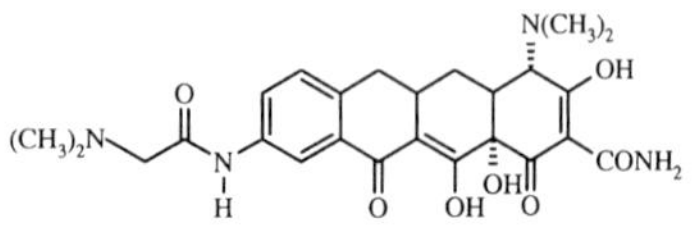

DMG-MINO **29**

Figure 9. Structure of DMG-DMDOT and DMG-MINO, two novel tetracyclines with high potency against resistant gram-positive and gram-negative microorganisms.

Decreased accumulation

Every antibiotic, in order to be effective, must reach a certain minimum concentration at the site of action. Since the biological targets of most antibiotics are located inside the bacterium, all the biological factors that affect the concentration of the compound inside the bacterial cell play an important role in the efficacy of the antibiotic.

The concentration of the antibiotic inside the bacteria depends on the diffusion rate of the compound through the outer membrane (OM) and/or on the presence of transmembrane efflux proteins.

Cell envelopes of gram-positive bacteria consist of a porous peptidoglycan layer which does not exclude most antimicrobial agents, whereas gram-negative organisms are surrounded by a peptidoglycan layer and a low-permeability OM. Nowadays, it is well known that the OM – composed of lipopolysaccharide – of gram-negative microorganisms acts as a permeation barrier, and that most hydrophilic solutes cross it through water-filled porin channels [44]. Porins are OM proteins that constitute the major route of uptake for hydrophilic antibiotics.

Although the permeability barrier alone is not enough to produce a high level of resistance, in certain cases, it was found to be an important factor when combined with other resistance mechanisms. This is particularly noticeable in *P. aeruginosa,* a gram-negative pathogen that possesses an OM 100–500 times less permeable than that of *E. coli.* This extremely low rate of penetration and the high production of β-lactamase act synergistically, rendering a number of cephalosporins ineffective [45].

At present, resistance caused by a permeation barrier can certainly be overcome by designing molecules that are likely to use specific routes of uptake. The observation that carbapenem antibiotics bearing certain basic groups in C-2 position are likely to use D2 porin channels as the main route of uptake, has allowed the synthesis of a series of potent antipseudomonal carbapenems such as imipenem **30**, meropenem **31**, biapenem **32** and panipenem **33** [46]. It has been shown that the high potency displayed by these compounds against resistant *P. aeruginosa* strains is due to their high penetration rates mainly through D2 porin channels (Fig. 10).

The rational exploitation of iron transport mechanisms has provided a novel approach to design antibiotics that are able to enter into the bacteria by using iron-specific receptors, therefore circumventing the usual porin route of uptake [47]. Siderophores are high-affinity iron-binding compounds that the bacteria release in their surroundings. These bacterial products that are capable of chelating ferric ion are then actively transported back into the organism by using the so-called *tonB* transport system. It has been shown that β-Lactam antibiotics containing siderophore-mimic moieties use the *tonB* protein to get into the bacteria [48]. For example, GR-69153 **34** [49] and KP-736 **35** [50] are two semisynthetic cephalosporins that have been shown to be actively transported into gram-negative cells via iron-regulated OM proteins regulated by

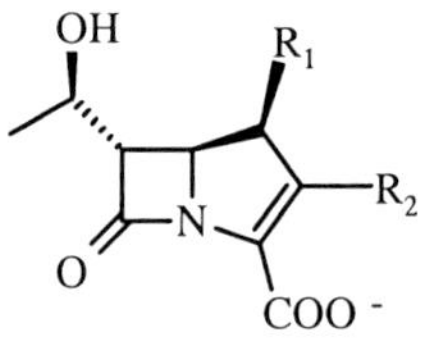

Carbapenem	R_1	R_2
Imipenem **30**	H	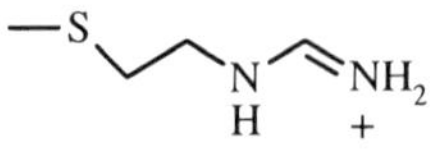
Meropenem **31**	CH_3	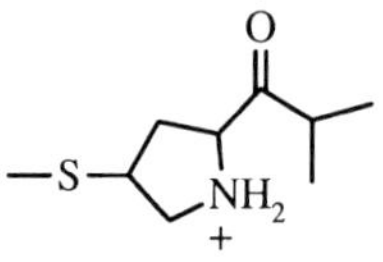
Biapenem **32**	CH_3	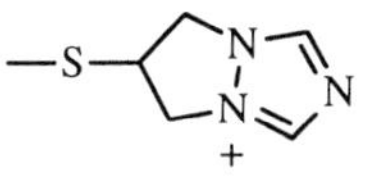
Panipenem **33**	H	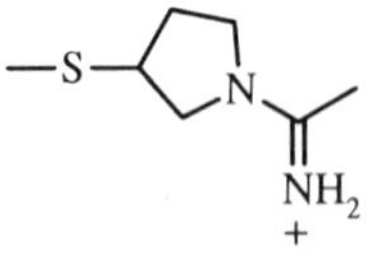

Figure 10. Structure of carbapenem antibiotics that use D2 porin channel to get into gram-negative bacteria.

the *tonB* gene product. As expected, these two antibiotics are highly potent against gram-negative microorganisms, especially against *P. aeruginosa* strains (Fig. 11).

Another mechanism of bacterial resistance related to permeability barriers is the formation of biofilms. Extracellular slime or glycocalyx production was found to play an important role in bacterial adherence and to be a bacterial strategy to resist the host defense mechanism, including anti-infective treatment [51]. The current approaches to eradicate bacterial biofilms rely on the use of combination therapies. The combinations of tobramycin-piperacillin, tosufloxacin with erythromycin, clindamycin and clarithromycin have been

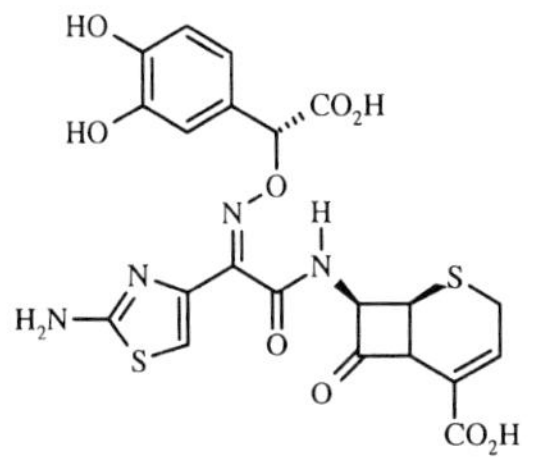

GR-69153 **34**

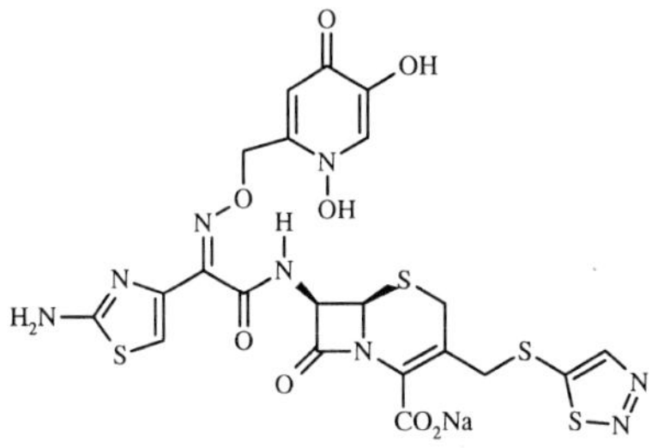

KP-736 **35**

Figure 11. Structure of GR-69153 and KP-736, two novel cephalosporins that have been shown to use *ton*B transport system to cross the outer membrane of gram-negative bacteria.

reported to be effective in the eradication of biofilm-associated bacteria. It has also been shown that certain fluoroquinolones, when given at sub-MICs, are able to inhibit *P. aeruginosa* glycocalyx formation and to prevent its adherence to plastic surfaces [52].

It has been reported that the use of polycations at sub-MICs can increase the permeability of the OM of gram-negative pathogens, rendering various microorganisms susceptible to many hydrophobic antibiotics. Polymyxin B nonapeptide – the most studied polycation – has been shown to lower the MIC of novobiocin against different gram-negative microogranisms and that this combination, when administered in multiple doses, protects mice challenged with a lethal dose of gram-negative bacteria [53].

The second major obstacle that may prevent the accumulation of the drug inside the bacterial cell is the presence of antibacterial efflux proteins. Resistance due to active efflux has been recognized as playing an important role in the resistance of many microorganisms to tetracycline, fluoroquinolones, macrolides and β-Lactam antibiotics [54].

The use of efflux pump inhibitors as antibiotic potentiators is an approach that is currently being considered. In a study carried out at Lederle Laboratories, various structural analogs of tetracycline were tested in an everted membrane vesicle system for their ability to inhibit the tetracycline efflux pump [55]. Compound **36** was identified as a potent inhibitor of bacterial efflux and exhibited potentiation *in vitro* against various tetracycline-resistant microorganisms. Recently, it was shown that combination of two proton-pump inhibitors – lansoprazole and omeprazole – with three different macrolides antibiotics – roxithromycin, flurithromycin and azythromycin – acted synergistically on most *H. pylori* strains [11] (Fig. 12).

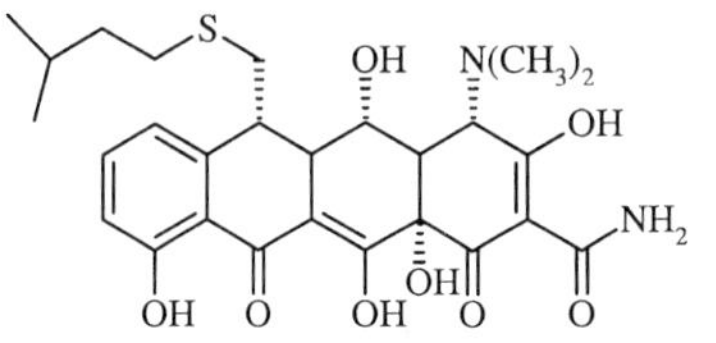

36

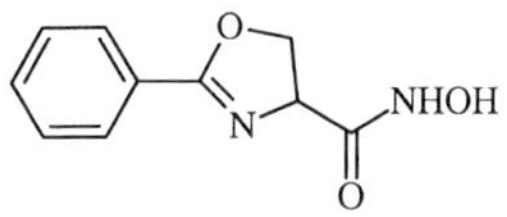

37

Figure 12. Compound **36** is a potent inhibitor of bacterial efflux. Compound **37** was shown to interfere with the second step in lipid A biosynthesis.

Novel antimicrobial agents with new or unexploited mechanisms of action

Antimicrobial agents currently in use owe their bactericidal/bacteriostatic activities to their ability to interfere with DNA, RNA, folic acid, bacterial protein and cell wall syntheses or to disrupt the bacterial cell membrane. Each of the enzymes involved in the metabolic pathway of these bacterial components is itself a potential biochemical target for new anti-infective agents. For instance, β-Lactam antibiotics and glycopeptides are two structurally unrelated antimicrobial classes of compounds that have a similar mechanism of action – inhibit cell wall synthesis – but at different biochemical targets.

Bacterial cell wall inhibitors

Despite the large number of antibiotics that inhibit bacterial cell wall synthesis, the search for new molecular entities that are able to inhibit unique biochemical steps in the metabolic pathway of cell wall formation is still being actively pursued. This is due to the fact that the enzymes involved in this specific process are not found in mammalian cells, implying that such inhibitors should display selective toxicity towards bacterial cells.

Lantibiotic Mersacidin is a new antibacterial agent that interferes with peptidoglycan synthesis in a different fashion to that of β-lactam and vancomycin [56]. Recent studies have shown that mersacidin inhibits the formation of lipid II – the ultimate monomeric peptidoglycan precursor – by binding to a site which is not targeted by any antibiotic currently used in clinical settings [57]. *In vitro* studies have shown that this antibiotic, when compared with vancomycin, teicoplanin and daptomycin, was 4- to 8-fold less active against *S. aureus* strains. However, *in vivo* results indicated that mersacidin was more effective than vancomycin in a mouse model of septicaemia.

Ramoplanin is a new lipoglycopeptide antibiotic that also inhibits cell wall peptidoglycan biosynthesis in gram-positive bacteria. Studies indicated that this antibiotic probably inhibits the N-acetylglucosaminyltransferase-catalyzed conversion of the lipid intermediate I to lipid intermediate II. Ramoplanin is a complex of three components designated as A1, A2 and A3 whose potential therapeutic application is severely limited because of its non-specific membrane-disrupting activity at high concentrations [58].

Recently, as a part of a broad screening program, scientists at Merck were able to identify inhibitors of UDP-3-O{R-3-hydroxymyristoyl}-GlcNAc deacetylase, the enzyme involved in the second step in lipid A biosynthesis. For instance, the lead compound L-573,655 **37** is a reversible inhibitor of this enzyme with an IC_{50} of about 30 nM and was found to have a modest antibacterial activity against a wild-type strain of *E. coli* [59] (Fig. 12).

Tunicamycin **38**, liposidomycin **39** and mureidomycin A **40** belong to the group of peptidylnucleoside antibiotics that act on the initial stage of peptidoglycan synthesis by inhibiting translocase [60]. Unlike tunicamycin, mureidomycin A and liposidomycin B are selective inhibitors of bacterial peptidoglycan biosynthesis [61]. The interesting antipseudomonal activities displayed by mureidomycins and their selective mode of inhibition constitute the basis for a rational design of novel antipseudomonal agents (Fig. 13).

Bacterial membrane-disrupting antibacterial agents

Peptide antibiotics or cationic peptides belong to a group of naturally occurring antibacterial agents with a broad spectrum of action that kill bacteria by forming large pores in bacterial cell membranes. These antibacterial peptides have some certain common features such as the possession of a net positive charge of at least +2 and are folded in three dimensions in such a way that they have both a hydrophobic and hydrophilic face [62].

Most cationic peptides do not have exceptionally high *in vitro* antimicrobial activities; however, it was found that they are able to kill bacteria much more quickly than conventional antibiotics and do not induce resistant mutants even after many passages on an antibiotic concentration close to the MIC.

Magainins consist of a mixture of two closely related, naturally occurring cationic peptides with a broad-spectrum antimicrobial activity, isolated from the skin of the African frog *Xenopus laevis* [63]. For instance, MSI-78 is a 22-residue magainin analog that is being developed for use as a topical agent. Two phase-III clinical trials of MSI-78 have been completed for the treatment of infected diabetic foot ulcers [64].

Like other cationic peptides (such as polymyxin B nonapeptide), magainins, acting as permeaters, were shown to be synergistic with cefpirome in a mouse infection model and to have endotoxin-neutralizing activity [65].

Cecropins are 35–39 amino acid peptides isolated from the *Hyalophora cecropia*, a giant silk moth [66]. They are active against gram-positive and gram-negative bacteria at 5–15 μg/ml and are active against *M. tuberculosis* [67]. A cecropin-metillin chimaera comprised of 30 amino acids, currently in

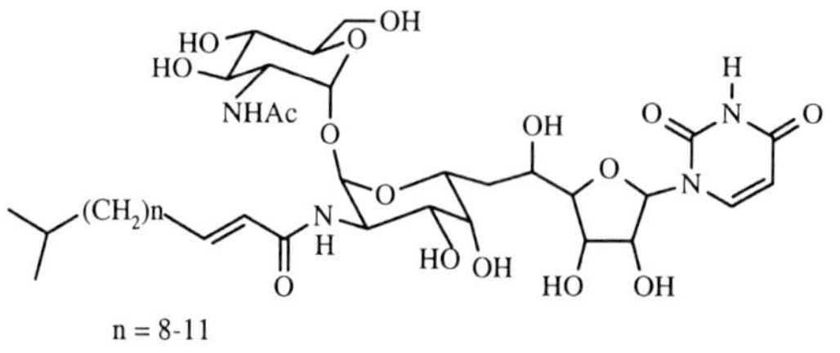

Tunicamycin **38**

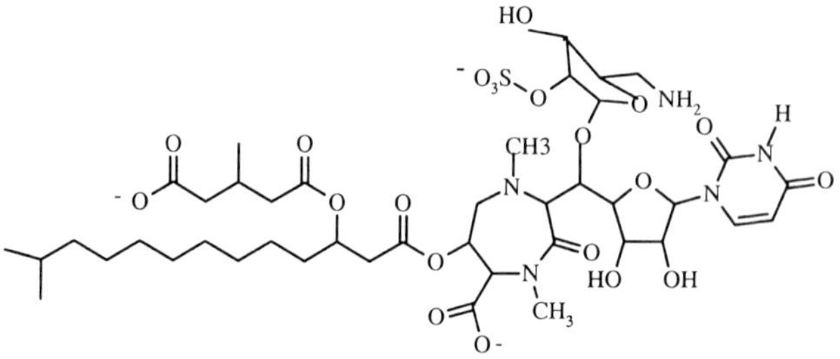

Liposidomycin B **39**

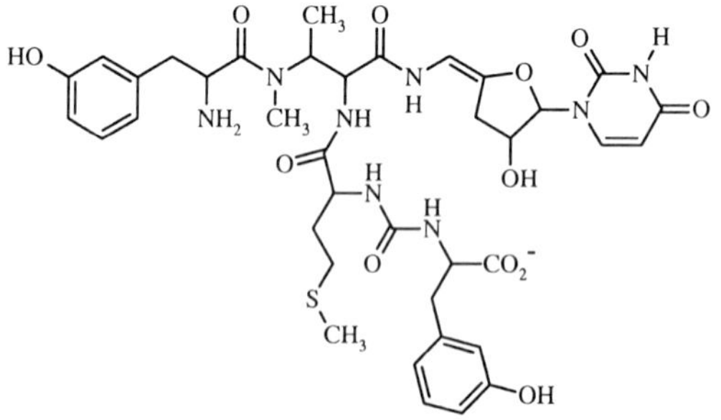

Mureidomycin A **40**

Figure 13. Structure of tunicamycin, liposidomycin B and mureidomycin A. These compounds inhibit the peptidoglycan synthesis by targeting translocase.

the developmental stage, has been shown to be active against various strains of *P. aeruginosa* with a 3 log kill at concentrations less than 1.5 μg/ml [68].

Protegrins are a group of cystein-rich peptides isolated from porcine leukocytes that have demonstrated to have a broad range of antimicrobial activity with a fast bactericidal action against either MRSA or resistant *P. aeruginosa* strains. Interestingly, MRSA strains treated with PG-1, an 18 amino acid β-sheet peptide belonging to the protegrin family, did not develop resistance after a serial transfer of the microorganism under conditions of cultivation which increased the MIC for norfloxacin by a factor of 85 [69]. *In vivo* results in mice demonstrated that protegrins have the potential in the treatment of systemic infection caused by clinically relevant pathogens such as MRSA, VRE and *P. aeruginosa*.

Bacterial protein synthesis inhibitors

Inhibition of different steps of the biological pathway leading to the synthesis of bacterial protein, continues to provide new biological targets to develop antimicrobial agents with novel modes of action.

Oxazolidinones are a group of new synthetic antimicrobial agents that exert their bactericidal or bacteriostatic action by acting at a very early stage of bacterial protein synthesis [70]. This new class of antibiotics is mainly active against gram-positive bacteria, including various resistant strains. For example, U-100592 **41** and U-100766 **42** are two orally available oxazolidinones, highly active against β-Lactam antibiotic-susceptible and –resistant gram-positive pathogens that are in clinical studies for the treatment of infections caused by gram-positive pathogens [71] (Fig. 14).

Pristinamycin is a mixture of microcyclic lactone (pristinamycin II **43**) and a cyclic depsipeptide antibiotic (pristinamycin I **44**) that belongs to the streptogramin family. These two related components (I_A and II_B) inhibit protein synthesis by binding to different sites on the 50S bacterial ribosomal subunit to form a stable ternary complex [72]. RP 59500 is a semisynthetic derivative of pristinamycin that is active against a wide range of gram-positive bacteria including MRSA, glycopeptide-resistant *E. faecium* and penicillin-resistant pneumococci [73] (Fig. 15).

GE2270 A **45** is a novel antibiotic isolated from the fermentation broth of *Planobispora rosea* that inhibits the bacterial protein synthesis by targeting the

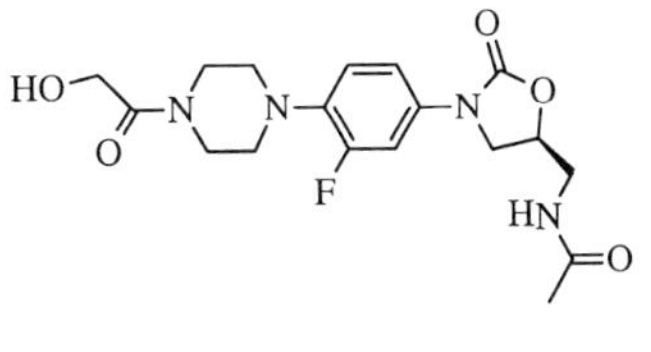

U-100592 **41**

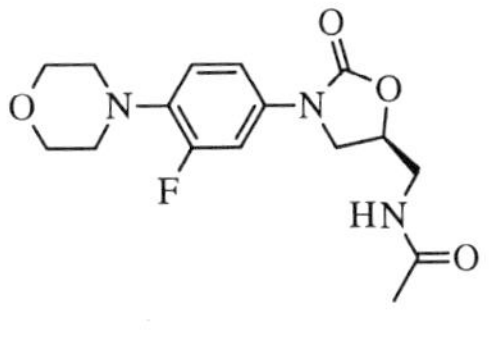

U-100766 **42**

Figure 14. Structure of compounds U-100592 and U-100766, two new oxazolidinones highly active against a wide variety of gram-positive antibiotic-resistant pathogens.

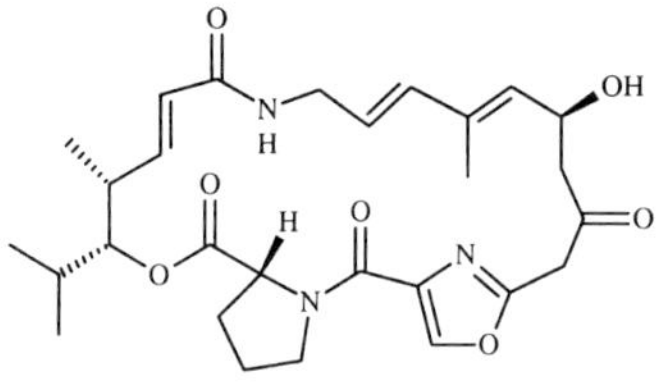

Pristinamycin II **43**

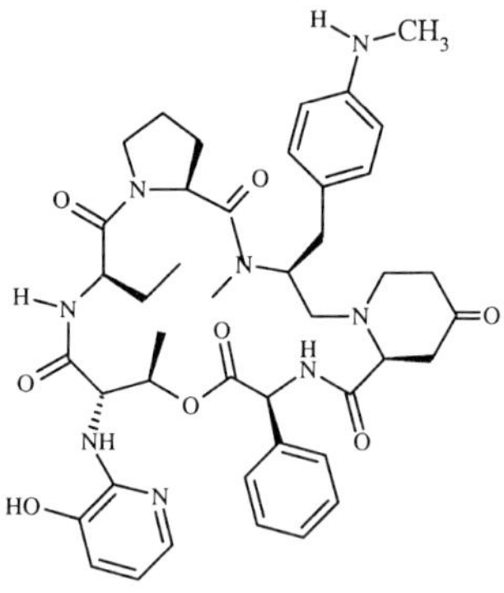

Pristinamycin I **44**

Figure 15. Structure of pristinamycin I and II. These two compounds inhibit protein synthesis by binding to different sites on the 50S bacterial ribosomal subunit.

elongation factor-Tu, which is the factor required for the binding of aminoacyl-tRNA to the A site of the ribosome. This natural product, structurally related to thiazolyl peptide antibiotics, is highly active against staphylococci, streptococci, enterococci and *M. tuberculosis* [74]. Similarly, compound GE-37468 A **46** is a relatively new thiazolyl peptide antibiotic that is functionally related to the GE2270 class of elongation factor-Tu inhibitors with activity against gram-positive bacteria, including multidrug-resistant strains [75] (Fig. 16).

Mupirocin or pseudomonic acid A **47** is a naturally occurring antibiotic that prevents bacterial protein synthesis by inhibiting the enzyme isoleucyl-tRNA synthetase [76]. It has potent antibacterial activity against *S. aureus*, *Staphylococcus* spp., and *S. pyogenes* and is effective in topical treatment of staphylococcal and streptococcal infections.

By applying rational drug design approaches, scientists at Cubist Pharmaceutical were able to identify CB-432 **48**, an inhibitor of the enzyme t-RNA synthetase. This compound showed *in vivo* efficacy against gram-positive microorganisms, especially *S. aureus* [77].

Conclusion

Rapid emergence and spread of drug-resistant bacteria have led to an increased demand for newer and more effective antibiotics or strategies to overcome resistance. Recent developments in an understanding of virulence mechanisms and advances in genomics are leading to the identification of novel bacterial targets and the means to design new and more effective antibiotics which may overcome problems associated with resistance. There is now an increased

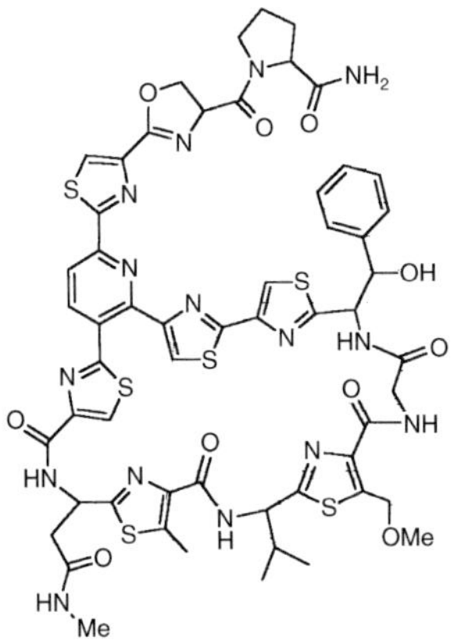

GE 2270 A **45**

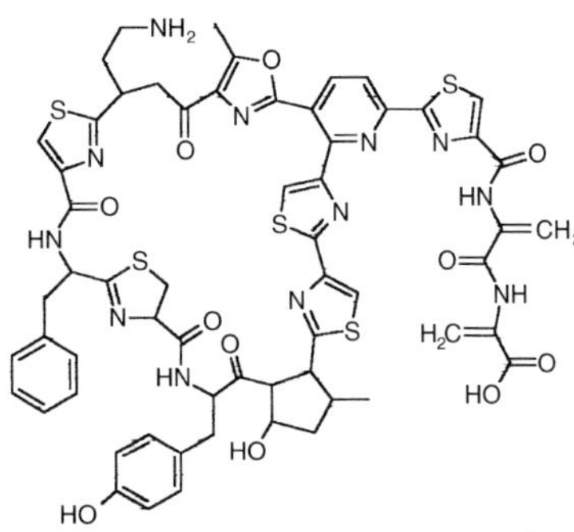

GE 37468 A **46**

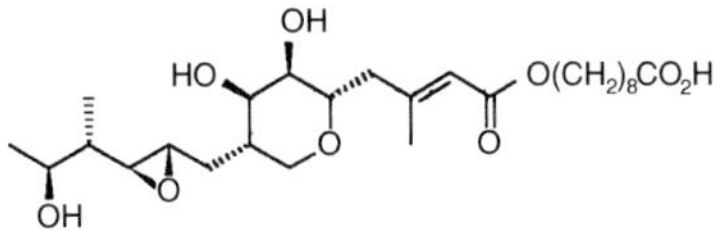

Mupirocin **47**

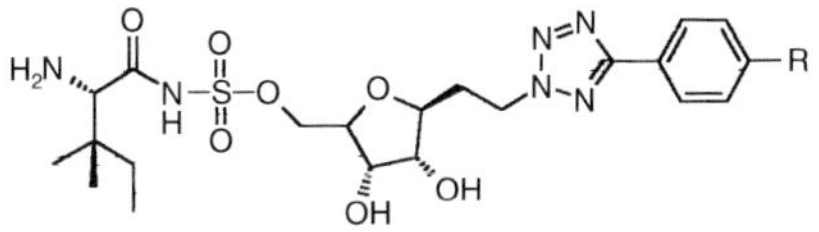

CB-432 **48**

Figure 16. Structure of GE2270A and GE37468 A, two natural occurring antibiotics that inhibit bacterial protein synthesis by targeting the elongation-factor-Tu. Mupirocin and CB-432 inhibit protein synthesis by targeting enzyme t-RNA synthetase and are active mainly against *S. aureus* strains.

awareness of the need for new and more effective antibiotics and it is evident that the research effort in this field will begin to grow.

Abbreviations

AAC: aminoglycoside N-acetyltransferase
AAD: aminoglycoside adenyltransferase
AG: aminoglycoside
AME: aminoglycoside-modifying enzyme
APH: aminoglycoside O-phosphotransferase
ATN: aminoglycoside O-nucleotidyltransferase
MIC: minimum-inhibitory-concentration
MRSA: Methicillin-resistant *Staphylococcus aureus*
OM: outer membrane
PBP: penicillin-binding proteins
TCS: two compnent systems
VF: virulence factors
VRE: vancomycin-resistant enterococci

Acknowledgement
We are grateful to Lina Quattrocchio Setti for her valuable help in the preparation of this manuscript.

References

1 Neu HC (1992) The crisis of antibiotic resistance. *Science* 257: 1064–1072
2 Setti EL, Micetich RG (1998) New trends in antimicrobial development. *Curr Med Chem* 5: 101–113
3 Mekalanos JJ (1992) Environmental signals controlling expression of virulence determinants in bacteria. *J Bacteriol* 174: 1–7
4 Dziejman M, Mekalanos JJ (1995) Two-component signal transduction and its role in the expression of bacterial virulence factors. *In*: JA Hoch, TJ Silhavy (eds): *Two component signal transduction.* AMS Press, Washington DC, 305–317
5 Nakao M, Tada M, Tsuchimori K, Uekata M (1995) Antibacterial properties of lansoprazole alone and in combination with antimicrobial agents against *Helicobacter pylori. Eur J Clin Microbiol Infect Dis* 14: 391–399
6 Roychoudhury S, Sakai K, Chakrabarty AM (1992) AlgR2 is an ATP/GTP-dependent protein kinase involved in alginate synthesis by *Pseudomonas aeruginosa. Proc Natl Acad Sci USA* 89: 2659–2663
7 Roychoudhury S, Zielinski NA, Ninfa AJ, Allen NE, Jungheim LN, Nikas TI, Chakrabarty AM (1993) Inhibition of two-component signal transduction systems: inhibition of alginate gene activation in *Pseudomonas aeruginosa. Proc Natl Acad Sci USA* 90: 965–969
8 Barrett JF, Hoch JA (1998) Two-component signal transduction as a target for microbial anti-infective therapy. *Antimicrob Agents Chemother* 42: 1529–1536
9 Sharon N, Lis H (1993) Carbohydrates in cell recognition. *Sci Amer* 268: 82–89
10 Aronson M, Medalia O, Schori L, Mirelman D, Sharon N, Ofek I (1973) Prevention of colonization of the urinary tract of mice with a *Escherichia coli* by blocking of bacterial adherence with methyl α-D-mannopyranoside. *J Infect Dis* 139: 329–332
11 Goldschmidt RM, Macielag MJ, Hlasta DJ, Barrett JF (1997) Inhibition of virulence factors in bacteria. *Curr Pharm Design* 3: 125–142
12 Davies J (1994) Inactivation of antibiotics and the dissemination of resistance genes. *Science* 264: 375–381
13 Silver LL, Bostian KA (1993) Discovery and development of new antibiotics: the problem of antibiotics resistance. *Antimicrob Agents Chemother* 37: 377–383
14 Frère JM, Nguyen-Distèche M, Coyotte J, Joris B (1992) Mode of action: interaction with the penicillin binding proteins. *In*: MI Page (ed.): *The chemistry of β-lactams.* Blackie Academic and Professional, New York, 148–197
15 Reading C, Cole M (1977) Clavulanic acid: a β-lactamase-inhibiting β-lactam from *Streptomyces clavuligerus. Antimicrob Agents Chemother* 11: 852–857
16 Aronoff SC, Jacobs MR, Johenning S, Yarrabo S (1986) Comparative activities of the β-lactamase inhibitor YTR830, clavulanate and sulbactam combined with extended-spectrum penicillins against ticarcillin-resistant Enterobacteriaceae and *Pseudomonas. J Antimicrob Chemother* 18: 177–184
17 Maiti SN, Phillips OA, Setti EL, Reddy AVN, Goyal J, Czajkowski DP, Luu HT, Micetich RG, Singh R, Catena R (1997) Syn-2190: Synthesis and structure-activity relationships of a novel series of 2-oxo-1-azetidine sulfonic acid as potent β-lactamase inhibitors. *37th Intersci Conf Antimicrob Agents Chemother* F188, Toronto, 178
18 Heinze-Krauss I, Angehrn P, Charnas RL, Gubernator K, Gutknecht EM, Hubschwerlen C, Kania M, Oefner C, Page MGP, Sogabe S et al (1998) Structure-based design of β-lactamase inhibitors.1. Synthesis and evaluation of bridged monobactams. *J Med Chem* 41: 3961–3971
19 Singh KV, Coque TM, Murray BE (1996) *In vitro* activity of the trinems sanfetrinem (GV104326) against gram-positive organisms. *Antimicrob Agents Chemother* 40: 2142–2146
20 Miller GH, Sabatelli FJ, Naples L, Hare RS, Shaw KJ (1995) The changing nature of aminoglycoside resistant mechanisms and the role of the isepamicin – a new broad-spectrum aminoglycoside. *J Chemotherapy* 7 (suppl 2): 31–44
21 Allen NE, Alborn WEJr Hobbs JNJr Kirst HA (1982) 7-Hydroxytropolone: an inhibitor of amino-

glycoside-2''-O-adenylyltransferase. *Antimicrob Agents Chemother* 22: 824–831
22 Roestamadji J, Grapsas I, Mobashery S (1995) Mechanism-based in activation of bacterial aminoglycoside 3'-phosphotransferases. *J Amer Chem Soc* 117: 80–84
23 Lambert T, Gerbaud G, Galimand M, Courvalin P (1993) Characterization of *Acinetobacter haemolyticus* AAC(2')-Ig gene encoding an aminoglycoside 6''-N-acetyltransferase which modifies amikacin. *Antimicrob Agents Chemother* 37: 2093–2100
24 Jones RN (1995) Isepamicin (SCH21420, 1-N-HAPA gentamicin B): microbiological characteristics including antimicrobial potency and spectrum of activity. *J Chemotherapy* 7 (suppl 2): 7–16
25 Arthur M, Courvalin P (1993) Genetics and mechanism of glycopeptide resistance in enterococci. *Antimicrob Agents Chemother* 37: 1563–1571
26 Wu Z, Walsh CT (1995) Phosphinate analogs of D-, D-dipeptides: slow-binding inhibition and proteolysis protection of VanX, a D-, D-dipeptidase required for vancomycin resistance in *Enterococcus faecium. Proc Natl Acad Sci USA* 92: 11 603–11 607
27 Wu Z, Walsh CT (1996) Dithiol compounds: potent time-dependant inhibitors of VanX, a zinc-dependant D-, D-dipeptidase required for vancomycin resistance in *Enterococcus faecium. J Amer Chem Soc* 118: 1785–1786
28 Spratt BG (1994) Resistance to antibiotics mediated by target alterations. *Science* 264: 388–393
29 Berger-Bächi B (1995) Factors affecting methicillin resistance in *Staphylococcus aureus. Int J Antimicrob Agents* 6: 13–21
30 Fontana R, Adegheri M, Ligozzi M, Lopez H, Sucari A, Satta G (1994) Overproduction of low-affinity penicillin-binding protein and high-level ampicillin resistance in *Staphylococcus aureus. Antimicrob Agents Chemother* 39: 910–916
31 Sumita Y, Nouda H, Kanazawa K, Fukasawa M (1995) Antimicrobial activity of SM-17466, a novel carbapenem antibiotic with potent activity against methicillin-resistant *Staphylococcus aureus. Antimicrob Agents Chemother* 39: 910–916
32 Malanoski G, Collins L, Eliopoulos CT, Moellering RCJr Eliopoulos GM (1995) Comparative *in vitro* activities of L-695,256, a novel carbapenem against gram-positive bacteria. *Antimicrob Agents Chemother* 39: 990–995
33 (a) Allen NE, Hobbs JN Jr, Nicas TI (1996) Inhibition of peptidoglycan biosynthesis in vancomycin-susceptible and –resistant bacteria by a semisynthetic glycopeptide antibiotic. *Antimicrob Agents Chemother* 40: 2356–2362. (b) Allen NE, LeTourneau DL, Hobbs JN Jr (1997) Molecular interactions of semisynthetic glycopeptide antibiotic with D-alanyl-D-alanine and D-alanyl-D-lactate residues. *Antimicrob Agents Chemother* 41: 66–71
34 (a) Nicas TI, Mullen DL, Flokowitsch JE, Preston DA, Snyder NJ, Zweifel MJ, Wilkie SC, Rodriguez MJ, Thompson RC, Cooper RDG (1996) Semisynthetic glycopeptide antibiotics derived from LY264826 active against vancomycin-resistant enterococci. *Antimicrob Agents Chemother* 40: 2194–2199. (b) Rodriguez MJ, Snyder NJ, Zweifel MJ, Wilkie SC, Stack DR, Cooper RDG, Nicas TI, Mullen DL, Butler TF, Thompson RC (1998) Novel glycopeptide antibiotics: N-alkylated derivatives active against vancomycin-resistant enterococci. *J Antibiot* 51: 560–569
35 Goldman RC, Kadam SK (1989) Binding of novel macrolide structures to macrolides-lincosamides-streptogramin B-resistant ribosomes inhibits protein synthesis and bacterial growth. *Antimicrob Agents Chemother* 33: 1058–1066
36 Fernandez PB, Baker WR, Freiberg LA, Hardy DJ, McDonald EJ (1989) New macrolides active against *Streptococcus pyogenes* with inducible or constitutive type of macrolide-lincosamide-streptogramin B resistance. *Antimicrob Agents Chemother* 33: 78–81
37 Griesgraber G, Kramer MJ, Elliott RL, Nilius AM, Ewing AM, Raney PM, Bui M-H, Flamm RK, Chu DTW, Plattner JJ et al (1998) Anhydrolide macrolide. 2. Synthesis and antibacterial activity of 2,3-anhydro-6-O-methyl 11,12-carbazate erythromycin A analogues. *J Med Chem* 41: 1660–1670
38 Pan X-S, Fisher LM (1997) Targeting of DNA gyrase in *Streptococcus pneumoniae* by sparfloxac*In*: selective targeting of gyrase or topoisomerase IV by quinolones. *Antimicrob Agents Chemother* 41: 471–474
39 Breines DM, Ouabdesselam S, Ng EY, Tankovic J, Shah S, Soussy CJ, Hooper DC (1997) Quinolone resistance locus nfxD of *Escherichia coli* is a mutant allele of the parE gene encoding a subunit of topoisomerase IV. *Antimcrob Agents Chemother* 41: 175–179
40 Kitamura A, Hoshino K, Kimura Y, Hayakawa I, Sato K (1995) Contribution of the C-8 substituent of DU-6859a, a new potent fluoroquinolone, to its activity against DNA gyrase mutants of

Pseudomonas aeruginosa. Antimicrob Agents Chemother 39: 1467–1471
41 Chopra I (1985) Mode of action of the tetracyclines and the nature of bacterial resistance to them. *In*: JJ Hlavka, JH Boothe (eds): *Handbook of experimental pharmacology. The tetracycline*. Springer-Verlag, Berlin, 317–390
42 Schnappinger D, Hillen W (1996) Tetracyclines: antibiotic action, uptake, and resistance mechanisms. *Arch Microbiol* 165: 359–369
43 Testa RT, Petersen PJ, Jacobus NV, Sum P-E, Lee VJ, Tally FP (1993) *In vitro* and *in vivo* antibacterial activities of the glycylcyclines, a new class of semisynthetic tetracyclines. *Antimicrob Agents Chemother* 37: 2270–2277
44 Nikaido H (1989) Outer membrane barrier as a mechanism of antimicrobial resistance. *Antimicrob Agents Chemother* 33: 1831–1836
45 Hancock REW, Woodruff WA (1988) Roles of porin and β-lactamase in β-lactam resistance of *Pseudomonas aeruginosa. Rev Infect Dis* 10: 770–775
46 Fung-Tomc JC, Huczko E, Banville J, Ménard M, Kolek B, Gradelski E, Kessler RE, Bonner DP (1995) Structure-activity relationships of carbapenems that determine their dependence on porin protein D2 for activity against *Pseudomonas aeruginosa. Antimicrob Agents Chemother* 39: 394–399
47 (a) Minnick AA, McKee JA, Dolence EK, Miller MJ (1992) Iron transport-mediated antibacterial activity of, development of resistance to hydroxamate, catechol siderophore-carbacephalosporin conjugates. *Antimicrob Agents Chemother* 36: 840–850. (b) Diarra MS, Lavoie MC, Jacques M, Darwish I, Dolence EK, Dolence JA, Ghosh A, Ghosh M, Miller MJ, Malouin F (1996) Species selectivity of new sideophore-drug conjugates that use specific iron uptake for entry into bacteria. *Antimicrob Agents Chemother* 40: 2610–2617
48 Curtis NAC, Eisenstadt RL, East SJ, Conford RJ, Walker LA, White AJ (1988) Iron-regulated outer membrane proteins of *Escherichia coli* K-12 and mechanism of action of catechol substituted cephalosporins. *Antimicrob Agents Chemother* 32: 1879–1886
49 Silley P, Griffiths JW, Monsey D, Harris AM (1990) Mode of action of GR69153, a novel catechol-substituted cephalosporin, and its interaction with the tonB-dependent iron transport system. *Antimicrob Agents Chemother* 34: 1806–1808
50 Tatsumi Y, Maejima T, Mitsuhashi S (1995) Mechanism of tonB-dependent transport of KP-736, a 1,5-dihydroxy-4-pyridone-substituted cephalosporin into *Escherichia coli* K-12 cells. *Antimicrob Agents Chemother* 39: 613–619
51 Anwar H, Strap JL, Costerton JW (1992) Establishment of aging biofilms: possible mechanism of bacterial resistance to antimicrobial therapy. *Antimcrob Agents Chemother* 36: 1347–1351
52 Yassien M, Khardori N, Ahmedy A, Toama M (1995) Modulation of biofilms of *Pseudomonas aeruginosa* by quinolones. *Antimcrob Agents Chemother* 39: 2262–2268
53 Ofek I, Cohen S, Rahmani R, Kabha K, Tamarkin D, Herzig Y, Rubinstein E (1994) Antibacterial synergism of polymyxin B nonapeptide and hydrophobic antibiotics in experimental gram-negative infections in mice. *Antimicrob Agents Chemother* 38: 374–377
54 Nikaido H (1994) Prevention of drug access to bacterial targets: permeability barriers and active efflux. *Science* 264: 382–388
55 Nelson ML, Park BH, Andrews JS, Georgian VA, Thomas RC, Levy SB (1993) Inhibition of the tetracycline efflux antiport protein by 13-thio-substituted 5-hydroxy-6-deoxy tetracyclines. *J Med Chem* 36: 370–377
56 Brötz H, Bierbaum G, Markus A, Molitor E, Sahl HG (1995) Mode of action of the lantibiotic mersacid*In*: inhibition of peptidoglycan biosynthesis via a novel mechanism? *Antimicrob Agents Chemother* 39: 714–719
57 Brötz H, Bierbaum G, Leopold K, Reynolds PE, Sahl HG (1998) The lantibiotic mersacidin inhibit peptidoglycan synthesis by targeting lipid II. *Antimicrob Agents Chemother* 42: 154–160
58 (a) Somner EA, Reynolds PE (1990) Inhibition of peptidoglycan biosynthesis by ramoplanin. *Antimicrob Agents Chemother* 34: 413–419. (b) Jones RN, Barry A (1989) *In vitro* evaluation of ramoplanin (A 16686 or MDL 62198), a new depsipeptide complex for potential topical use. *Diagn Microbiol Infect Dis* 12: 279–282
59 (a) Chen MH, Steiner MG, de Laszlo SE, Patchett AA, Anderson MS, Hyland SA, Onishi HR, Silver LL et al (1999) Carbohydroxamido-oxazolidines: antibacterial agents that target lipid A biosynthesis. *Bioorg Med Chem Lett* 9: 313–318. (b) Onishi HR, Pelak BA, Gerckens LS, Silver LL, Kahn FM, Chen MH, Patchett AA, Galloway SM, Hyland SA, Anderson MS, Raetz CRH (1996) Antibacterial agents that inhibit lipid A biosynthesis. *Science* 274: 980–982

60 Brandish PE, Kimura KI, Inukai M, Southgate R, Lonsdale JT, Bugg TDH (1996) Mode of action of tunicamycin, liposidomycin B, and mureidomycin A: inhibition of phospho-N-acetylmuramyl-pentapeptide translocase from *Escherichia coli*. *Antimicrob Agents Chemother* 40: 1640–1644

61 Inukai M, Isono F, Takatsuki A (1993) Selective inhibition of the bacterial translocase reaction in peptidoglycan synthesis by mureidomycins. *Antimicrob Agents Chemother* 37: 980–983

62 (a) Hancock REW (1997) Peptide antibiotics. *Lancet* 349: 418–422. (b) Larrick JW, Wright SC (1996) Cationic antimicrobial proteins. *Drugs Future* 21: 41–48

63 Zasloff M (1987) Magainins, a class of antimicrobial peptide from *Xenopus* skin: isolation, characterization of two active forms, and partial cDNA sequence of a precursor. *Proc Natl Acad Sci USA* 84: 5449–5453

64 Fuchs PC, Barry AL, Brown SD (1998) *In vitro* antimicrobial activity of MSI-78, a magainin analog. *Antimicrob Agents Chemother* 42: 1213–1216

65 Darveau RP, Cunningham MD, Seaford CL, Cassiano-Clough L, Cosand WL, Blake J, Watkins CS (1991) *Beta*-lactam antibiotics potentiate magainin II antimicrobial activity *in vitro* and *in vivo*. *Antimicrob Agents Chemother* 35: 1153–1159

66 Hultmark D, Steiner H, Rasmuson T, Boman HG (1980) Insect immunity. Purification and properties of three inducible bactericidal proteins from hemolymph of immunized pupae of *Hyalophora cecropia*. *Eur J Biochem* 106: 7–16

67 Moore AJ, Beazley WD, Bibby MC, Devine DA (1996) Antimicrobial activity of cecropins. *J Antimicrob Chemother* 37: 1077–1089

68 Piers KL, Brown MH, Hancock RE (1994) Improvement of outer membrane-permeabilizing and lipopolysaccharide-binding activities of an antimicrobial cationic peptide by C-terminal modification. *Antimicrob Agents Chemother* 38: 2311–2316

69 Fiddes JC, Loury DJ, Ho JF, Cheng FC (1996) Protegrin protects mice from systemic infection by antibiotic resistant pathogens. *36th Intersci Conf Antimicrob Agents Chemother* F244, New Orleans, 142

70 Eustice D, Feldman PA, Zajac I, Slee AM (1988) Mechanism of action of DuP721: inhibition of an early event in protein synthesis. *Antimicrob Agents Chemother* 32: 1218: 1222

71 (a) Mason EO Jr, Lamberth LB, Kaplan SL (1996) *In vitro* activities of oxazolidinones U-100592, U-100766 against penicillin resistant, cephalosporin-resistant strains of *Streptococcus pneumoniae*. *Antimicrob Agents Chemother* 40: 1039–1040. (b) Eliopoulos GM, Wennersten CB, Gold HS, Moellering RC Jr (1996) *In vitro* activities of a new oxazolidinone antimicrobial agent against enterococci. *Antimicrob Agents Chemother* 40: 1745–1747

72 Vannuffel P, Cocito C (1996) Mechanism of action of streptogramin and macrolides. *Drugs* 51 (suppl 1): 20–30

73 (a) Brumfitt W, Hamilton-Miller JMT, Shah S (1992) In vitro activity of RP59500, a new semi-synthetic streptogramin antibiotic, against gram-positive bacteria. *J Antimicrob Chemother* 30 (suppl A): 29–37. (b) Verbist L, Verhaegen J (1992) Comparative in vitro activity of RP59500. *J Antimicrob Chemother* 30 (suppl A) 39–44. (c) Goto S, Miyazaki S, Kaneko Y (1992) The *in vitro* activity of RP59500 against gram-positive cocci. *J Antimicrob Chemother* 30 (suppl A): 25–28

74 (a) Selva E, Beretta G, Montanini N, Saddler GS, Gastaldo L, Gastaldo L, Ferrari P, Lorenzetti R, Landini P, Ripamonti F et al (1991) Antibiotic GE2270 A: a novel inhibitor of bacterial protein synthesis. *J Antibiot* 44: 693–701. (b) Goldstein BP, Berti M, Ripamonti F, Resconi A, Scotti R, Denaro M (1993) *In vitro* antimicrobial activity of a new antibiotic, MLD62,879 (GE2270A). *Antimicrob Agents Chemother* 37: 741–745

75 Stella S, Montanini N, Le Monnier F, Ferrari P, Colombo L, Landini P, Ciciliato I, Goldstein BP, Selva E, Denaro M (1995) Antibiotic GE37468 A: a new inhibitor of bacterial protein synthesis. I. Isolation and characterization. *J Antibiot* 48: 780–786

76 Hughes J, Mellows G (1978) Interaction of pseudomonic acid A with *Escherichia coli* B isoleucyl-tRNA synthetase. *Biochem J* 191: 209–219

77 Hill JM, Yu G, Shue YK, McCarthy P, Wendler PA, Li T (Sept 1996) Synthesis and activity of a new class of antibacterial agents which target isoleucyl t-RNA synthetase. *36th Intersci Conf Antimicrob Agents Chemother* F235, New Orleans, 140

New Approaches to Drug Development
ed. by P. Jollès

Trends in antifungal research

Vorapann Mahaguna[1], Robert O. Williams III[1] and Thomas C. Hardin[2]

[1] *College of Pharmacy, The University of Texas at Austin, Austin, Texas, USA*
[2] *The University of Texas Health Science Center, San Antonio, TX 78284-6620, USA*

Summary. With the increasing use of aggressive immunosuppressive therapies in the management of a variety of patient populations, the continuing presence of the AIDS pandemic and the therapeutic advances employed in critical care settings, an increasing number of serious fungal infections are being encountered by today's practicing clinicians. Traditionally, antifungal drug therapy has been delivered by means of intravenous infusion, oral administration, or topical application. Recently, a number of alternative routes of antifungal drug delivery have been developed and investigated, and the traditional means of antifungal administration have been improved to facilitate the therapeutic use of new and reformulated antifungal agents. Organized based on the route of administration, this chapter reviews these advances in antifungal drug delivery.

Introduction

With the advent of the increased use of aggressive immunosuppressive therapies associated with solid organ and bone marrow transplantation programs, the widespread prevalence of the AIDS pandemic and the advances in technology used in intensive care settings, clinicians are requiring safe and effective treatments for many opportunistic infections. Many of these infections are caused by fungal pathogens, and antifungal drug development has become an important aspect of contemporary medical research. The number of antifungal agents currently available and in clinical trials is unfortunately limited, therefore efforts to enhance our drug delivery capabilities of these agents through innovative technologies have become important areas of antifungal research.

Parenteral delivery

Emulsion delivery system

Amphotericin B (AmB) is the most effective and widely used antifungal agent for treating systemic mycoses caused by opportunistic fungi. The standard intravenous formulation for over 30 years has been AmB deoxycholate (Fungizone®, Bristol Myers Squibb, Princeton, NJ, USA). While usually very effective, AmB deoxycholate therapy is associated with several concerns, including acute infusion-related reactions, nephrotoxicity and occasional clinical failure or relapse of fungal infection [1–3]. In most cases, the clinical use of AmB is limited by dose-related nephrotoxicity, which reportedly occurs in between 15% to 80% of patients [2, 4, 5]. Although this nephrotoxicity is largely reversible, many patients who complete a full course of AmB therapy can have some residual impairment of kidney function. Therefore, there have been many attempts to evaluate alternative formulations of AmB with the primary aim of reducing associated side-effects while maintaining antifungal activity.

AmB is an amphiphilic drug that binds to sterols and intercalates into lipid bilayers [6]. This suggests that AmB would be particularly suitable for use with lipid-based drug delivery systems. Several studies have shown that formulating AmB into a lipid-emulsion carrier does not alter its antifungal activity; however, there is often a significant reduction in observed toxicities [7–10].

Kirsh and colleagues introduced an alternative emulsion-based delivery system using a lipid solution commercially available for parenteral nutrition [8]. The formulation was prepared with AmB diluted in sodium deoxycholate and dimethylacetamide (DMA), filtered, and then added to a vial containing 20% Intralipid® (Kabi-Pharmacia, Stockholm, Sweden). The efficacy and reduced toxicity of this AmB emulsion was demonstrated in an experimental model of murine candidiasis. A similar formulation was reported by other investigators to have less nephrotoxicity and be better tolerated clinically than regular AmB deoxycholate when given to neutropenic or HIV-infected patients with candidiasis [11]. However, an *in vitro* study reported that AmB deoxycholate mixed in the lipid emulsion formulation was not stable [12]. A clear, yellow precipitate and an increase in size distribution of the particles were observed. This formulation is, therefore, not recommended for clinical use since the formation of the larger lipid particles may be associated with pulmonary microembolism.

Egito and coworkers investigated the injectable AmB emulsion-based delivery system containing a mixture of saturated medium-chain fatty acids as the oil phase, egg lecithin as the lipophilic surfactant, and using methanol as a solvent for AmB [13]. The structure of the system was evaluated using spectroscopic methods at various AmB concentrations and the results suggested that this AmB emulsion-based delivery system has potential applications meriting further evaluation. The *in vitro* and *in vivo* toxicity and activity of this system compared with those of AmB deoxycholate were investigated [14]. The AmB emulsion system was less toxic than AmB deoxycholate against human red blood cells and equally active against *Candida albicans*. Additionally, the AmB emulsion system appeared more effective than AmB deoxycholate in the treatment of systemic mycoses at the maximum tolerated doses [14].

Micellar delivery system

AmB has been incorporated into mixed micelles composed of egg lecithin and bile salt (glycolate, deoxycholate, or taurocholate); these formulations were compared with AmB deoxycholate as treatment for murine models of systemic fungal infections [15]. It was found that the micelle formulations were nontoxic to mice at a certain dose and were therapeutically more active than the maximally tolerated dose of AmB deoxycholate in murine infections. However, at an equivalent dose of AmB, AmB-mixed micelles were more active in treating murine candidiasis, whereas AmB deoxycholate was more active in treating murine crytococcosis [15].

Liposomal delivery system

As previously described, AmB would be particularly suitable for use with lipid-based delivery systems since it can bind to sterols and intercalate into lipid-bilayers [6]. Attempts to investigate various preparations of AmB with the aim of reducing side-effects while maintaining its antifungal activity led to the successful incorporation of AmB into liposomes. Since liposomes do not readily concentrate in organs such as the kidney, drugs such as AmB, that are intended to have primary therapeutic activity in other body tissues, produced little or no nephrotoxicity when encapsulated into liposomes [16].

Significant research has confirmed that incorporating AmB into liposomes significantly reduces the systemic toxicity of this agent without a concomitant loss of efficacy against fungal infections [17–28]. Additionally, the therapeutic index can be dramatically improved at least 10-fold [25].

Kretschmar et al. reported *in vivo* results that even when given in the same dosage as conventional AmB deoxycholate or AmB deoxycholate dissolved in a lipid emulsion, liposomal AmB was more effective in the treatment of murine candidiasis [29]. Recently, results of clinical trials with liposomal formulations of AmB have confirmed these findings.

Comprehensive reviews of the liposomal formulation of AmB, including its pharmacokinetics, experimental antifungal activity, toxicity and clinical trials have recently been published [30, 31].

A liposome-incorporated nystatin delivery system for treating disseminated fungal infection in an animal has been developed. Nystatin is incorporated into a stable multilamellar vesicle comprising one or more phospholipids, such as phosphomonoglycerides, phosphatidic acid, and sphingolipid. It is administered by the parenteral route in most instances, but also by the oral or topical routes as indicated [32].

Other lipid-based delivery systems

Studies with AmB prepared with various types of phospholipids, sterols or sterol derivatives, including cholesteryl hemisuccinate and cholesteryl sulfate, have indicated that both the efficacy and toxicity of a drug can be altered by the lipid composition, size and type of the lipid carriers. Stable AmB-cholesteryl sulfate discs have been developed [33]. The cholesteryl sodium sulfate forms a stable colloidal lipid complex with AmB at a 1:1 drug-to-lipid molar ratio. The disc is different from conventional phospholipid liposomes in its ultrastructure and has remarkably reduced toxicity *in vivo* compared to a micellar formulation of AmB deoxycholate. The unique structure and tight association of the drug with the complex seems to prevent the incorporation of AmB into the host tissues and cells. This may be responsible for the reduced AmB distribution to tissues, such as in the kidney, where side-effects commonly occur.

Nanosphere/microsphere delivery systems

Griseofulvin has a very low solubility in water, a slow dissolution rate in the gastrointestinal fluids and has been shown to be incompletely and irregularly absorbed after oral administration of the micronized drug [34–36]. Its physicochemical and pharmacokinetic parameters make it a good candidate for a controlled release dosage form. Vojnovic et al. prepared regular microspheres of 200–260 μm average size from vinylpyrrolidone/vinylacetate (PVP/VA) copolymer using a solvent evaporation method. Griseofulvin was incorporated into these microspheres and its physical characteristics evaluated. PVP/VA copolymers are used primarily in film-coating tablets and capsules because of their solubility in relatively non-toxic solvents, low hygroscopicity and very low chronic oral toxicity. The release rate of the drug incorporated into the microspheres was shown to be biphasic and dependent on the penetration of water into the microspheres, hydration and dissolution of the polymer and finally dissolution of the drug [37].

The stability of lyophilized itraconazole-loaded nanosphere suspensions and the cryopreservation conditions were investigated using differential scanning calorimetry (DSC) and zeta potential measurements [38]. Poly-epsilon-caprolactone nanosphere formulations containing Pluronic F68 as a nonionic surfactant and different cryoprotective agents (i.e., glucose, sucrose, dextran, trehalose and mannitol) were evaluated. Results indicate that the addition of carbohydrates leads to a partial protection of the colloidal suspension, with leakage of 30% of the drug under the best cryopreservation conditions (10% of glucose or sucrose). Zeta potential measurements reveal that the main destabilization mechanism during freeze-drying was surface modifications of the nanospheres, and particularly drug desorption. Replacing Pluronic F68 by the anionic surfactant, sodium deoxycholate, results in a complete stabilization of nanospheres after freeze-drying in the presence of 10% sucrose with no drug desorption, but this could not be reproduced using glucose.

In another study, the association of itraconazole with intravenously compatible colloidal drug carriers (liposomes, cholesterol complexes and nanospheres) was evaluated [39]. The results showed that the different lipid-based drug carriers yielded low association efficiencies (0.6%), whereas itraconazole loading into chemically modified beta-cyclodextrin nanospheres reached 6.8%. The highest association yields were obtained with nanospheres composed of the most hydrophobic polymer tested, poly-epsilon-caprolactone, and a negatively charged surfactant, sodium deoxycholate. The drug was included in the matrix (40%) and adsorbed at the surface of the nanospheres (60%). This may explain the nanosphere instability with time because of continuous drug desorption from the nanospheres, although the mean size of the nanospheres remained unchanged.

Tolnaftate microcapsules and microspheres have been prepared by gelatin-acacia coacervation and by modified emulsion-solvent evaporation methods. The physical state of the drug in these formulations was determined by using scanning electron microscopy (SEM), X-ray diffraction and DSC. Stability determination by high performance liquid chromatography (HPLC) indicated that these formulations are stable for at least 6 months. In addition, the Fourier Transform Infrared Spectroscopy (FTIR) results indicated the absence of any drug interaction with the polymeric matrix materials during preparation of the dosage form [40].

A new stable AmB nanosphere-based delivery system was obtained by a solvent displacement process [41]. A reproducible and monodispersed size distribution of 220 nm was obtained when AmB and poly(epsilon-caprolactone) were dissolved in an appropriate solvent mixture. Zeta potential measurements indicated that amphiphilic AmB could be adsorbed onto the nanospheres. This system may be an alternative way to reduce acute toxicity of AmB by association of the drug with nanospheres. However, the antifungal activity of these systems needs further investigation.

Recently, a low-dose therapeutic system for AmB, NS-718, has been developed [42]. NS-718, which incorporates AmB into the Lipid Nano-Sphere (LNS®, a fine lipid emulsion with a particle size of 25–50 nm), is a lyophilized dosage form containing equal amounts of soybean oil and egg lecithin. NS-718 yields higher concentrations of AmB in the plasma and body tissues than did other dosage forms of AmB, while maintaining the potent activity of AmB against fungal pathogens. Consequently, it has the potential to become a low-dose therapeutic system for AmB, so that most of the severe side-effects of AmB can be minimized by decreasing the total dose of AmB administered.

In addition, a system utilizing thermosetting gels has been developed to deliver various antifungal agents by injection. The thermosetting gel formulation is a liquid at room temperature and becomes a semi-solid or gel at body temperature [43]. Preclinical information on this delivery system is incomplete at this time.

Topical delivery

Studies of the solubilization of antifungal agents (clotrimazole, ciclopirox, olamine, and econazole nitrate) with topical therapeutic activity in pharmaceutically acceptable nonionic surfactant systems (POE(20) sorbitan monooleate) has been reported. Ternary water/nonionic surfactant/oil formulations of microemulsions were used to study drug solubilization in creams or lotions. The rheological properties were investigated to determine the suitability of selected compositions as topical formulations. The results showed that it was possible to solubilize 1% w/w of antifungal agent in suitable topical formulations with a water content greater than 50% w/w [44].

A delivery system utilizing thermosetting gels was developed to deliver antifungal agents for application to the skin. The thermosetting gel formulation is a liquid at room temperature and becomes a semi-solid or gel when administered topically and warmed by the body [45].

A stick delivery system intended for topical application of an antifungal agent has been developed and patented. The system includes a solvent as a polyhydric alcohol, a gelling agent that can be an alkali metal stearate, an alkali metal palmitate or mixtures, and a treatment agent including antifungal agents [46].

Recently, another topical formulation containing sulconazole and natifine hydrochloride as antifungal agents, combined with an acetate penetration-enhancing compound, has been developed. It is intended for use in the treatment of toenail and fingernail onychomycoses [47].

Oral delivery

Itraconazole, a triazole derivative available for oral administration as either capsules or solution, is more effective than other azoles against *Aspergillus* species, thus making it an important therapeutic entity [48, 49]. Additionally, in doses used clinically, itraconazole lacks the adverse effects on testosterone and cortisol synthesis previously observed with ketoconazole. Itraconazole oral liquid solution in cyclodextrin is intended for patients who have difficulty swallowing capsules or who require aggressive dosing considerations [50]. In addition, the optimal bioavailability of the solution is not dependent on an acidic intragastric environment and is best when the solution is administered on an empty stomach. The oral capsule formulation of itraconazole requires an acidic intragastric pH for absorption; the bioavailability may be improved by administration of the dose with a fatty meal.

The absorption and bioavailability of fluconazole tablets after oral administration appears to be more predictable than that of itraconazole and ketoconazole, and the pharmacokinetic profile of fluconazole suggests significant advantages, such as good penetration into cerebrospinal fluid, prolonged half-life, and low protein binding [51]. Moreover, fluconazole is water-soluble and is commercially available for intravenous administration.

Flucytosine (5-fluorocytosine) is a water-soluble antifungal agent (1.5 g/100 ml at 25°C) that is rapidly and completely absorbed after oral administration. It has been used in different types of controlled-release capsules: capsules of melting granules containing 80% w/w flucytosine and 20% w/w hydrogenated castor oil, capsules of hydroxypropyl-methyl cellulose (HPMC) granules containing 50% w/w flucytosine and 50% w/w HPMC 60SH 4000 (Shin-Etsu, Japan); and controlled-release capsules containing 50% w/w flucytosine and 50% w/w hydrogenated castor oil. *In vitro-in vivo* correlation studies of these dosage forms suggest little value over the commercially available product [52].

A new triazole antifungal agent, MFB-1041 [(+)-2-(2, 4-difluorophenyl)-3-methyl-1-(1H-1, 2, 4-triazol-1-yl)-3-(6-(1H-1, 2, 4-triazol-1-yl)pyridazin-3-ylthio)butan-2-ol], showed poor oral absorption and was practically insoluble in water (1.2 μg/ml). A solid dispersion technique using an enteric polymer, such as HPMC phthalate and carboxymethylethylcellulose, and the nonenteric polymer, hydroxypropylmethylcellulose, were used to improve drug absorption and solubility. The oral bioavailabilities of these solid dispersions in beagle dogs were over 6 times greater than that of a suspension system with increasing drug solubility in an alkaline medium. These results suggest that the oral preparation of MFB-1041 has a superior absorption profile and a higher solubility could be obtained through a phase change from a crystalline to an amorphous state, especially for the spray-drying method using enteric polymers [53].

Oral transmucosal delivery

Buccal gels containing miconazole nitrate are currently used for the treatment of topical fungal infections, such as oropharyngeal candidiasis. A disadvantage of these systems is that they must be applied several times a day. The maximal salivary concentration of the drug is found immediately after application, after which the drug is rapidly cleared from the oral cavity [54–56]. A higher and sustained level would be preferred in order to reduce the duration of treatment. A chewing gum formulation containing miconazole for topical treatment of oral candidiasis has been developed by using a solid dispersion technique. It was shown that the addition of lecithin and the application of a miconazole-PEG 6000 solid dispersion increased the release of miconazole from a chewing gum formulation, yielding therapeutically active concentrations of the drug in saliva [57]. While during the chewing period the local drug concentration was elevated, it cleared rapidly when the chewing stopped. A clinical study performed to examine the effect and tolerance of miconazole chewing gum in comparison to the miconazole gel for the treatment of oral candidiasis indicated that the miconazole released from chewing gum was as effective as the miconazole gel. The chewing gum reduced the dosage of miconazole for treatment of oral candidiasis and the patients described the chewing gum as an acceptable delivery system [58].

To increase the buccal residence time of miconazole in cases of oral candidiasis, a bioadhesive buccal tablet with slow-release properties was developed [59]. The main advantage of this system is a reduction in the frequency of administration, possibly improving patient compliance. The bioadhesive tablet formulation containing miconazole nitrate and a mixture of thermally modified starch and 5% polyacrylic acid was compared with a buccal gel (Daktarin Buccal Gel, Janssen) in terms of salivary miconazole concentrations after administration. Although the amount of drug administered via the bioadhesive tablet was six-fold lower than the dose in the gel, the salivary miconazole concentrations were higher and remained above the minimum inhibitory concentration of *Candida albicans* for more than 10 h. The bioadhesive tablet appeared to be a promising drug delivery system for buccal administration of antifungal drugs for local therapy.

The Mucosal Oral Therapeutic System (MOTS) containing 200 000 units of nystatin (ALZA Corporation, Palo Alto, CA) is a controlled-release osmotic system for oral cavity therapy [60]. It is comprised of a two-layer core containing the drug and osmotic excipients and is coated with a water-permeable membrane having an orifice on the drug layer side. In a moist environment, water enters both layers at a controlled rate, expanding the osmotic compartment and forming a drug suspension in the other compartment. As the osmotic compartment expands, it pushes the drug suspension out of the other compartment through the orifice at the same rate that water enters. It is designed to deliver nystatin over several hours. The salivary nystatin concentrations after administration of the osmotic system were compared with commercial nystatin pastilles (Mycostatin, Bristol-Myers Squibb) and it was demonstrated that the MOTS maintained high salivary nystatin concentrations throughout a 2-h dosing interval [60].

A frozen oral medication delivery system containing nystatin for treatment of oral candidiasis has been patented [61]. The medication is stored, either in a liquid suspension or in a dry powder form with a delivery liquid, until dispensed for use by a patient. The medication is mixed as required and frozen in the storage container until hard, when it is administered to the patient in the form of a frozen popsicle. One embodiment of the system includes a flexible plastic container containing the active agent mixed with a delivery liquid. Another has the medication in powder form and a delivery liquid is stored in a chambered container separated by a membrane that is ruptured to mix the container contents prior to freezing. This technique is particularly well suited for delivering nystatin to the very young, the elderly and those unable to take medications by traditional administration methods [61].

Other mucosal adhesive devices for long-acting delivery of pharmaceutical combinations in the oral cavity have been described [62]. These devices consist of either a dosage unit wherein a mucoadhesive polymer layer having intermixed pharmaceuticals is adhered to a protective backing layer, or a tablet comprising a mucosal adhesive polymer mixture having intermixed pharmaceuticals. These yield a gradual and relatively constant release of the antifungal agent over at least a 12-h period [62].

A mouthrinse formulation containing antifungal agents (i.e., nystatin, AmB), hexetidine and sodium bicarbonate has also been developed. The stability and efficacy of nystatin and AmB contained in the mouthrinses have been recently reviewed [63]. Chlorhexidine gluconate, which is known to reduce *Candida albicans* adhesion to oral mucosal cells, was formulated with chitosan for local oral mucosal delivery [64]. Chitosan was shown to prolong the adhesion time on the oral mucosa as well as drug release from the gel. In addition, it inhibited the adhesion of *Candida albicans* cells to human buccal cells, acted non-specifically on fungal and epithelial cells and prevented the development of infection [65]. Chitosan containing chlorhexidine gluconate in a gel and in films was found to be applicable for oral mucosa application as an antifungal preparation. *In vitro* release from chitosan films also showed an increase in the residence time of chlorhexidine gluconate for local therapy in the oral cavity [64].

A comprehensive review of the delivery of both old and new antifungal agents to the oral cavity, including advantages and disadvantages of the delivery systems, is available [59].

Nasal delivery

Intranasal administration of AmB deoxycholate has been investigated for the prevention of invasive aspergillosis. A suspension of AmB deoxycholate (5 mg/ml in sterile distilled water) was instilled intranasallly using an atomizer. This approach was well tolerated, safe and reduced the incidence of invasive aspergillosis [66].

Pulmonary delivery

Aerosolized AmB-liposomes for the treatment of systemic *Candida* infections in mice was reported. Small-particle aerosols containing AmB liposomes increased the survival time of infected mice [67].

For prophylaxis against *Aspergillus* infections for patients receiving immunosuppressive therapy and/or organ transplantation, AmB may be a useful drug when inhaled as an aerosol. The aerosolization of AmB has been investigated using different medical nebulizers under various operating conditions and with different AmB concentrations. The evaluation of dosages of the drug delivered by the various types of nebulizers to different regions of the respiratory tract yielded a semi-empirical deposition model based on experimental aerosol deposition data from over 20 normal adults. Results suggested that AmB can be delivered by nebulization in the treatment of invasive pulmonary aspergillosis [68].

Ocular delivery

An intraocular drug delivery vehicle composed of an oil-in-water submicron emulsion was developed and patented. The system can be used to deliver a variety of compounds, including antifungal drugs. The administered vesicles reduced eye irritation and provided increased local drug concentrations [69].

Miyamoto et al. have evaluated the feasibility of using a biodegradable polymeric scleral implant containing fluconazole as a potential intravitreal-controlled release drug delivery system. Different drug loadings were prepared with biodegradable polymers of poly(DL-lactide-co-glycolide) and the *in-vivo* studies were conducted in rabbits. Results showed that the sclera implant gradually biodegraded and disappeared by 4 months after implantation. No substantial toxic reactions were found in the ocular tissues. This system provided a promising intravitreal drug delivery system to treat fungal endophthalmitis [70].

Vaginal delivery

Vaginal delivery systems described for the treatment of fungal infections contain an active imidazole compound released in a controlled manner. The systems are composed of at least two phases: an external lipoidal phase and an internal nonlipoidal phase. These are claimed to allow reduction of the overall treatment period or quantity of imidazole administered to treat vaginal fungal infections to be reduced by at least 25% [71].

Rectal delivery

Fluconazole suppositories claim to offer a novel route to administer this systemically active agent. Bioequivalence was demonstrated between 200 mg rectal suppositories and 200 mg oral capsules, and between 25 mg rectal suppositories and 25 mg fluconazole dosed in an oral suspension. Even though absorption of the drug after rectal administration of the suppositories was slower than following oral ingestion of the capsules or suspension, good systemic blood concentrations and acceptable local tolerance of the fluconazole suppositories were observed [72].

Miscellaneous

A novel peptide carrier system involving peptidase-activated prodrugs as antimicrobial agents, based on the concept of chemical instability of alpha-substituted glycine analogues, was developed with the explicit intent of actively transporting therapeutically useful compounds into microbial cells. Peptides containing 5-fluorouracil (5-FU) linked to the peptide backbone were selected to test the feasibility of the new delivery system. These peptide conjugates were designed to be substrates for both the microbial peptide permeases and peptidases. After entry into the cells, enzymatic hydrolysis of the peptide generates an unstable alpha-(5-FU)-glycine that spontaneously decomposed to release 5-FU. The 5-FU-peptide conjugates were found to have antifungal activities against *Candida albicans* comparable to free 5-FU [73].

Some antimycotic imidazole derivatives (econazole and miconazole) and beta-cyclodextrin complexes, isolated by freeze-drying, have been patented [74, 75]. These preparations containing beta-cyclodextrin appear superior to the pure drugs with respect to effectiveness, such as vulvovaginal candidosis [76]. The effect of pH on the complex formation between each drug and beta-cyclodextrin and the antimycotic activity of the complexes indicated that the antimycotic effect of the econazole-beta-cyclodextrin complex against a strain of *Candida albicans* was superior to the effect of a physical mixture of the two compounds.

References

1 Gallis HA, Drew RH, Pickard WW (1990) Amphotericin B: 30 years of clinical experience. *Rev Infect Dis* 12: 308–329
2 Clements JSJr Peacock JE Jr (1990) Amphotericin B revisited: reassessment of toxicity. *Amer J Med* 88(Suppl 5N): 22N–7N
3 Krick JA, Remington JS (1976) Opportunistic invasive fungal infections in patients with leukemia and lymphoma. *Clin Lab Haematol* 5: 249–310
4 Pathak A, Pien FD, Carvalho L (1998) Amphotericin B use in a community hospital, with special emphasis on side effects. *Clin Infect Dis* 26: 334–8
5 Perfect JR, Pickard WW, Hunt DL, Palmer B, Schell WA (1991) The use of amphotericin B in nosocomial fungal infection. *Rev Infect Dis* 13: 474–479
6 De Kruijff B, Demel RA (1974) Polyene antibiotic-sterol interactions in membranes of *Acholeplasma laidlawii* cells and lecithin liposomes. III-Molecular structure of the polyene antibiotic-cholesterol complex. *Biochim Biophys Acta* 339: 57–70
7 Kirsh R, Goldstein R, Tarloff J, Parris D, Hook J, Hanna N, Bugelski P, Poste G (1988) An emulsion formulation of amphotericin B improves the therapeutic index when treating systemic murine candidiasis. *J Infect Dis* 158: 1065–70
8 Smith TM, Strozyk WR (1994) Lack of nephrotoxicity with the administration of amphotericin B in a lipid emulsion. *Ann Pharmacother* 28(11): 1307–1308
9 Walsh TJ, Yeldani V, McEvey M, Gonzalez C, Chanock S, Freifeld A (1998) Safety, tolerance, and pharmacokinetics of a small unilamellar liposomal formulation of amphotericin B (AmBisome) in neutropenic patients. *Antimicrob Agents Chemother* 42(9): 2391–8
10 Souza LC, Maranhao RC, Schreier S, Campa A (1993) *In-vitro* and *in-vivo* studies of the decrease of amphotericin B toxicity upon association with a triglyceride-rich emulsion. *J Antimicrob Chemother* 32: 123–132
11 Caillot D, Reny G, Solary E, Casasnovas O, Chavanet P, Bonnott B (1994) A controlled trial of the tolerance of amphotericin B infused in dextrose or in Intralipid in patients with hematological malignancies. *J Antimicrob Chemother* 33: 603–613
12 Ranchere JY, Latour J, Fuhrmann C, Lagallarde C, Loreuil F (1996) Amphotericin B Intralipid formulation: Stability and particle size. *J Antimicrob Chemother* 37(6): 1165–1169
13 Egito ESTD, Fessi H, Appel M, Barratt G, Legrand P, Bolard J, Devissaguet JP (1996) A morphological study of an amphotericin B emulsion-based delivery system. *Int J Pharm* 145: 17–27
14 Egito ESTD, Appel M, Fessi H, Barratt G, Puisieux F, Devissaguet JP (1996) *In vitro* and *in vivo* evaluation of a new amphotericin B emulsion-based delivery system. *J Antimicrob Chemother* 3893: 485–497
15 Brajtburg J, Elberg S, Travis SJ, Kobayashi GS (1994) Treatment of murine candidiasis and crytococcosis with amphotericin B incorporated into egg lecithin-bile salt mixed micelles. *Antimicrob Agents Chemother* 38(2): 294–299

16 Gokhale PC, Kshirsagar NA, Pandya SK (1993) Development and therapeutic application of liposomal amphotericin B. *Current Sci* 65(6): 448–454
17 Graybill JR, Craven PC, Taylor RL, Williams DM, Magee WE (1982) Treatment of murine Crytococcosis with liposomal-associated amphotericin B. *J Infect Dis* 145: 748–752
18 Lopez-Berestein G, Mehta R, Hopfer RL, Mills K, Kasi L, Mehta K, Fainstein V, Luna M, Hersh EM, Juliano R (1983) Treatment and prophylaxis of disseminated infection due to *Candida albicans* in mice with liposome-encapsulated amphotericin B. *J Infect Dis* 147: 939–945
19 Lopez-Berestein G, Hopfer RL, Mehta R, Mehta K, Hersh EM, Juliano RL (1984) Liposome-encapsulated amphotericin B for treatment of disseminated *Candida* in neutropenic mice. *J Infect Dis* 150: 278–283
20 Lopez-Berestein G (1987) Liposomes as carriers of antimicrobial agents. *Antimicrob Agents Chemother* 31: 675–678
21 Mehta R, Lopez-Berestein G, Hopfer R, Mills K, Juliano RL (1984) Liposomal amphotericin B is toxic to fungal cells but not to mammalian cells. *Biochim Biophys Acta* 770: 230–234
22 Wiebe VJ, DeGregorio MW (1988) Liposome-encapsulated amphotericin B: A promising new treatment for disseminated fungal infections. *Rev Infect Dis* 10(6): 1097–1010
23 Janoff AS, Boni LT, Popescu MC, Minchey SR, Cullis PR, Madden TD, Taraschi T, Gruner SM, Shyamsunder E, Tate MW et al (1988) Unusual lipid structures selectively reduce the toxicity of amphotericin B. *Proc Natl Acad Sci USA* 85: 6122–6126
24 Gondal JA, Swartz RP, Ranhman A (1989) Therapeutic evaluation of free and liposome-encapsulated amphotericin B in the treatment of systemic Candidiasis in mice. *Amtimicrob Agents Chemother* 33: 1544–1548
25 Szoka FC (1990) The future of liposomal drug delivery. *Biotech Appl* 12(5): 496–500
26 Juliano RL, Grant CWM, Barber KR, Kalp MA (1987) Mechanism of the selective toxicity of amphotericin B incorporated into liposomes. *Mol Pharmacol* 31: 1–11
27 Szoka Jr FC, Milholland D, Barza M (1987) Effect of lipid composition and liposome size on toxicity and *in vitro* fungicidal activity of liposome-intercalated amphotericin B. *Antimicrob Agents Chemother* 31: 421–429
28 Jullien S, Comtrepois A, Sligh JE, Domart Y, Yeni P, Brajtburg J, Medoff G, Bolard J (1989) Study of the effects of liposomal amphotericin B on *Candida albicans, Cryptococcus neoformans,* and erythrocytes by using small unilamellar vesicles prepared from saturated phospholipids. *Antimicrob Agents Chemother* 33: 345–349
29 Kretschmar M, Nichterlein T, Hannak D, Hof H (1996) Effects of amphotericin B incorporated into liposomes and in lipid suspensions in the treatment of murine candidiasis. *Arzneim Forsch-Drug Res* 46(7): 711–715
30 Wong-Beringer A, Jacobs RA, Guglielmo BJ (1998) Lipid formulations of amphotericin B: clinical efficacy and toxicities. *Clin Infect Dis* 27(3): 603–18
31 Coukell AJ, Brogden RN (1998) Liposomal amphotericin B: therapeutic use in the management of fungal infections and visceral leishmaniasis. *Drugs* 55(4): 585–612
32 Lopez-Berestein G, Mehta R, Hopfer R, Juliano RL (1989) Liposome-incorporated nystatin. *US Patent* 4812312
33 Guo LSS, Fielding RM, Lasic DD, Hamiton RL (1991) Novel antifungal drug delivery: stable amphotericin B-cholesteryl sulfate discs. *Int J Pharm* 75: 45–54
34 Kadir S, Murakami T, Higashi Y, Yata N (1986) Gastrointestinal absorption of griseofulvin from liquid organic acids and esters in rats. *Int J Pharm* 33: 235–242
35 Chiou WL, Riegelman S (1971) Absorption characteristics of solid dispersed and micronized griseofulvin in man. *J Pharm Sci* 60: 1376–1379
36 Canal T, Lovrecich M, Rubessa F (1988) Solid dispersions of griseofulvin. *CHIMICAoggi* 1: 61–64
37 Vojnovic D, Rubessa F, Bogataj M, Mrhar A (1993) Formulation and evaluation of vinylpyrrolidone/vinylacetate copolymer microspheres with griseofulvin. *J Microencapsul* 10 (1): 89–99
38 De Chasteigner S, Cave G, Fessi H, Devissaguet JP, Puisieux F (1996) Freeze-drying of itraconazole-loaded nanosphere suspensions: a feasible study. *Drug Dev Res* 38(2): 116–124
39 De Chasteigner S, Fessi H, Devissaguet JP, Puisieux F (1996) Comparative study of the association of itraconazole with colloidal drug carriers. *Drug Dev Res* 38(2): 125–133
40 Dash AK (1997) Determination of the physical state of drug in microcapsule and microsphere formulations. *J Microencapsul* 14(1): 101–112
41 Espuelas MS, Legrand P, Irache JM, Gamazo C, Orecchioni AM, Devissaguet JP, Ygartua P

(1997) Poly(omega-caprolacton) nanospheres as an alternative way to reduce amphotericin B toxicity. *Int J Pharm* 158(1): 19–27
42 Fukui H, Koike T, Saheki A, Sonoke S, Yoshikawa H (1998) Characteristics of NS-718, a low-dose therapeutic system for amphotericin B with lipid nano-sphere (LNS®). *In*: *Proceedings of the 25th International Symposium on Controlled Release of Bioactive Materials.* Controlled Release Society, Las Vegas, Nevada, 388–389
43 Haslem JL, Higuchi T, Mlodozeniec AR (1984) Drug delivery system utilizing thermosetting gels. *US Patent* 4474752
44 Garcia-Celma MJ, Azemar N, Pes MA, Solans C (1994) Solubilization of antifungal drugs in water-POE920) sorbitan monooleate-oil systems. *Int J Pharm* 105(1): 77–81
45 Haslem JL, Higuchi T, Mlodozeniec AR (1984) Topical drug delivery system utilizing thermosetting gels. *US Patent* 4474753
46 Lapidus H (1996) Stick delivery system for topical application of a treatment agent. *US Patent* 5543148
47 Knowles WR (1997) Topical composition for fungal treatment. *US Patent* 5652256
48 Graybill JR, Ahrens J (1985) Itraconazole treatment of murine aspergillosis. *Sabouraudia* 23: 219–23
49 Jacobson PA, Johnson CE, Walters JR (1995) Stability of itraconazole in an extemporaneously compounded oral liquid. *Amer J Health-Syst Pharm* 52: 189–191
50 Como JA, Dismukes WE (1994) Oral azole drugs as systemic antifungal therapy. *N Engl J Med* 330: 263–72
51 Jacobson PA, Johnson CE (1995) Bioavailability of itraconazole from oral liquids in questions: 2. *Amer J Health-Syst Pharm* 52:1708
52 Bonny JD, Kyowa M (1995) Use of *in vitro* release tests for the prediction of the *in vivo* behavior and the development of flucytosine controlled-release capsules. *J Pharm Sci* 84(5): 619–623
53 Kai T, Akiyama Y, Nomura S, Sato M (1996) Oral absorption improvement of a poorly soluble drug using solid dispersion technique. *Chem Pharm Bull Tokyo* 44(3): 568–571
54 Odds FC (1981) Persistence of miconazole in saliva after a single oral dose. *Clin Res Rev* 1: 231–232
55 Turner A, Warnock DW (1982) Determination of miconazole in human saliva using high-performance liquid chromatography. *J Chromatogr* 227: 229–232
56 Bouckaert S, Schautteet H, Lefebvre RA, Remon JP, van Clooster R (1992) Comparison of salivary miconazole concentrations after administration of a bioadhesive slow-release buccal tablet and an oral gel. *Eur J Clin Pharmacol* 43: 137–140
57 Pederson M, Rassing MR (1990) Miconazole chewing gum as a drug delivery system application of a solid dispersion technique and lecithin. *Drug Dev Ind Pharm* 16: 2015–2030
58 Rindum JL, Holmstrup P, Pedersen M, Rassing MR, Stoltze K (1993) Miconazole chewing gum for treatment of chronic oral candidosis. *Scand J Dent Res* 101(6): 386–390
59 Samaranayake LP, Ferguson MM (1994) Delivery of antifungal agents to the oral cavity. *Adv Drug Deliv Rev* 13(Jan–Feb): 161–179
60 Encarnacion M, Chin I (1994) Salivary nystatin concentrations after administration of an osmotic controlled release tablet and a pastille. *Eur J Clin Pharmacol* 46: 533–535
61 Harvey BM, Knight PM (1995) Frozen oral medication delivery system and method. *US Patent* 5431915
62 Chien YW, Nair M (1996) Mucosal adhesive device for long-acting delivery of pharmaceutical combinations in oral cavity. *US Patent* 5578315
63 Vincent I, Schmit B, Houze S, Benhaiem N, Taburet AM (1995) Stability of antifungal mouthrinses. *J Pharmacie Clinique* 14(2): 106–110
64 Senel S, Ikinci G, Kas S, Yousefi-Rad A, Hincal AA (1998) Chitosan films and hydrogels of chlorhexidine gluconate for oramucosal delivery. *In*: *Proceedings of the 25th International Symposium on Controlled Release of Bioactive Materials.* Controlled Release Society, Las Vegas, Nevada, 790–791
65 Knapczyk J, Macura AB, Pawlik B (1992) Simple tests demonstrating the antimycotic effect of chitosan. *Int J Pharm* 80: 33–38
66 Meunier F (1989) New methods for delivery of antifungal agents. *Rev Infect Dis* 11(Suppl 7): S1605–12
67 Gilbert BE, Wyde PR, Lopez-Berestein G, Wilson SZ (1994) Aerosolized amphotericin B-liposomes for treatment of systemic *Candida* infections in mice. *Antimicrob Agents Chemother* 38(2):

356–359
68 Roth C, Gebhart J, Nuebling GJ, Rothe BV, Reeb IB (1996) Characterization of amphotericin B aerosols for inhalation treatment of pulmonary aspergillosis. *Infection* 24(5): 354–360
69 Aviv H, Friedman D, Bar-Ilan A, Vered M (1996) Submicron emulsions as ocular drug delivery vehicles. *US Patent* 5496811
70 Miyamoto H, Ogura Y, Hashizoe M, Kunou N, Honda Y, Ikada Y (1997) Biodegradable scleral implant for intravitreal controlled release of fluconazole. *Curr Eye Res* 16(9): 930–935
71 Riley TC (1993) Vaginal delivery system. *US Patent* 5266329
72 Pfaff G, Zimmermann T, Lach P, Yeates R, Simon G, Wildfeuer A (1993) Pharmacokinetics and tolerance of fluconazole suppositories in healthy volunteers. *Arzneim Forsch-Drug Res* 43(3): 391–395
73 Kingsbury WD, Boehm JC, Mehta RJ, Grappel SF, Gilvarg C (1984) A novel peptide delivery system involving peptidase activated prodrugs as antimicrobial agents. Synthesis and biological activity of peptidyl derivatives of 5-fluorouracil. *J Med Chem* 27(11): 1447–1451
74 Doorne HV, Bosch EH, Lerk CF (1988) Formation and antimicrobial activity of complexes of beta-cyclodextrin and some antimycotic imidazole derivatives. *Pharm Weekbl Sci Ed* 10: 80–85
75 Doorne HV, Bosch EH, Lerk CF (1988) Interactions between cyclodextrins and some antimycotic imidazole derivatives: Studies on solubility and antimicrobial activity. *In*: O Huber, J Szejtli (eds): *Proceedings of the Fourth International Symposium on Cyclodextrins,* Kluwer, Dordrecht, 285–291
76 Pedersen M, Edelsten M, Nielsen VF, Scarpellini A, Skytte S, Slot C (1993) Formation and antimycotic effect of cyclodextrin inclusion complexes of econazole and miconazole. *Int J Pharm* 90: 247–254

New Approaches to Drug Development
ed. by P. Jollès

Vaccines and infectious disease

Mark A. Fletcher and Pierre Saliou

Aventis Pasteur SA, F-69367 Lyon, France

Summary. The exponential growth in vaccine research over the last decade, in which many infectious diseases now appear to be amenable to prevention through immunization, is built upon three factors: first, a richer understanding of the immune response (in particular, cellular immunity), second, a greater finesse in understanding the molecular biology of pathogenicity, and third, an expanding use of genetic engineering techniques either to create micro-organisms of greatly attenuated virulence that may be used as vaccines, or to sequence, and express, potential vaccine antigens. With respect to vaccines composed of purified antigens, parallel work is underway to develop immuno-modulating agents (adjuvants) that will selectively and safely induce the necessary immune response. Finally, within this plethora of vaccine candidates, vaccinologists are devoting much effort to alternatives to immunization via injection, such as administration of a vaccine through the mucosal route (e.g., oral, intranasal, intravaginal, etc.), through the transcutaneous route, and even by expression of vaccine antigens in edible fruits and vegetables.

A vaccinologist's classification of micro-organisms

A micro-organism's view of the human body

The human body has been studied, and its parts classified and categorized, by many savants over many centuries: clinical practitioners learned to palpate and to regard its surface and contours attentively; surgeons entered the body, uncovering layers, attachments and organs; finally, physiologists discovered the activity of interlacing systems, directing particular body functions. Each micro-organism (i.e., virus, bacterium, fungus, or parasite) that causes disease has a "vision" of the human body, as well, which determines how it infects, the type of immune response it induces and which types of vaccines it "fears" might be most effective. Through an appreciation of these views at the microscopic level, pathogens may be classified in a manner that will allow a better understanding of the types of vaccines that have proven to be effective, as well as to make predictions about the approaches most likely to succeed for new vaccines.

Three predilections of pathogens

Each pathogenic agent that is a target of immunization may be described by its predilection: micro-organisms that prefer body surfaces ("strictly surface"), those that invade the bloodstream ("surface to bloodstream" and "strictly bloodstream") and those that favor immunologically privileged regions, where they are relatively safe from the immune system ("surface to immunologically privileged sites" and "bloodstream to immunologically privileged sites").

The human body as a doughnut

For a micro-organism, the surface of the human body is continuous–rather disjointed from the common sense of an "exterior", such as the skin, and an "interior", such as the bowel. Much as the mathematician likes to prove that a coffee cup and a doughnut are, topologically speaking, the same object, following a micro-organism through its (hypothetical) surface exploration leads to an expanded view of the surface of the body.

A micro-organism beginning outside the body, at the forehead, for example, and traveling into the ear canal, still remains on the surface. The surface has not been breached if this same organism slides down the Eustachian tube into the nasopharynx and then the trachea; the micro-organism remains outside of the body even while exploring the intricate surfaces of the lung. As the pulmonary tract and the gastrointestinal tract are contiguous for the micro-organism, its surface journey may continue along the wall of the intestine (i.e., esophagus, stomach, small intestine, large intestine and rectum) until reaching the anus. At this juncture, a brief climb into the urinary tract (i.e., urethra, bladder and ureters) still does not allow the micro-organism to access the interior of the body. Neither does an excursion along the vaginal wall or the lining of the uterus grant the micro-organism entry.

To enter the body, a micro-organism is obliged to break through the layer of epithelial cells, whether these be at the epidermis of the skin, the mucosa of the intestines (or the mucosa of the nasopharynx, the lungs, or the genito-urinary tract), or the urothelium of the bladder. Histologists have given many different names to this contiguous epithelial layer.

"Superficial infections"

Most micro-organisms found on the skin and mucosa are non-pathogenic. Nevertheless, some of the most important diseases of mankind, such as diarrhoea and respiratory infections, may be caused by micro-organisms that remain on the surface of the body.

Each year, for example, there are about 4 billion episodes of diarrhoeal disease in the world, causing as many as 3 million deaths. Important diarrhoea pathogens operating on an intestinal mucosal level (and for which there are vaccines available) include *Vibrio cholerae* [1] and rotavirus [2] (Tab. 1); other diarrhoeal disease vaccine candidates under development through clinical trials in human volunteers include those against enterotoxigenic *E. coli* [3], *Shigella* spp. [4], *Helicobacter pylori* [5] and Norwalk virus [6] (Tab. 2).

Table 1. Some vaccines for human use (year first available)

	Live attenuated	Killed	Protein	Polysacharide
Strictly surface	– *Vibrio cholera* (1996)† – Rotavirus (1998)	– *Vibrio cholera* (1896)† ** – *Bordetella pertussis* (1914) – Influenza virus (1943)	– *Corynebacterium diphtheriae* (1923) – Influenza virus (1967) – *Bordetella pertussis* (1981)	
Surface to bloodstream	– Smallpox virus (1798)** – *Bacillus anthracis* (1940)† ** – Measles virus (1963) – Mumps virus (1967) – Rubella virus (1969) – *Salmonella typhi* (1980) – Junin virus (1998)†	– *Salmonella typhi* (1896)** – *Bacillus anthracis* (1950)† ** – Measles virus (1963)** – *Coxiella burnetti* (1976)† – Hepatitis A virus (1996)	– *Bacillus anthracis* (1970)	– *Neisseria meningitidis* (1972) – *Streptococcus pneumoniae* (1977) – *Haemophilus influenzae* type b (1985) – *Salmonella typhi* (1993)
Surface to immunologically privileged sites*	– *Mycobacterium tuberculosis* (1927) – Poliovirus (1961) – Adenovirus† ** (1980) – Varicella-zoster virus (1983)	– Poliovirus (1955)	– Hepatitis B virus (1981)	
Strictly bloodstream	– Yellow fever virus† (1936) – *Francisella tularensis* (1956)†	– *Yersinia pestis* (1897)† – *Francisella tularensis* (1956)†	– *Clostridium tetani* (1926)	
Bloodstream to immunologically privileged sites*	– Venezuelan equine encephalomyelitis virus (1974)†	– Rabies virus (1885)† – Rift Valley fever virus (1962)† – Eastern equine encephalomyelitis virus (1970)† – Western equine encephalomyelitis virus (1971)† – Japanese encephalitis virus (1974)† – Tick-borne encephalitis (1976)†	– *Borrelia burgdorferi* (1998)†	

* "Immunological privilege" includes latency and sanctuary sites (i.e., brain, epididymis, kidney, neurons, gametes and red blood cells).
† Vaccines for special risk groups or selected geographical areas; ** vaccine no longer used.

Table 2. Some human vaccines under clinical trial

	Live	Live-vectors	Killed	Protein	Polysacharide	DNA
Strictly surface	– ETEC	– ETEC	– ETEC	– ETEC		
				– *H. pylori*		
		– HPV		– HPV		
	– Flu virus†		– Flu virus†	– Flu virus†		
			– *M. pneumoniae*			
	– PIV					
				– Norwalk virus		
	– RSV			– RSV		
	– *Shigella* spp.				– *Shigella* spp.	
			– *V. cholera*†			
Surface to bloodstream					– *Streptococcus* B	
	– Hantaan virus	– Hantaan virus				
	– *L. major*		– *L. major*			
			– *P. aeruginosa*	– *P. aeruginosa*		
					– *S. aureus*	
				– *S. pneumoniae*†		
Surface to immunologically privileged sites*			– *C. immitis*			
	– CMV	– CMV		– CMV		
		– EBV		– EBV		
	– HSV			– HSV		
		– HIV	– HIV	– HIV		– HIV
	– *M. leprae*		– *M. leprae*			
	– *M. tuberculosis*†					
				– *N. meningitidis* B		

Table 2. (continued)

	Live	Live-vectors	Killed	Protein	Polysacharide	DNA
Strictly bloodstream	– Chikungunya virus – RVF virus		– RVF virus			
Bloodstream to immunologically privileged sites*	– Dengue fever virus – JEV†	– JEV† – *P. falciparum*		– *P. falciparum*		– *P. falciparum*

* "Immunological privilege" includes latency and sanctuary sites (i.e., brain, epididymis, kidney, neurons, gametes and red blood cells).
† Improvement of a vaccine that is already available for human use
(Abbreviations: CMV, cytomegalovirus; EBV, Epstein-Barr virus; ETEC, Enterotoxigenic *E. coli*; HSV, *Herpes simplex* virus; HIV, Human immunodeficiency virus; HPV, Human papillomavirus; JEV, Japanese encephalitis virus; PIV, Parainfluenza virus; RVF virus, Rift Valley fever virus; RSV, Respiratory syncytial virus)

An equally important cause of human suffering is respiratory infections [7]. Overall, it has been estimated that there are about 400 million episodes of acute lower respiratory tract infections yearly, and over 4 million subsequent deaths. One potentially devastating upper respiratory tract infection in children, whooping cough, which is caused by the nasopharyngeal mucosal pathogen *Bordetella pertussis*, accounts for 40 million cases per year and 350 000 deaths yearly, in part because of inadequate application of the vaccines available. As well as a vaccine for *Bordetella pertussis*, vaccines exist against the influenza virus [8], *Corynebacterium diphtheriae* [9] and *Coxiella burnetii* [10] (Tab. 1). Vaccines that are under development against respiratory infections include ones against parainfluenza virus [11], respiratory syncytial virus [12] and *Mycoplasma pneumoniae* [13] (Tab. 2).

As for inducing immunity to protect the mucosal surfaces of the genital tract, clinical trials have begun for a vaccine against human papillomavirus, a cause of cancer of the cervix [14] (Tab. 2).

A bit of immunology

When a micro-organism attempts to invade the surface or the bloodstream, it finds the way blocked by the innate immune system and by the adaptive immune system (humoral and cellular immunity). Components of the innate immune system include pyrogenic cytokines, neutrophils, natural killer cells, monocytes and macrophages, and the complement cascade. As for the rather complex adaptive immune system, certain generalizations may be made. First, antibodies are the defensive agent of the humoral immune response: sIgA for the mucosal sites and IgG, IgM and IgE for the bloodstream [15]. One type of cellular immunity, via a class of T cells identified as CD4+ Th2, assists in the humoral immune response [16, 17]. Second, a different type of cellular immunity is directed at clearing cells that have been invaded by a pathogenic micro-organism; these CD8+ Tc1 cells often function as cytotoxic T-lymphocytes [CTL] that recognize and lyse infected target cells [18] and they are helped in this function by CD4+ Th1 cells. Third, a set of cytotoxic T cells are present at the mucosa, as well; these CD8+ intraepithelial lymphocytes destroy infected cells of the mucosal layer [19]. The protective immune responses predominant for each infectious pathway (i.e., surface, bloodstream, or immunologically privileged) are provided in Table 3.

The heart of the matter

Generally speaking, pathogens exploit one of two methods to enter the body: the micro-organism might use its own virulence factors to invade directly through the mucosa, or the micro-organism might hitch-hike a ride with, for example, a biting insect or a contaminated hypodermic needle. In either case,

Table 3. A classification of pathogenic micro-organisms that cause vaccine-preventable diseases

	Region(s) of the body infected by the micro-organism			
Region of the body targeted by the micro-organism	Surface	Bloodstream	Immunologically privileged sites	Some protective immune response(s)
Surface				
Strictly surface				sIgA, intraepithelial lymphocytes (CD8+)
Surface to bloodstream				Inate, IgG / IgM / IgE, T helper cells (CD4+ Th2)
Surface to immunologically privileged site(s)				Cytotoxic T lymphocytes (CD8+), T helper cells (CD4+ Th1)
Bloodstream				
Strictly bloodstream				Inate, IgG / IgM / IgE, T helper cells (CD4+ Th2)
Bloodstream to immunologically privileged site(s)				Cytotoxic T lymphocytes (CD8+), T helper cells (CD4+ Th1)

the micro-organism enters the bloodstream, finding a route via the "bloodstream" (the cardio-vascular system or the lymphatic system), to destinations like the musculo-skeletal system or the major organs of the abdominal cavity.

Effective vaccines have been developed for pathogens that invade the mucosal surfaces to ultimately infect the bloodstream and cause systemic disease. These include smallpox virus [20], *Streptococcus pneumoniae* [21], measles virus [22], rubella virus [23], *Haemophilus influenzae* type b [24], *Salmonella typhi* [25, 26], hepatitis A virus [27] and Junin virus [28] (Tab. 1). Vaccines targeted against *Leishmania major* [29], Hantaan virus [30, 31] and *Pseudomonas aeruginosa* [32, 33] and *Staphylococcus aureus* [34] are undergoing clinical trials (Tab. 2).

Pathogens that exploit ways to bypass the cutaneous barrier of the skin also have been subjects of successful vaccine development, examples of which include *Yersinia pestis* [35], *Clostridium tetani* [36], yellow fever virus [37] and *Francisella tularensis* [38, 39] (Tab. 1). Work that continues on vaccines to protect against diseases caused by biting insects focuses on Chikungunya virus [40] and Rift Valley fever virus [41] (Tab. 2).

Safe from harm's way

In the course of a pathogen's invasion of the bloodstream, some micro-organisms are able to surmount the immune response and may cause a persistent infection through establishment within "sanctuary sites" that are inaccessible to immune surveillance. Examples of such regions include the brain, epididymis and kidney; particular cells that escape immune surveillance include the neurons, gametes and red blood cells. A pathogen may circumvent the immune system as well, through tactics such as becoming latent (and therefore hidden from immune surveillance), interfering with the cellular immune response, or destroying components of the immune system.

Underlining the importance of persistent infectious diseases are the examples of *Mycobacterium tuberculosis* (which remains dormant within pulmonary macrophages) [42] and hepatitis B virus (which replicates within hepatocytes) [43]. It has been estimated that nearly 2 billion people now carry the tuberculosis bacilli, with 9 million new cases of tuberculosis diagnosed each year [44]. The hepatitis B virus, as well, is burdensome to public health; there are believed to be about 350 million chronic carriers of this virus throughout the world, who are at risk of developing hepatocellular carcinoma [45].

"Sanctuary site" infections form a large category of vaccine-preventable diseases: some examples are rabies virus [46], *Mycobacterium tuberculosis* [47], poliovirus [48], Japanese encephalitis virus [49], tick-borne encephalitis [50], adenovirus [51], hepatitis B virus [52], varicella-zoster virus [53] and *Borrelia burgdorferi* [54] (Tab. 1).

Possible "sanctuary site" infection vaccines targeted for clinical trial development are equally vast, including cytomegalovirus [55], Herpes simplex virus

[56, 57], *Mycobacterium leprae* [58], Epstein-Barr virus [59], human immunodeficiency virus [60], *Coccidioides immitis* [61], *Neisseria meningitis* [Group B] [62], Dengue fever virus [63] and *Plasmodium falciparum* [64, 65] (Tab. 2).

Vaccinology's traditional armamentarium

Vaccines against infectious disease have been developed over the past century using variations on two crucial techniques, those that create "live" vaccines and those that create "inactivated" vaccines.

Methods of attenuation

A live vaccine may be defined as an attenuated micro-organism that replicates in the vaccinee to a varying degree, but usually without causing a systemic infection. Occasionally, a naturally attenuated variant has been uncovered that was pathogenic for species other than man; such was the case for the cowpox vaccine used by Edward Jenner to inoculate against smallpox (Tab. 1). More often, however, laboratory methods must be applied to attenuate sufficiently the virulence of a live organism to become an appropriate vaccine candidate. Traditional laboratory techniques include (a) attenuation in animals (b) attenuation by adaptation in cell culture and (c) attenuation by cold-adaptation or by development of a temperature-sensitive mutant.

Benefits and risks of live vaccines

Live attenuated vaccines have the advantages of being able to stimulate antibodies at mucosal surfaces (e.g., oral poliovirus vaccine) and in the bloodstream (e.g., live measles virus vaccine), while producing cellular immunity (e.g., bacillus Calmette-Guérin, BCG, to prevent tuberculosis). In addition, the induced immunity persists.

On the other hand, live attenuated vaccines may revert to virulence, they may invade immunocompromised vaccinees and they may spread to others who have contact with the vaccinees.

Methods of inactivation

It is not always possible to coax a living organism to give up its virulence so that it may serve as a vaccine, which led to the efforts to create inactivated vaccines that began at the turn of the century with plague, typhoid fever and cholera vaccines (Tab. 1). An inactivated vaccine may be an entire organism killed by heat or chemicals, or may involve chemical inactivation or directed mutagenesis of parts of organisms or of single proteins, including toxins. Capsular polysaccharides also have proven to be effective vaccines, in particular when the polysaccharide is chemically conjugated to a carrier protein [66]. Examples of vaccines made with each of these techniques are provided (Tab. 4).

Table 4. Techniques for the production of inactivated vaccines

Type of vaccine	Technique	Examples of licensed vaccines
Killed whole-cell organisms		Poliovirus, *Bordetella pertussis*, Japanese encephalitis virus
Protein-based	Inactivated toxin(s) or excreted antigen(s)	*Corynebacterium diphtheriae*, *Clostridium tetani*
	Native subunit protein	Influenza virus, *Bordetella pertussis*
	Soluble protein(s) produced *in vitro*	Hepatitis B virus, *Borrelia burgdorferi*
Polysaccharide-based	Capsular polysaccharide(s)	*Neisseria meningiditis*, *Salmonella typhi*, *Streptococcus pneumoniae*
	Protein conjugated polysaccharide(s)	*Haemophilus influenzae* type b

Advantages and drawbacks of inactivated vaccines

An inactivated vaccine is capable of stimulating, in a dose-dependent manner, serum antibodies. Furthermore, booster injections may be used to overcome poor immunity or the limited persistence of immunity.

Nonetheless, production of inactivated vaccines at an industrial level may be a complex affair, requiring steps to purify and concentrate the protective antigen, as well as to eliminate any toxins. In addition, inactivated vaccines often require adjuvants to boost their immunogenicity. Finally, inactivated vaccines are often not successful at inducing mucosal or cellular immunity.

New tricks for the vaccinologist

Vaccine development efforts throughout the world have accelerated, driven by the need to improve currently used vaccines, the possibility of exploiting novel vaccine technologies and the uncovering of new vaccine targets.

Sprucing up the old car

Work continues to improve some of the most widely-used vaccines, particularly with the goals of improved immunogenicity, diminished reactogenicity and enhanced ease of immunization. A decade of extensive clinical trials have been completed, for instance, which will allow eventual replacement of the killed whole-cell *Bordetella pertussis* vaccine by less reactogenic, protein-based vaccines that make use of a pertussis toxoid and one or more of the surface antigens [67]. As elaborated in Table 2, other contemporary vaccines undergoing improvement include those against *Mycobacterium tuberculosis* (to enhance protective efficacy) [42], *Neisseria meningitidis* (to allow effective childhood vaccination) [68], *Vibrio cholerae* (for oral administration) [69], influenza virus (for nasal administration) [70] and Japanese encephalitis virus (to decrease reactogenicity) [71].

New technologies are everywhere!

Increasingly, genetic engineering techniques are used to diminish the virulence of a micro-organism that has been chosen for development of a potential attenuated vaccine; methods include (a) reassortment of genes to create hybrids (b) creation of mutants by addition or deletion of DNA [for DNA viruses] (c) creation of mutants through "reverse genetics" [for RNA viruses] and (d) development of replication-defective mutants. In addition, genes that express vaccine antigens are inserted into an organism of another species that then serves as the vector.

Novel vaccine technologies for the creation of inactivated vaccines include (a) subunit antigens produced in recombinant organisms (b) "pseudo-particles" produced *in vitro* that have the structure of a virus but lack replicative capacity (c) synthetic peptides linking B-cell epitopes and T-cell epitopes (to induce both specific humoral and cellular immune responses) and (d) polysaccharides that are chemically conjugated to proteins. The use of "naked DNA" vaccines (in fact, DNA coding for protein antigens inserted into a plasmid) has attracted enthusiasm recently, and clinical trials in humans for DNA vaccines are in the early stages for human immunodeficiency virus [72] and *Plasmodium falciparum* [73, 74].

(New) infectious diseases and new vaccines

A very busy generation

As detailed in Table 2, a remarkable number of vaccines under clinical development are directed against "new" infectious diseases, whether they be age-old scourges of mankind against which no licensed vaccine yet exists (such as *Plasmodium falciparum* or one of the *Shigella* spp.), a pest that is "reemerging" (such as *Mycobacterium tuberculosis*) or an "emerging" infectious disease, as is exemplified by the human immunodeficiency virus. With respect to "emerging" infectious diseases, at least a dozen new pathogens have been recognized in the past 30 years, with the rate of discovery accelerating (Tab. 5).

Table 5. Some "emerging" infectious disease micro-organisms

Infectious agent	Disease	First identified
LaCrosse virus	Encephalitis	1960
Rotavirus	Dehydrating infantile diarrhoea	1973
Helicobacter pylori	Peptic ulcer disease	1984
HHV-6	"Sixth disease"	1986
Chlamydia pneumoniae	Atypical pneumonia	1986
Hepatitis C virus	Non-A, non-B hepatitis	1989
Vancomycin-resistant enterococci	Nosocomial infections	1990
Tropheryma whippelii	Whipple's disease	1991
Bartonella henselae	Cat scratch disease	1992
Sin nombre virus	Hantavirus pulmonary syndrome	1993
HHV-8	Kaposi's sarcoma	1994
TT virus	Post-transfusion hepatitis	1996

The pace quickens since the Golden Age

The history of vaccines may be represented graphically (Fig. 1). Years before the present ("years ago") are displayed on the abscissa–log scale–against the stage of vaccine development ("disease", "germ", "vaccine", "vaccination" and "disease control") on the ordinate–linear scale. For diseases of Antiquity until the Golden Age of Bacteriology (in the late Nineteenth Century), the time between identification of a disease ("disease") and isolation of its infectious aetiology ("germ") could be measured in centuries; roughly 50 years were needed to create a vaccine ("vaccine") once the germ was located, and another 25 years were required to produce and then distribute the vaccine on a wide

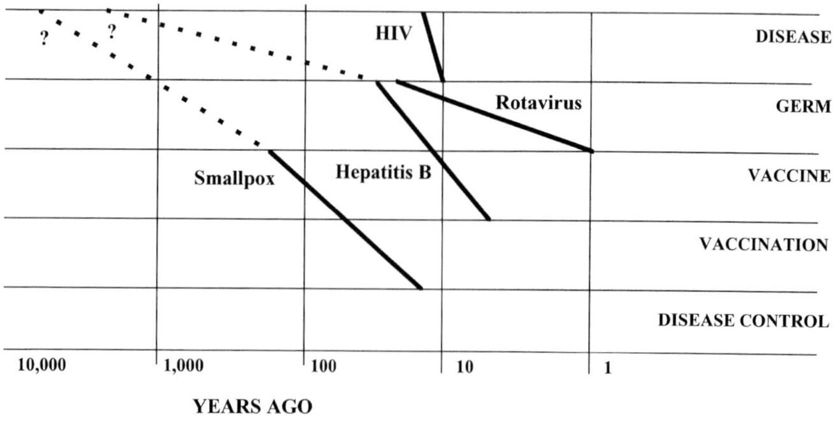

Figure 1.

scale ("vaccination"). In one case, so far, successful application of this progression has led to the elimination of an infectious disease, smallpox ("disease control"), which was certified in 1980, and soon promises to lead to the eradication of a second, poliomyelitis.

For vaccines focused on "new" infectious diseases, as exemplified by hepatitis B virus, rotavirus, human immunodeficiency virus and others (see Tab. 2), steps from disease to germ to vaccine are shorter, and may be measured in decades or even years.

Setting priorities

Vaccinologists find themselves with an over-abundance of possibilities, leading to the "problems" of best choosing appropriate vaccine targets and of creating imaginative ways of distributing vaccines to those who most need them [75, 76]. A proposal for selection of appropriate vaccine targets is provided: as indicated, the complex task of establishing priorities could be based upon the characteristics of the disease, the experience that infectious disease experts have gathered from other vaccines and the potential suitability of vaccination within established healthcare systems (Fig. 2). In addition to these medical issues, scientific criteria, intellectual property rights, formulation processes and delivery systems, production steps and marketing aspects each play a role in the creation of new vaccines.

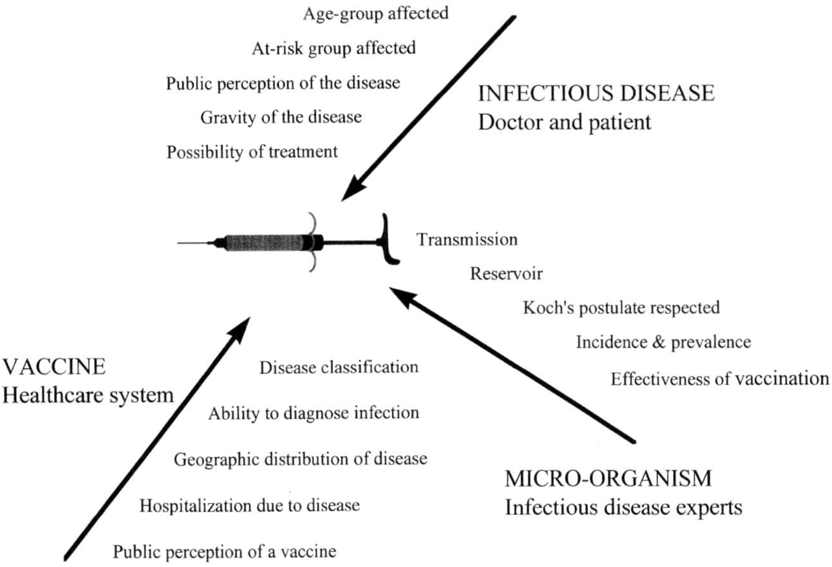

Figure 2.

As a reflection of these different demands, it is important to note that only 2 of the 12 possible new vaccine targets presented in Table 5 have been pushed through research and development to the stage of clinical trials. Furthermore, the recent licensing of a rotavirus vaccine required 25 years subsequent to the virus's discovery, much of this a reflection of the intense labor needed to go from "germ" to "vaccine". The next potential success might be a *Helicobacter pylori* vaccine [5], which has begun initial clinical trials (i.e., Phase I) with one candidate [77], but could require at least another 10 years of clinical development before licensing of a final vaccine.

Other trends to watch for

A sampling of major changes in vaccinology includes

- replacement of whole-cell *Bordetella pertussis* vaccine by acellular (protein sub-unit) vaccines and extension of *Bordetella pertussis* vaccination to adults [78, 79];
- presentation of paediatric vaccines containing multiple vaccines in a single injection (such as a combination of vaccines against *Corynebacterium diphtheriae* – *Clostridium tetani* – *Bordetella pertussis* – poliovirus – *Haemophilus influenzae* type b – hepatitis B virus in one vaccine), which could improve compliance and broaden coverage [80];
- microencapsulation of antigens for parenteral or oral immunization, to limit the need for booster injections [81];
- administration via non-parenteral, mucosal routes (e.g., oral, intranasal, intravaginal, or intrarectal) to reach regional immune responses or a common mucosal immune response [82, 83];
- creation of edible vaccines expressed by genetically modified plants, or by plant viruses that infect the plants [84];
- augmentation of the immunogenicity of recombinant proteins and peptides with new adjuvants [85] and
- induction of cellular immune responses through targeting to antigen-presenting cells [86–88].

The big picture

Multiple lessons can be drawn from the century of vaccinology summarized in Table 1, "Some vaccines for human use", as well as in Figure 1.

First, a greater number of vaccines have been developed against pathogens that begin at the surface of the body (particularly respiratory tract infections) as against pathogens that are introduced directly into the bloodstream (via biting insect vectors).

Second, for infections that begin at the surface of the body, live vaccines have proven to be most effective.

Third, immunization with inactivated vaccines (i.e., killed, protein, or polysaccharide) confers effective protection against micro-organisms that have a bloodstream phase in their disease pathogenesis (even if the infection begins at the surface or concludes in an immunologically privileged site).

Fourth, for the group of "surface to bloodstream" bacterial pathogens that are encapsulated, which includes *Streptococcus pneumoniae*, *Haemophilus influenzae* type b and *Salmonella typhi*, polysaccharide vaccines have been remarkably successful.

Future trends

Following these hard-earned lessons, future trends may be surmised from the work in progress, Table 2, "Some human vaccines in clinical trials".

Surface pathogens will continue to predominate as targets for candidate vaccines. Currently, pathogens at the surface of the body represent most of the vaccines in the final phases of clinical development. In contrast, there are a limited number of vaccine candidates for micro-organisms that enter directly through the bloodstream (e.g., *Plasmodium falciparum*, Dengue fever virus, Rift Valley fever virus, Chikungunya virus and Japanese encephalitis virus).

Inactivated vaccines plus mucosal adjuvants may be developed for surface pathogens. While live vaccines continue to be the preferred choice against surface pathogens, there is a tendency for the parallel development of inactivated vaccines against the same targets. Often the inactivated vaccine candidate is given by the oral or intranasal route in the presence of an adjuvant that enhances mucosal immune responses, in particular, production of sIgA and activation of intraepithelial lymphocytes (CD8+).

Polysaccharide-protein conjugates will continue to be used for "surface to bloodstream" pathogens. To protect against "surface to bloodstream" pathogens, such as *Pseudomonas aeruginosa*, *Staphylococcus aureus*, or *Streptococcus pneumoniae*, the auspicious combination of a polysaccharide antigen conjugated to a protein carrier continues to be exploited, and the induction of specific functional serum antibodies is accepted as a surrogate for a protective immune response.

Inactivated vaccines plus cellular immune response adjuvants may be used for "privileged site" infections. Surmounting the problem of persistent infections, or of protecting against pathogens that remain within immunologically privileged sites, both the live vaccine and the inactivated vaccine approaches continue to be tested. Live vaccine candidates against cytomegalovirus, Herpes simplex virus, or *Mycobacterium leprae*, for example, have a counter-part each in a protein-based vaccine candidate, which may be given with a adjuvant that promotes a cytotoxic T lymphocyte (CD8+) response.

DNA vaccines may be used for "privileged site" infections. Of particular note, the only DNA vaccines to have been tested in man so far are against micro-organisms that favor immunologically privileged sites, i.e., *Plasmodium falciparum* (red blood cells) and human immunodeficiency virus (T lymphocytes). It appears that DNA vaccines could be most suited for induction of a cellular immune response when a live attenuated vaccine approach is impractical.

Therapeutic vaccination will be used for persistent infections. Finally, some exciting studies into therapeutic vaccination are underway, where–so far–inactivated vaccines have been used to treat a chronic infection, the best examples being hepatitis B virus [89] and *Mycobacterium leprae* [90]. Multiple injections are given over a long enough period to break down the state of immunological privilege from which the pathogen benefited.

Parting thoughts

A few months before he died, an ailing Louis Pasteur wrote a letter from Marnes-la-Coquette to the *Président du Conseil Municipal de Paris*. In declining an invitation to attend a ceremony in his honor, Pasteur took the occasion to express his wishes for the future for

> *l'oeuvre que j'ai rêvée: laboratoires de recherches et laboratoires d'enseignement, travaux de longue patience pour arriver soit aux découvertes, soit aux développements de certaines méthodes, et leçons professées pour faire connaître une science nouvelle* [91].[1]

In the century since his death, immunization on a public health scale has eradicated one disease, smallpox, and may soon eliminate a second dreaded disease, polio [92]. Over the past two decades, children throughout the world have been vaccinated against many of the formerly common diseases of childhood, such as whooping cough, diphtheria and tuberculosis, through the tireless efforts of international organizations [76, 93]. At the same time, a remarkable number of new vaccines are becoming available to prevent (and even treat) infectious diseases in all age groups [94]. Each of these successes owes a debt of gratitude to the *découvertes* made and the *méthodes* developed by and subsequent to Pasteur.

[1] ...the work of which I have dreamed: laboratories for research and laboratories for training, painstaking work either to achieve discoveries or to achieve the development of particular methods, and lectures to promulgate a new science (authors' translation).

References

1 Taylor DN, Sanchez JL, Castro JM, Lebron C, Parrado CM, Johnson DE, Tacket CO, Losonsky GA, Wasserman SS, Levine MM et al (1999) Expanded safety and immunogenicity of a bivalent, oral, attenuated cholera vaccine, CVD 103-HgR plus CVD 111, in United States military personnel stationed in Panama. *Infect Immunity* 67: 2030–2034

2 Jacobson RM (1999) The current status of the rotavirus vaccine. *Vaccine* 17: 1690–1699

3 Savarino SJ, Hall ER, Bassily S, Brown FM, Youssef F, Wierzba TF, Peruski L, El-Masry NA, Safwat M, Rao M et al (1999) Oral, inactivated, whole cell enterotoxigenic *Escherichia coli* plus cholera toxin b subunit vaccine: results of the initial evaluation in children. *J Infect Dis* 179: 107–114

4 Sansonetti PJ (1998) Pathogenesis of shigellosis: from molecular and cellular biology of epithelial cell invasion to tissue inflammation and vaccine development. *Jpn J Med Sci Biol* 51: S69–S80

5 Michetti P (1998) Oral immunization against *Helicobacter pylori* – a future concept. *J Gastroenterol* 33: 66–68

6 Ball JM, Estes MK, Hardy ME, Conner ME, Opekun AR, Graham DY (1996) Recombinant Norwalk virus-like particles as an oral vaccine. *Arch Virol* 12: 243–249

7 Monto AS, Lehmann D (1998) Acute respiratory infections (ARI) in children: prospects for prevention. *Vaccine* 16: 1582–1588

8 Webster RG (1998) Influenza: an emerging disease. *Emerg Infect Dis* 4: 436–441

9 Vitek CR, Brennan MB, Gotway CA, Bragina VY, Govorukina NV, Kravtsova ON, Rhodes PH, Bisgard KM, Strebel PM (1999) Risk of diphtheria among schoolchildren in the Russian Federation in relation to time since last vaccination. *Lancet* 353: 355–358

10 Ackland JR, Worswick DA, Marmion BP (1994) Vaccine prophylaxis of Q fever – A follow-up study of the efficacy of Q-Vax (CSL) 1985–1990. *Med J Australia* 160: 704–708

11 Ewasyshyn M, Cates G, Jackson G, Scollard N, Symington A, Klein M (1997) Prospects for a parainfluenza virus vaccine. *Pediat Pulmonol* 16: 280–281

12 Dudas RA, Karron RA (1998) Respiratory syncytial virus vaccines. *Clin Microbiol Rev* 11: 430–439

13 Cimolai N, Cheong AC, Morrison BJ, Taylor GP (1996) *Mycoplasma pneumoniae* reinfection and vaccination: protective oral vaccination and harmful immunoreactivity after re-infection and parenteral immunization *Vaccine* 14: 1479–1483

14 Rowen D, Lacey C (1998) Toward a human papillomavirus vaccine. *Dermatol Clin* 16: 835–838

15 Slifka MK, Ahmed R (1998) Long-lived plasma cells: a mechanism for maintaining persistent antibody production. *Curr Opin Immunol* 10: 252–258

16 Alexander J, Fikes J, Hoffman S, Franke E, Sacci J, Appella E, Chisari FV, Guidotti LG, Chesnut RW, Livingston B et al (1998) The optimization of helper T lymphocyte (HTL) function in vaccine development. *Immunol Res* 18: 79–92

17 Maloy KJ, Burkhart C, Freer G, Rulicke T, Pircher H, Kono DH, Theofilopoulos AN, Ludewig B, Hoffmann-Rohrer U, Zinkernagel RM et al (1999) Qualitative and quantitative requirements for CD4(+) T cell- mediated antiviral protection. *J Immunol* 162: 2867–2874

18 Harty JT, Bevan MJ (1999) Responses of CD8(+) T cells to intracellular bacteria. *Curr Opin Immunol* 11: 89–93

19 De Panfilis G (1998) CD8+ cytolytic T lymphocytes and the skin. *Exp Dermatol* 7: 121–131

20 Dunn PM (1996) Dr Edward Jenner (1749–1823) of Berkeley, and vaccination against smallpox. *Arch Dis Child* 74: F77–F78

21 Poland GA (1999) The burden of pneumococcal disease: the role of conjugate vaccines. *Vaccine* 17: 1674–1679

22 Wild TF (1999) Measles vaccines, new developments and immunization strategies. *Vaccine* 17: 1726–1729

23 Rosa C (1998) Rubella and rubeola. *Semin Perinatol* 22: 318–322

24 Bisgard KM, Kao A, Leake J, Strebel PM, Perkins BA, Wharton M (1998) *Haemophilus influenzae* invasive disease in the United States, 1994–1995: near disappearance of a vaccine-preventable childhood disease. *Emerg. Infect. Dis.* 4: 229–237

25 Zenilman JM (1997) Typhoid fever. *JAMA* 278: 847–850

26 Engels EA, Falagas ME, Lau J, Bennish ML (1998) Typhoid fever vaccines: a meta-analysis of studies on efficacy and toxicity. *BMJ* 316: 110–116

27 Levy MJ, Herrera JL, DiPalma JA (1998) Immune globulin and vaccine therapy to prevent hepatitis A infection *Amer J Med* 105: 416–423
28 Maiztegui JI, McKee KT, Oro JGB, Harrison LH, Gibbs PH, Feuillade MR, Enria DA, Briggiler AM, Levis SC, Ambrosio AM et al (1998) Protective efficacy of a live attenuated vaccine against Argentine hemorrhagic fever. *J Infect Dis* 177: 277–283
29 Momeni AZ, Jalayer T, Emamjomeh M, Khamesipour A, Zicker F, Ghassemi RL, Dowlati Y, Sharifi I, Aminjavaheri M, Shafiei A et al (1999) A randomised, double-blind, controlled trial of a killed L-major vaccine plus BCG against zoonotic cutaneous leishmaniasis in Iran. *Vaccine* 17: 466–472
30 Lu Q, Zhu Z, Weng J (1996) Immune responses to inactivated vaccine in people naturally infected with hantaviruses. *J Med Virol* 49: 333–335
31 Astakhova TI, Slonova RA, Kompanets GG (1997) [Approaches to selecting a strain of hantavirus for developing an inactivated brain vaccine]. [Russian] *Vopr Virusol* 42: 105–108
32 Campbell WN, Hendrix E, Cryz S, Cross AS (1996) Immunogenicity of a 24-valent *Klebsiella* capsular polysaccharide vaccine and an eight-valent *Pseudomonas* O-polysaccharide conjugate vaccine administered to victims of acute trauma. *Clin Infect Dis* 23: 179–181
33 Von Specht BU, Lücking HC, Blum B, Schmitt A, Hungerer KD, Domdey H (1996) Safety and immunogenicity of a *Pseudomonas aeruginosa* outer membran protein I vaccine in human volunteers. *Vaccine* 14: 1111–1117
34 Welch PG, Fattom A, Moore J, Schneerson R, Shiloach J, Bryla DA, Li XR, Robbins JB (1996) Safety and immunogenicity of *Staphylococcus aureus* type 5 capsular polysaccharide-*Pseudomonas aeruginosa* recombinant exoprotein A conjugate vaccine in patients on hemodialysis. *J Amer Soc Nephrol* 7: 247–253
35 Knobloch J (1994) Indication for plague vaccine and plague chemoprophylaxis. *Dtsch Med Wochenschr* 119: 1714–1715
36 Dietz V, Galazka A, van Loon F, Cochi S (1997) Factors affecting the immunogenicity and potency of tetanus toxoid: implications for the elimination of neonatal and non-neonatal tetanus as public health problems. *Bull WHO* 75: 81–93
37 Poland JD, Calisher CH, Monath TP, Downs WG, Murphy K (1981) Persistence of neutralizing antibody 30–35 years after immunization with 17D yellow fever vaccine. *Bull WHO* 59: 895–900
38 Waag DM, Sandstrom G, England MJ, Williams JC (1996) Immunogenicity of a new lot of *Francisella tularensis* live vaccine strain in human volunteers. *FEMS Immunol Med Microbiol* 13: 205–209
39 Kormilitsyna MI, Meshcheryakova IS (1996) The new vaccine strains (or variants) of *Francisella tularensis. FEMS Immunol. Med. Microbiol.* 13: 215–219
40 McClain DJ, Pittman PR, Ramsburg HH, Nelson GO, Rossi CA, Mangiafico JA, Schmaljohn AL, Malinoski FJ (1998) Immunologic interference from sequential administration of live attenuated alphavirus vaccines. *J Infect Dis* 177: 634–641
41 Morrill JC, Mebus CA, Peters CJ (1997) Safety and efficacy of a mutagen-attenuated Rift Valley fever virus vaccine in cattle. *Amer J Vet Res* 58: 1104–1109
42 Ehlers S (1999) Immunity to tuberculosis: a delicate balance between protection and pathology. *FEMS Immunol. Med. Microbiol.* 23: 149–158
43 Grob PJ (1998) Hepatitis B: virus, pathogenesis and treatment. *Vaccine* 16: S11–S16
44 Raviglione MC, Snider DE, Kochi A (1995) Global epidemiology of tuberculosis: Morbidity and mortality of a worldwide epidemic. *JAMA* 273: 220–226
45 Buendia MA (1998) Hepatitis B viruses and cancerogenesis. *Biomed Pharmacother* 52: 34–43
46 Dreesen DW (1997) A global review of rabies vaccines for human use. *Vaccine* 15: S2–S26
47 Ortona L, De Luca A (1998) BCG and prospects for new vaccines against tuberculosis. *Rays* 23: 225–230
48 Zimmerman RK, Spann SJ (1999) Poliovirus vaccine options. *Amer Fam Physician* 59: 113–118
49 Tsai TF (1993) Inactivated Japanese encephalitis virus vaccine recommendations of the Advisory Committee on Immunization Practices (ACIP). *MMWR Morb Mortal Wkly Rep* 42(RR-1): 1–15
50 Muller A (1994)[Active vaccination against early summer meningoencephalitis]. [German] *Schweiz Med Wochenschr* 128: 1110–1116
51 Ludwig SL, Brundage JF, Kelley PW, Nang R, Towle C, Schnurr DP, Crawford-Miksza L, Gaydos JC (1998) Prevalence of antibodies to adenovirus serotypes 4 and 7 among unimmunized US Army trainees: Results of a retrospective nationwide seroprevalence survey. *J Infect Dis* 178: 1776–1778

52 Kane MA (1998) Status of hepatitis B immunization programmes in 1998. *Vaccine* 16: S104–S108
53 White CJ, Stinson D, Staehle B, Cho I, Matthews H, Ngai A, Keller P, Eiden J, Kuter B (1997) Measles, mumps, rubella, and varicella combination vaccine: safety and immunogenicity alone and in combination with other vaccines given to children. Measles, Mumps, Rubella, Varicella Vaccine Study Group. *Clin Infect Dis* 24: 925–931
54 Steere AC, Sikand VK, Meurice F, Parenti DL, Fikrig E, Schoen RT, Nowakowski J, Schmid CH, Laukamp S, Buscarino C et al (1998) Vaccination against Lyme disease with recombinant *Borrelia burgdorferi* outer-surface lipoprotein A with adjuvant. Lyme Disease Vaccine Study Group. *N Engl J Med* 339: 209–215
55 Plotkin SA, Higgins R, Kurtz JB, Morris PJ, Campbell DA Jr, Shope TC, Spector SA, Dankner WM (1994) Multicenter trial of Towne strain attenuated virus vaccine in seronegative renal transplant recipients. *Transplantation* 58: 1176–1178
56 Bernstein DI, Stanberry LR (1999) Herpes simplex virus vaccines. *Vaccine* 17: 1681–1689
57 Stanberry LR (1998) Herpes vaccines for HSV. *Dermatol Clin* 16: 811–816
58 Jacobson RR, Krahenbuhl JL (1999) Leprosy. *Lancet* 353: 655–660
59 Khanna R, Sherritt M, Burrows SR (1999) EBV structural antigens, gp350 and gp85, as targets for *ex vivo* virus-specific CTL during acute infectious mononucleosis: Potential use of gp350/gp85 CTL epitopes for vaccine design. *J Immunol* 162: 3063–3069
60 Wagner R, Shao YM, Wolf H (1999) Correlates of protection, antigen delivery and molecular epidemiology: basics for designing an HIV vaccine. *Vaccine* 17: 17–1710
61 Galgiani JN (1999) Coccidioidomycosis: a regional disease of national importance. Rethinking approaches for control. *Ann Intern Med* 130: 293–300
62 Tappero JW, Lagos R, Ballesteros AM, Plikaytis B, Williams D, Dykes J, Gheesling LL, Carlone GM, Hoiby EA, Holst J et al (1999) Immunogenicity of 2 serogroup B outer-membrane protein meningococcal vaccines – A randomized controlled trial in Chile. *JAMA* 281: 1520–1527
63 Cardosa MJ (1998) Dengue vaccine design: issues and challenges. *Brit Med Bull* 54: 395–405
64 Kumar S, Kaslow DC, Hoffman SL (1999) An overview of malaria vaccine development efforts. *In*: P Perlmann, H Wigzell (eds): *Vaccines* (Handbook of experimental pharmacology; v. 133). Springer-Verlag, Berlin, 397–442
65 Stoute JA, Ballou WR (1998) The current status of malaria vaccines. *Biodrugs* 10: 123–136
66 Goldblatt D (1998) Recent developments in bacterial conjugate vaccines. *J Med Microbiol* 47: 563–567
67 Loosmore S, Zealey G, Yacoob R, Klein M (1993) The new generations of acellular and recombinant pertussis vaccines. *Vaccine Res* 2: 173–181
68 Macdonald NE, Halperin SA, Law BJ, Forrest B, Danzig LE, Granoff DM (1998) Induction of immunologic memory by conjugated vs plain meningococcal C polysaccharide vaccine in toddlers: a randomized controlled trial. *JAMA* 280: 1685–1689
69 Levine MM, Noriega F (1993) Vaccines to prevent bacterial enteric infections in children. *Pediat Ann* 22: 719–725
70 Belshe RB, Mendelman PM, Treanor J, King J, Gruber WC, Piedra P, Bernstein DI, Hayden FG, Kotloff K, Zangwill K et al (1998) The efficacy of live attenuated, cold-adapted, trivalent, intranasal influenzavirus vaccine in children. *N Engl J Med* 338: 1405–1412
71 Hennessy S, Liu Z, Tsai TF, Strom BL, Wan CM, Liu HL, Wu TX, Yu HJ, Liu QM, Karabatsos N et al (1996) Effectiveness of live-attenuated Japanese encephalitis vaccine (SA14-14-2): a case-control study. *Lancet* 347: 1583–1586
72 Boyer JD, Chattergoon M, Shah A, Ginsberg R, Macgregor RR, Weiner DB (1998) HIV-1 DNA based vaccine induces a CD8 mediated cross-clade CTL response. *Dev Biol Stand* 95: 147–153
73 Wang R, Doolan DL, Charoenvit Y, Hedstrom RC, Gardner MJ, Hobart P, Tine J, Sedegah M, Fallarme V, Sacci JB Jr et al (1998) Simultaneous induction of multiple antigen-specific cytotoxic T lymphocytes in nonhuman primates by immunization with a mixture of four *Plasmodium falciparum* DNA plasmids. *Infect Immunity* 66: 4193–4202
74 Wang RB, Doolan DL, Le TP, Hedstrom RC, Coonan KM, Charoenvit YP, Jones TR, Hobart P, Margalit M, Ng J et al (1998) Induction of antigen-specific cytotoxic T lymphocytes in humans by a malaria DNA vaccine. *Science* 282: 476–480
75 Dupuy JM, Freidel L (1990) Lag between discovery and production of new vaccines for the developing world. *Lancet* 336: 733–734
76 Bart KJ, Lin KF (1990) Vaccine-preventable disease and immunization in the developing world. *Pediat Clin N Amer* 37: 735–756

77 Michetti P, Kreiss C, Kotloff KL, Porta N, Blanco JL, Bachmann D, Herranz M, Saldinger PF, Corthesy-Theulaz I, Losonsky G et al (1999) Oral immunization with urease and *Escherichia coli* heat-labile enterotoxin is safe and immunogenic in *Helicobacter pylori*-infected adults. *Gastroenterology* 116: 804–812
78 Cherry JD (1997) The role of *Bordetella pertussis* infections in adults in the epidemiology of pertussis Dev Biol Stand 89: 181–186
79 Edwards KM (1997) Pertussis in older children and adults. *Adv Pediatr Infect Dis* 13: 49–77
80 Pines E, Barrand M, Fabre P, Salomon H, Blondeau C, Wood SC, Hoffenbach A (1999) New acellular pertussis-containing paediatric combined vaccines. *Vaccine* 17: 1650–1656
81 Cleland JL (1999) Single-administration vaccines: controlled-release technology to mimic repeated immunizations. *Trends Biotech* 17: 25–29
82 Nugent J, Po AL, Scott EM (1998) Design and delivery of non-parenteral vaccines. *J Clin Pharm Ther* 23: 257–285
83 Kaul D, Ogra PL (1998) Mucosal responses to parenteral and mucosal vaccines. *Dev Biol Stand* 95: 141–146
84 Mor TS, Gomez-Lim MA, Palmer KE (1998) Perspective: edible vaccines – a concept coming of age. *Trends Microbiol* 6: 449–453
85 Velin D, Hopkins S, Kraehenbuhl JP (1998) Delivery systems and adjuvants for vaccination against HIV. *Pathobiology* 66: 170–175
86 Murphy BR, Hall SL, Kulkarni AB, Crowe JE, Collins PL, Connors M, Karron RA, Chanock RM (1994) An update on approaches to the developement of respiratory syncytial virus (RSV) and parainfluenza virus type 3 (PIV3) vaccines. *Virus Res* 32: 13–30
87 Pamer EG (1999) Antigen presentation in the immune response to infectious diseases *Clin Infect Dis* 28: 714–716
88 Sigal LJ, Crotty S, Andino R, Rock KL (1999) Cytotoxic T-cell immunity to virus-infected nonhaematopoietic cells requires presentation of exogenous antigen. *Nature* 398: 77–80
89 Pol S, Couillin I, Michel ML, Driss F, Nalpas B, Carnot F, Berthelot P, Brechot C (1998) Immunotherapy of chronic hepatitis B by anti HBV vaccine. *Acta Gastro-Enterol Belg* 61: 228–233
90 Chaudhury S, Hajra SK, Mukerjee A, Saha B, Majumdar V, Chattapadhya D, Saha K (1997) Immunotherapy of lepromin-negative borderline leprosy patients with low-dose Convit vaccine as an adjunct to multidrug therapy; a six-year follow-up study in Calcutta. *Int J Lepr Mycobact Dis* 65: 56–62
91 Pasteur L (1951) L'étape des maladies virulentes (suite); vaccination de l'homme contre la rage; dernières années 1885–1895. *In*: R Vallery-Radot (ed.): *Correspondance de Pasteur (1840–1895).* Flammarion, Paris, 360
92 Feery BJ, Boughton CR (1994) The evidence in favour of immunisation – a world without smallpox – a world without polio. *Med J Australia* 160: 459–460
93 Grossman M (1995) Immunization: Past successes, future challenges. *Infect Dis Clin N Amer* 9: 325–333
94 Eskola J (1994) Epidemiological views into possible components of paediatric combined vaccines in 2015. *Biologicals* 22: 323–327

New Approaches to Drug Development
ed. by P. Jollès

Manipulation of polyketide biosynthesis for new drug discovery

Christopher W. Carreras[1] and Gary W. Ashley[2]

Departments of [1]New Technologies and [2]Chemistry, Kosan Biosciences, Inc., Hayward, CA 94545, USA

Summary. Modular polyketide synthases (PKS) are large multifunctional proteins which direct the condensation of activated short chain carboxylic acids into products of defined length and functionality using a dedicated set of active sites, or module, for each step in the polymerization. The structure of the product is directly related to the number, content and sequence of modules in a PKS. Technology is described which allows the rational manipulation of the biosynthesis of these compounds and enables the generation of specific novel polyketide structures. Examples of polyketide drugs whose structures may be manipulated using this technology are given.

Introduction – what are polyketides?

The facile generation of complex polymers of specific utility is a hallmark of biological systems. While the preservation, recombination and expression of genetic material is directed by nucleic acid polymerases and ribosomes, polymers of nucleotides and amino acids, the harbor for these reactions is contained within membranes consisting mainly of fatty acids, polymers of acetate. Many organisms (including actinomycetes, myxobacteria, fungi and some marine animals) have developed defense mechanisms based upon polyketide metabolites, which are biosynthesized using a mechanism that is likely to have evolved from fatty acid synthesis. These polyketide secondary metabolites are typically secreted after the organism has completed the rapid phase of its growth cycle, when growth is limited by the lack of necessary nutrients or space, as a way of defending an organism's needs from competing organisms. Because of the potent and specific bioactivities of many of these molecules, polyketide metabolites have become an important source of pharmaceutical agents and include numerous antibiotics, anticancer agents immunosuppressants, antiparasitic agents, antifungals, cardiovascular agents and veterinary products [1] (Tab. 1). Technology is rapidly becoming available which allows the rational manipulation of the biosynthesis of these compounds and enables the generation of specific novel polyketide structures [2, 3]. The goal of this review is to acquaint the reader with the biochemistry behind novel polyketide biosynthesis using established examples of genetic and precursor-directed strategies.

Table 1. Polyketide-Based Pharmaceuticals

Product	Theraputic area	Company
Azithromycin	Antibacterial	Pfizer
Clarithromycin	Antibacterial	Abbott
Erythromycin	Antibacterial	Abbott, others
Josamycin	Antibacterial	Yamanouchi
Minocycline (Dynacil)	Antibacterial	Wyeth-Ayerst
Miokamycin	Antibacterial	Meiji Seika
Mycinamicin	Antibacterial	Asahi
Oleandomycin	Antibacterial	Pfizer
Pseudomonic acid	Antibacterial	SmithKline Beecham
Rifamycins (Rifampin)	Antibacterial	Novartis, Lepetit
Rokitamycin (Ricamycin)	Antibacterial	Asahi
Tetracyclines	Antibacterial	Pfizer, Weth-Ayerst
Aclarubicin (Aclacinomycin)	Anticancer	Bristol-Myers Squibb
Adriamycin (Doxorubicin)	Anticancer	Pharmacia-Upjohn
Chromomycin	Anticancer	Takeda
Daunorubicin	Anticancer	Astra, Chiron
Enediynes	Anticancer	Wyeth-Ayerst
Idarubicin (Idamycin)	Anticancer	Pharmacia-Upjohn
Amphotericin B	Antifungal	Bristol-Myers Squibb
Candicidin	Antifungal	Hoechst Marion Roussel
Griseofulvin	Antifungal	Schering, Wyeth-Ayerst, Ortho
Nystatin/Mycostatin	Antifungal	Bristol-Myers Squibb, others
Spiramycin	Antirickettsial	Rhône-Poulenc
Mevacor (Lovastatin)	Cholesterol-lowering	Merck
Mevastatin (compactin)	Cholesterol-lowering	Sankyo
Pravastatin	Cholesterol-lowering	Sankyo, Bristol-Myers Squibb
Zocor	Cholesterol-lowering	Merck
Zearalenone	Estrogenic activity	
Ascomycin (Immunomycin)	Immunosuppressant	Merck
FK506	Immunosuppressant	Fujisawa
Rapamycin	Immunosuppressant	Wyeth-Ayerst
Spinosad	Insecticide	Dow Elanco
Avermectin	Veterinary product	Merck
Lasalocid A	Veterinary product	Hoffman LaRoche
Milbemycin	Veterinary product	Sankyo
Monensin	Veterinary product	Lilly
Tylosin	Veterinary product	Lilly

Description of polyketide biosynthesis

Polyketide synthase (PKS) pathways direct the condensation of activated short-chain carboxylic acids into polymers of defined length and functionality in a manner that is highly analogous to fatty acid biosynthesis. Reviews have appeared summarizing key aspects of polyketide biosynthesis [2–6]. The starting metabolites for both polyketide and fatty acid biosyntheses are short-chain acyl-coenzyme A thioesters used as chain priming units, and malonyl- or α-substituted-malonyl-coenzyme A thioesters used as chain extender units.

Three enzyme activities are responsible for polyketide and fatty acid chain elongation [6](and references therein): (1) an acyl carrier protein (ACP) carries the growing chain attached to its phosphopantetheine prosthetic group through a thioester (2) an acyltransferase (AT) selects the malonyl-coenzyme A extender unit and (3) a β-ketoacylsynthase (KS) catalyzes the decarboxylative condensation of the extender unit from the AT onto the ACP-bound growing chain (Fig. 1). Following each cycle of chain elongation, enzyme activities may reduce the β-carbonyl that is formed after each cycle of chain elongation. A ketoreductase (KR) domain uses NADPH to stereospecifically reduce the carbonyl to a hydroxyl group. A dehydratase (DH) activity may then act to convert the alcohol to an olefin by elimination of water. The olefin may then be reduced to the saturated hydrocarbon by an NADPH-dependent enoylreductase (ER). Fatty acid synthase (FAS) systems utilize KR, DH and ER activities following each step of chain elongation, yielding saturated fatty acids, while PKS systems specify the level of modification that occurs at each step of the polymerization process and therefore yield much more diverse structures.

Polyketide synthase protein architecture

PKS architectures mirror those of FAS systems, leading to the suggestion that polyketide biosynthesis may have evolved from fatty acid biosynthesis. FAS pathways are ancient, and bacteria, fungi and animals have found increasingly elegant solutions to the problem of coordinating the necessary enzyme activities. Bacteria encode these activities on several individual polypeptides (6000–50 000 daltons) [7]. Yeasts encode the majority of their FAS activities on two multifunctional polypeptides, which form an $\alpha_6\beta_6$ active complex [8]. Animals encode the seven necessary FAS activities on a single polypeptide, which is active as a dimer [9].

PKS responsible for the biosynthesis of bacterial aromatic polyketides ("aromatic PKS") such as actinorhodin, granaticin and tetracenomycin are comprised of a relatively small set of active sites, each of which is encoded as a distinct protein [10–12]. These proteins are likely to form a higher order complex, and subunits obtained from various PKS systems may be coexpressed, resulting in the production of novel aromatic polyketides (cf. [13, 14]). The function of the aromatic PKS appears limited to elaboration of a true polyketide chain, i.e., poly(1,3-dicarbonyls), of defined length. Subsequent ketoreductions, cyclizations and aromatizations are performed by auxiliary enzymes in the pathway. The genetic manipulation of aromatic polyketide pathways to date has thus focused upon alterations in chain length, cyclizations, and post-PKS modifications such as oxidation and glycosylation.

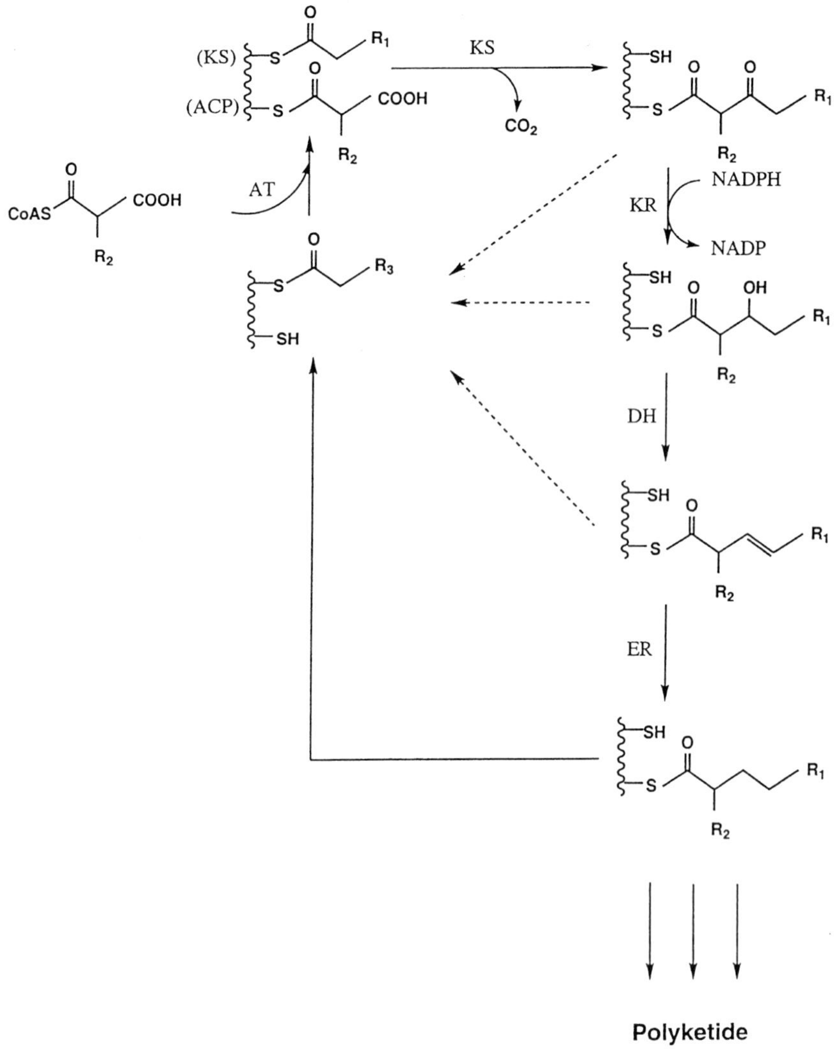

Figure 1. The overall catalytic cycle of a polyketide synthase. Different polyketide synthases show variability with regard to the length of the polyketide chain, the choice of monomer incorporated at each step, the degree of reduction of each β-keto group, and the stereochemistry at each chiral center. Dashed arrows illustrate how the degree of β-ketoreduction can vary at any given carbonyl.

The strategy used for biosynthesis of bacterial aromatic polyketides can be contrasted with that of the complex or "modular" PKSs, which direct the biosynthesis of macrolides such as erythromycin [15, 16], avermectin [17] and rapamycin [18]. The modular PKSs described to date are complexes of large multifunctional proteins with dedicated domains for each enzyme-catalyzed step in polyketide carbon chain assembly and modification (Fig. 2). Active

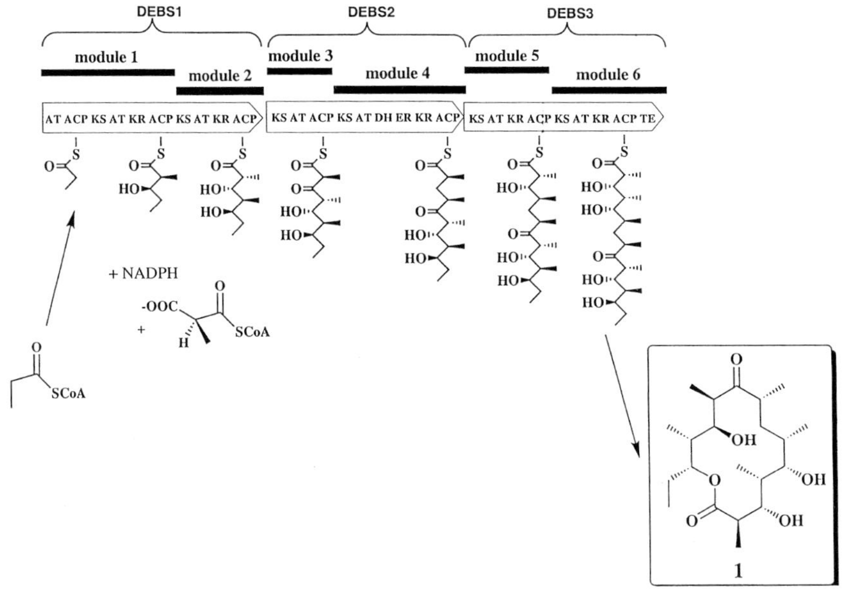

Figure 2. Assembly-line nature of 6-deoxyerythronylide B synthase (DEBS), which is responsible for biosynthesis of the erythromycin aglycone. [16]. Each of the 6 modules accounts for one cycle of polyketide chain extension and β-ketoreduction. The active sites are designated as follows: acyltransferase (AT), β-ketoacyl-ACP transferase (KS), acyl carrier protein (ACP), β-ketoreductase (KR), dehydratase (DH), enoylreductase (ER), and thioesterase (TE). Thus, the AT domain in module four is referred to as AT4, the KS domain in module 6 is referred to as KS6, and so on.

sites are clustered into modules, with each module containing a full complement of sites required for one condensation and associated reduction cycle. The genetic clustering of modules is highly variable, with several modules often encoded by a single gene, and with several large genes encoding the entire PKS. The overall organization of the modules within the genes is the same for all known PKSs, however, and there is always a direct correspondence between the genetic and the functional order of the modules. Modular PKSs appear to act in an assembly-line fashion in which the growing polyketide chain is passed from module to module as polymerization occurs. The erythromycin PKS from *Saccharopolyspora erythraea*, known as 6-deoxyerythronolide B synthase (DEBS), is the only modular PKS to date on which there is available structural information, and this is rather limited. It is clear that it is dimeric, and that the dimeric structure of the erythromycin PKS is essential for its activity as both subunits participate in the synthesis of each polyketide product [19, 20].

The use of individual enzyme activities for each step of modular polyketide biosynthesis gives rise to the possibility of altering selected or multiple sites in the polyketide product, suggesting a staggering potential for genetic manipulation. This review thus focuses on polyketides synthesized by modular PKSs because of their current and future therapeutic value and their potential for structural alteration through genetic manipulation.

Mechanism of modular polyketide biosynthesis

Because of the availability of facile systems for the genetic manipulation of DEBS, the proof-of-principle experiments summarized below were performed using these systems [16, 21]. Other modular PKSs seem analogous to DEBS; they contain genetically-encoded modules whose sequence and content may be used to predict the structure of the resulting polyketide product, and portions of some of these systems may be grafted into DEBS to create functional hybrid enzymes. It seems likely that the results obtained with DEBS will be readily extrapolated to other modular PKSs.

Modular polyketide structures are "encoded" by the number and content of modules in a PKS [16, 18]. Unlike nucleic acid and protein biosynthesis, which are template-directed, polyketide biosynthesis uses a series of covalently-tethered enzymes to specify product structure. The linear polyketide chain is sequentially elongated as the reaction proceeds from the amino to carboxy end of the PKS. The domains of a modular PKS are generally segregated from amino to carboxy termini into a "loading domain", multiple chain-extention "modules" and a "releasing domain". The loading domain is responsible for priming the PKS with a starter unit, and in DEBS consists of AT and ACP subdomains. An alternate loading domain structure is seen in the rapamycin PKS, which consists of a coenzyme A-ligase, an enoylreductase and an ACP domain.

Each subsequent chain-extension module of the PKS contains the essential KS, AT and ACP domains responsible for the polymerization activity as well as optional KR, DH or ER enzyme domains which modify the β-carbon of a growing polyketide chain. The AT encoded by the module is responsible for selecting the required extender unit, malonyl- or α-substituted malonyl, and together the complement of encoded reductive activities determines the structural unit introduced into the growing polyketide chain.

After the polyketide chain is complete, the product is released from the PKS. In DEBS this is accomplished through macrolactonization by a C-terminal thioesterase activity. In rapamycin, an auxiliary enzyme transfers the polyketide chain to an amino acid acceptor, pipecolic acid, and performs a macrolactonization.

The molecular mechanism of modular polyketide biosynthesis has benefited greatly from analogies drawn from studies of FASs [6–9]. Polyketide synthesis is initiated when the phosphopantetheine prosthetic group of the ACP in the loading domain is primed with the starting acyl unit through acylation either by a loading AT, as in DEBS, or through an acyl-coenzyme A ligase, as in rapamycin synthase. The AT of each chain-extension module concurrently recognizes a particular malonyl-CoA (malonyl or α-substituted malonyl) and transfers this malonyl group to the ACP of that same module. Once the PKS is primed with acyl- and malonyl-ACPs, the acyl group of the loading domain migrates to form a thiol ester with a cysteine residue of the ketosynthase of the first module (KS1). The KS1 domain then catalyzes decarboxylation of the malonyl-ACP1 to generate an intermediate which adds to the acyl group present on KS1, generating a new acyl-ACP1 with a backbone 2-carbons longer than the loading unit. The growing polyketide chain is transferred from the ACP1 to the KS of the next module, KS2, and the process continues. The polyketide chain, growing by two carbons at each module, is sequentially passed from module to module as covalently bound thiol esters, in an assembly line-like process.

In addition to the condensing activities just described, there may be KR, DH and ER active sites which modify the resulting β-carbonyl group following each cycle of chain elongation prior to its transfer to the next module. These β-carbonyl-modifying enzymes, together with the specificity of the ATs in each module, are largely responsible for the diversity of the polyketide chains and the resultant polyketide products. Once a polyketide chain traverses the final module of a PKS, it encounters the releasing activity. Here, the polyketide is cleaved from the enzyme (and often cyclized) by thioesterase (DEBS) or amide synthase (rapamycin, rifamycin) activities. The parent polyketide is often further modified by "tailoring enzymes," which add carbohydrate groups, methyl groups, or make other modifications on the polyketide core molecule [5, 22].

Manipulation of polyketide biosynthesis

The striking similarities between polyketide and fatty acid biosyntheses suggest a potential evolutionary relationship between these biochemical pathways. Gene fusion and duplication events could result in a multi-module synthase, and mutagenic inactivation of the KR, DH and/or ER domains could generate novel modules catalyzing varied levels of β-carbonyl modification. Further genetic drift could result in varied specificity in the AT domains, leading to increased structural variation. Once several multi-module PKSs had evolved, genetic recombination of various domains or modules could hasten the evolution of additional synthases. The direct correspondence between polyketide structure and modular PKS gene organization and genetic engineering provides a short-cut to "directed evolution" of novel PKSs for the production of new polyketides in a predictable and reproducible manner. Several strategies to engineer modular PKSs will be discussed in this section, including manipulation of polyketide chain length, choice of extender units, oxidation levels and stereochemistry. Precursor-directed strategies which circumvent PKS specificity to some degree will also be presented.

Manipulation of polyketide chain length

Since a polyketide chain lengthens by 2 carbon atoms as it traverses each module of a PKS, it follows that the number of modules in a PKS should dictate the size of the resulting polyketide. Protein-engineering experiments in which modules were removed from DEBS, a 6-module PKS which normally produces a heptaketide product, show appropriately smaller polyketide products, and support the general notion that polyketide size can be altered simply by altering the number of modules in a PKS (Fig. 3). Repositioning of the chain-terminating thioesterase (TE) domain from the C-terminus of module 6 of DEBS to the C-terminus of the second module (DEBS1+TE) results in biosynthesis of the expected triketide product [23, 24]. DEBS1+TE has served as a useful model system in experiments described below. Similarly, repositioning of the TE domain to the C-terminus of module 3 or 5 results in biosynthesis of the expected tetra- and hexaketides [25, 26].

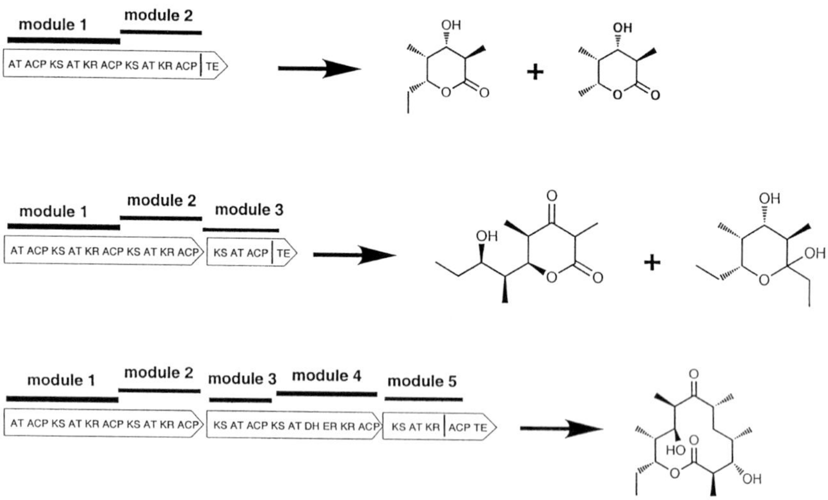

Figure 3. Polyketides produced by *S. coelicolor* strains expressing truncated forms of DEBS [23–26]. Structures were determined by isotope enrichment and NMR spectroscopy.

Manipulation of extender unit choice

The specificity of the AT in a module encodes the identity of the α-alkyl substituent of each 2-carbon unit, such that AT replacement could be a strategy for controlling the groups attached to alternate carbons along the chain. Such experiments were originally performed using the truncation mutant DEBS1+TE described above, which normally encodes a PKS elaborating a triketide lactone. A hybrid gene was made in which the sequence encoding the module 1 AT (specific for methylmalonyl-CoA) was replaced with that encoding the module 2 AT of the rapamycin PKS (specific for malonyl-CoA). The resulting PKS gene was expressed in *Streptomyces coelicolor* CH999, and this organism was found to produce a triketide product similar to that produced by DEBS1+TE but lacking a methyl group at C4, consistent with alteration of the module 1 AT specificity [27]. This also demonstrated that downstream modules could (at least in this case) process altered polyketides resulting from mutations earlier in the pathway.

In the first report of production of a complete novel macrolactone via an internal AT exchange, researchers at Kosan Biosciences substituted the AT domain from module 6 of DEBS with the AT from module 2 of rapamycin synthase to provide 2-nor-6-deoxyerythonolide B (dEB) in *Streptomyces coelicolor*, which was subsequently converted to 2-norerythromycins by *Saccharopolyspora erythraea* [28]. Katz and coworkers at Abbott Laboratories exchanged AT domains of modules 1 and 2 of chromosomal DEBS in *Saccharopolyspora erythraea* with AT domains from other PKSs specific for malonyl-CoA, which led to the production of 10 and 12-nor-erythromycin derivatives [29]. The successful introduction of an ethyl malonyl-CoA-specific AT into DEBS has also been demonstrated [30]. In this work, the AT5 domain from the niddamycin PKS was used to replace AT4 of DEBS in *S. erythraea*. Production of 6-ethyl-6-desmethyl-erythromycin by this strain required coexpression of crotonyl-CoA reductase, a primary metabolic enzyme involved in butyryl-CoA production, to supply precursors for biosynthesis of the ethyl malonyl-CoA substrate of the heterologous AT domain.

It is clear that there is substantial flexibility in the substitution of AT domains within modular PKSs, and that significant structural variation in polyketides can be achieved through engineering of this type.

Manipulation of oxidation level

A second strategy for manipulating modular PKSs involves alteration of the β-carbonyl reduction activities often present in modules. The variable occurrence of these activities within modules suggests that they may be particularly flexible sites for genetic manipulation. Experiments performed by the Abbott group on the chromosomal DEBS genes of *S. erythraea* confirmed this hypothesis. KR and ER domains in DEBS were inactivated by chromosomal mutagenesis, yielding the predicted incompletely reduced polyketides [16, 31]. In this manner, the 5-keto and 6,7 anhydro analogs of 6-dEB were produced by targeted knock-out mutations of the KR5 and ER4 of DEBS, respectively.

McDaniel et al. at Kosan Biosciences provided the first demonstration that entire domains may be predictably *added* to a modular PKS to *increase* the functionality of the enzyme. Here, using a three-module system, the KR domain of DEBS module 2 was replaced with the DH and KR domains module 4 of rapamycin synthase [32]. This resulted in the predicted substitution of a double bond for the hydroxyl group normally introduced by the KR in module 2 of the wild-type. In another study, substitution of the KR domain of DEBS module 2 with the complete set of DH, KR and ER domains from module 1 of rapamycin synthase resulted in complete reduction of the second ketide unit in the polyketide chain and formation of an 8-membered lactone [33].

Subsequent work has helped delimit the tolerance of downstream modules in DEBS towards oxidative alterations in the polyketide chain [40]. While there appears at present to be somewhat more stringency with respect to oxidation level (at least in DEBS), the degree of tolerance is still remarkably wide. Alteration of oxidation level is thus a second highly successful strategy for manipulation of polyketide structure.

Change of stereochemistry

The stereochemistry of polyketide hydroxyls can also be controlled through genetic manipulation. The KR domain of module 2 of DEBS1+TE, which normally reduces the polyketide intermediate to an alcohol with the (*S*)-configuration, was replaced by either the KR domain of rapamycin synthase module 2 or 4, both of which create alcohols with the (*R*)-configuration [34]. These replacements resulted in synthesis of a triketide lactone with an inverted stereocenter at C-3 relative to the triketide lactone normally produced by DEBS1+TE, demonstrating that control of alcohol stereochemistry in polyketide backbones resides in KR domains.

Precursor-directed strategies for novel polyketide production

Two approaches have been developed for the generation of polyketides containing altered starter units. The first approach relies on feeding desired carboxylic acids to a strain harboring an engineered PKS containing a promiscuous loading domain, while the second approach utilizes the feeding of activated synthetic diketide intermediates to a strain whose first module has been inactivated through mutagenesis.

In the first approach, the loading domain of DEBS was replaced with one obtained from the avermectin PKS, which has a broad specificity for branched chain carboxylic acids [35]. This resulted in a formation of a mixture of erythromycins D, B and A, along with corresponding analogs that had iso-propyl and sec-butyl at the C-13 position. Thus, the avermectin PKS loading domain functions in the context of the DEBS PKS, and the possibility of introducing unnatural starter units from carboxylic acids into erythromycin, seems feasible. However, since in the natural avermectin PKS the avermectin loading domain recognized only ~40 out of >800 potential precursors tested [36], it is expected that the hybrid gene will show similar limitations of specificity.

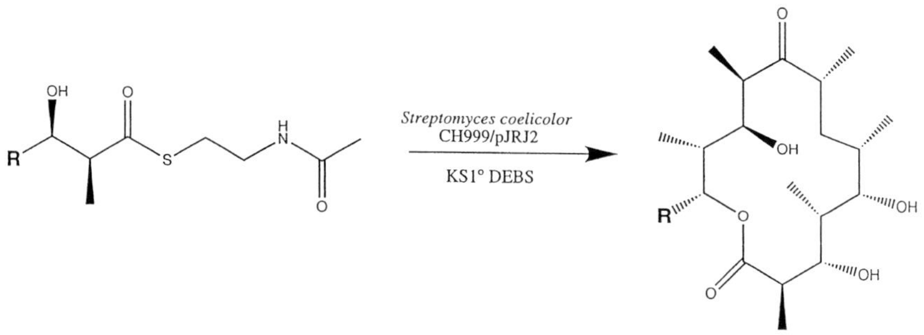

Figure 4. Precursor directed "chemobiosynthesis" [37]. Synthetic diketide intermediates, activated as N-acetylcysteamine thioesters, are incorporated into erythromycin aglycones by a mutant PKS containing an inactive module 1.

In the second approach, Jacobsen et al. inactivated the first KS domain (and thereby module 1) of the DEBS protein using site-directed mutagensis to block natural polyketide production. This blocked mutant was found to produce 6-dEB when provided with the expected product of module 1, a diketide thioester (Fig. 4) [37]. When provided with synthetic analogs of this diketide thioester, the analogs loaded directly onto KS2 of DEBS [38], and were incorporated into 6-dEB scaffolds to produce novel 13-substituted 6-dEB analogs in good yield. The 6-dEB analogs produced from these activated diketide precursors were good substrates for the *S. erythraea* "converter strains" which lack PKS activity but contain the necessary tailoring activities for the formation of biologically active erythromycin analogs [39]. Feeding of these 13-substituted 6-dEB analogs to *S. erythraea* converter strains resulted in formation of novel, biologically active 13-substituted erythromycins [37].

Combinatorial potential

Several recent accomplishments demonstrate the potential of hybrid modular PKSs for the generation of combinatorial libraries of novel macrolides. First, a library of more than 50 novel erythromycin aglycones was produced from a set of individually engineered DEBS enzymes, many of which contained more than one of the above-mentioned genetic alterations (Fig. 5) [40]. This work demonstrated the feasibility of creating large polyketide libraries through genetic manipulation of modular PKS genes and the productivity of a modular PKS containing more than one genetic modification. Since the strains were individually constructed with each PKS mutation individually grafted into an expression plasmid containing all three DEBS genes, generation of the library was labor-intensive. A similar library was subsequently constructed using a multi-plasmid expression system in which each of the three DEBS genes was encoded independently on a separate, compatible plasmid. This allowed com-

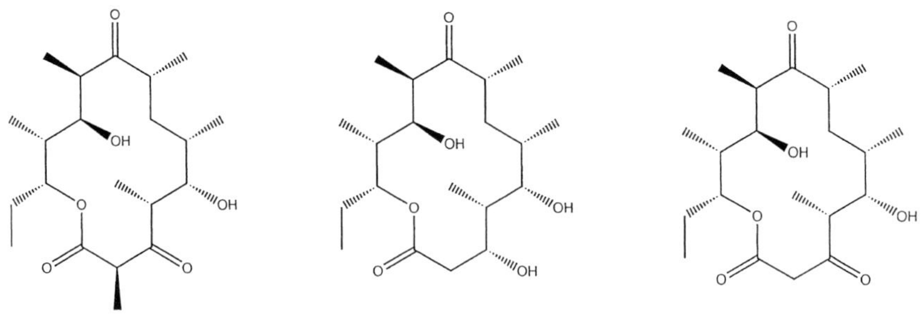

Figure 5. Examples of of erythromycin aglycones produced through genetic manipulation of DEBS [40]. A) 3-keto-6-dEB produced by inactivation of the KR activity in module 6 of DEBS; B) 2-nor-6-dEB produced by exchanging AT6 of DEBS for the AT2 of the Rapamycin PKS; C) 2-nor-3-keto-6-dEB, produced by a combination of the AT6 and KR6 mutations described above.

bination of mutations via combinatorial expression of mutated plasmids without the need to combine the mutations on a single plasmid. [41]. This work reproduced the individually constructed library, as well as identifying additional combinations of productive mutations and provides a critical technology for the rapid generation of extremely large polyketide libraries. In addition, the multi-plasmid system can be used for combinatorial expression of enzymes which further modify polyketide products.

A number of the possible polyketide products were not observed in the above experiments, which is likely to be due to the low productivity of many hybrid PKSs. Careful optimization of domain boundaries has not been completed, however, and recent work from the Khosla group at Stanford suggests that productive synthases may be obtained through appropriate engineering of the regions which link the active sites together [42]. These findings together indicate that extremely large libraries of novel polyketides may be produced rapidly via genetic manipulation of PKS genes.

To illustrate the vast potential of combinatorial polyketide libraries such as those described above, the "combinatorial potential" (CP) of a polyketide library created from modular building blocks may be calculated using the formula

$$CP = AT_L * [AT_E * 4]_M$$

where AT_L is the number of loading acyl transferases of differing specificity, AT_E is the number of extender α-substituted malonyl transferases of differing specificity, 4 represents the number of β-carbon modifiers and M is the number of modules in the gene cluster. Not included are the two possible AT stereochemistries, two possible KR stereochemistries and the possible tailoring enzymes, which would all increase the CP. For a 6-module PKS such as the one found in the erythromycin gene cluster, and 3 ATs (specific for malonyl-, methylmalonyl- or ethyl-malonyl-CoAs), the theoretical number of polyketides that could be made approaches 10 million. Note that PKSs are known that contain 10 or more modules [18]. While the potential diversity of such libraries is enormous, the immense power of molecules derived in this manner comes from the fact they are based upon molecules of known biological activity which are members of a class of compounds known for their pharmaceutical utility.

Therapeutic targets of polyketide drugs

With the major proof-of-principle experiments for the generation of large libraries of novel drug candidates accomplished, the technology can be brought to bear on relevant therapeutic problems. While it is likely that novel polyketides will be used to treat conditions for which there is currently no satisfactory drug, the immediate application of the technology is in the search for improvements over currently prescribed medicines such as those described below (Fig. 6).

Macrolide antibiotics

The classic therapeutic use associated with macrolide polyketides is as antibacterial agents. Since the discovery of erythromycin from an organism found in a Philippine soil sample in 1952, a variety of useful macrolide antibacterials have been discovered. An excellent review of the known macrolide antibiotics has been given [43].

Two "second-generation" erythromycin derivatives, clarithromycin (Biaxin, 6-O-methylerythromycin A) and azithromycin (Zithromax) are of particular importance. Both derivatives show reduced gastrointestinal disturbance due to their stability against acid-catalyzed degradation in the stomach. Both compounds show comparable efficacies in the treatment of pneumonia, acute bronchitis, pharyngitis and otitis media [44]. The next generation of macrolide antibacterials, the ketolides, are semisynthetic analogs of erythromycin, structurally related to picromycin. These compounds lack the 3-O-cladinose unit characteristic of the erythromycins, and thus bypass the induced resistance mechanism common to many pathogens [45].

Macrolide antiparasitics

The avermectins are 16-member macrolide polyketides possessing broad-range antiparasitic and antithelminthic activity. Avermectins and the closely allied milbemycins occur as mixtures of related components in the fermenta-

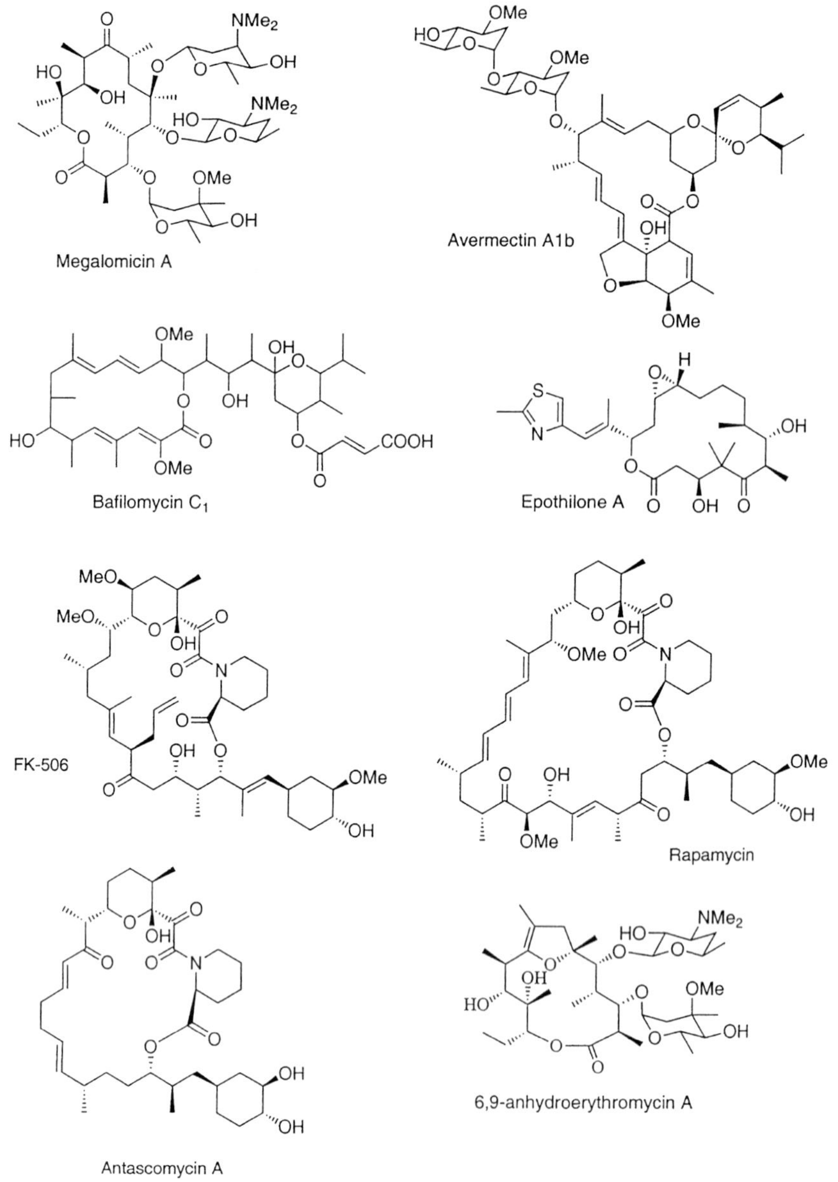

Figure 6. Structures of polyketide drugs and lead compounds for drug discovery. Reengineering the biosynthesis of these compounds using genetic methods may result in therapeutically useful derivatives.

tion broths of *Streptomyces avermitilis* and *Streptomyces hygroscopicus*, respectively. Various semisynthetic analogs (Ivermectin, moxidectin, abamectin) have been found to possess enhanced activities, and Ivermectin has found significant use in the treatment of human filarial worm infections like onchocerciasis (*Onchocerca volvulus*, River Blindness) as well as in veterinary medicine [46, 47]. The mechanism of action of the avermectins appears to be to activate GABA-gated chloride channels [48].

Several other macrolide polyketides show promising antiparasitic activity. The bafilomycins, 16-member macrolides produced by a variety of *Streptomyces*, show antithelminthic activity with no apparent cross-resistance to the avermectins [49]. The megalomicins (often misspelled as "megalomycins" in the literature) are derivatives of erythromycin having a third sugar group (megosamine) attached to the 6-hydroxyl group [50]. Recent studies have demonstrated substantial antiparasitic activity against *Trypanosoma cruzi* (the causative agent of Chagas' Disease), and lesser activity against *T. brucei*, as well as activity against *Leishmania donovani,* and *L. major* (agents in Leishmaniasis) [51]. Megalomicin also blocks intracellular replication of asexual *Plasmodium falciparum*-infected erythrocytes, and so may find use in the treatment of malaria.

Macrolide anticancer agents

The epothilones are macrolide polyketides isolated from the myxobacterium *Sorangium cellulosum* [52]. Originally isolated based on their rather unassuming antifungal activity, upon closer inspection they revealed astonishing microtubule-stabilizing activity [53]. This places them within the activity sphere of the taxols, and the epothilones indeed possess potent antitumour activity. Many facets of the epothilones make them more attractive antitumour agents than the taxols, such as increased activity against multi-drug resistant cells and increased water solubility.

Macrolide immunosuppressants

Two structurally similar but mechanistically unrelated macrolide immunosuppressants have been identified as clinically useful drugs. FK-506 (Tacrolimus) is produced by several strains of *Streptomyces*, and rapamycin (Sirolimus, Rapimmune) is produced by *S. hygroscopicus*. Both molecules are tight binders of the "FK-binding proteins" (FKBP) due to a common structural motif, and these drug-FKBP complexes are important modulators of the immune response. The FK-506/FKBP complex binds the protein phosphatase calcineurin and inhibits its activity, resulting in elevated levels of phosphorylated proteins such as the transcription factor NFAT [54]. The rapamycin/FKBP complex binds a different protein variously known as FRAP, RAFT or TOR, resulting in cell cycle arrest [55].

Macrolide neurotrophic agents

A side-effect of the immunosuppressive macrolides FK-506 and rapamycin, at least in cell culture, is to promote neurite outgrowth [56, 57]. This has led to a search for non-immunosuppressive analogs of these compounds which might be useful for the treatment of neurodegenerative diseases such as Alzheimer's Disease. It appears that the immunosuppressive and the neurotrophic activities can indeed be separated, as neurite outgrowth has been demonstrated with semisynthetic analogs of FK-506 [58]. Furthermore, naturally-occurring non-immunosuppressive analogs, the antascomycins, have been discovered in the culture broths of a *Micromonospora* strain [59].

Macrolide antiinflammatory agents

Several reports have pointed out the ability of macrolide polyketides to influence inflammation. Such antiinflammatory activities have been postulated to explain the effectiveness of erythromycin in the treatment of chronic sinusitis, asthma and diffuse panbronchiolitis, where antibacterial activity is apparently unimportant [60]. Despite substantial attention, the exact mechanism of action of macrolides in inflammation remains uncertain. Several reports demonstrate a modulatory effect of erythromycin on interleukin (IL)-8 expression [61–63], while others indicate an interaction with oxidative processes or nitric oxide (NO) release via cyclic adenosine monophosphate (cAMP)-dependent protein kinase [64–66].

Other macrolides have been shown to modulate intracellular cytokine levels. The polynactins, nonactin analogs, have been found to inhibit the transcription of the IL-5 gene in atopic patients [67]. Conceivably, such agents could find use in the treatment of chronic eosinophilic inflammation found in asthma and atopic dermatitis, especially if they showed fewer side-effects than glucocorticoids or immunosuppressants.

Macrolide prokinetic agents

Treatment of infections with erythromycin is often associated with gastrointestinal distress. The nature of this side-effect has been traced to the formation of 6,9-anhydroerythromycin in the acidic environment of the stomach. Examination of this compound revealed it to act as an agonist of the motilin receptor, and treatment of smooth muscle with this "motilide" results in initiation of type-III migratory motor complex [68]. The ideal motilide would be a safe and effective treatment for common heartburn, and is anticipated to have applications in the treatment of diabetic gastroparesis, postoperative ileus, scleroderma and as a quick and safe procedure to empty the stomach before emergency surgery [69]. Substantial medicinal chemistry has gone into developing clinical motilide candidates based on 6,9-anhydroerythromycin [70].

Conclusions

The modular nature of the genes and enzymes which direct the biosynthesis of the polyketide core of macrolides such as erythromycin is a relatively new discovery [16]. Excitement surrounding the possibility of genetically rearranging the modules to create new enzymes to direct the biosynthesis of novel products has sparked a furious pursuit of technologies to accomplish this goal. During the past eight years it has been shown that polyketide structures can be predictably modified by genetic engineering of the individual domains of modular PKS genes, and many of the major proofs of principle for generating large libraries of novel polyketides are now in hand. The immediate goal is to bring the technology to bear on practical therapeutic problems facing the world today. This requires the identification and isolation of relevant PKS gene clusters, and application of the technology to these systems. In cases where the organisms which harbor these activities are not easily manipulated, suitable genetic methods must be developed.

Although many diverse manipulations have been used to produce specific novel polyketides, in many cases the yield of these molecules has been low. Commercial development of a therapeutic candidate requires large amounts of compound. Industrial laboratories, such as Kosan Biosciences, are devoting resources to produce novel polyketides on a commercial scale. Since polyketides are produced by fermentation rather than via chemical syntheses, general, scalable solutions to the problems of strain improvement and commercial production of diverse polyketides should be accessible in the near term. With the current excitement surrounding novel polyketides, it is anticipated that many therapeutic areas will benefit from access to this rich source of pharmaceutical diversity in the next ten years.

References

1 O'Hagan D (1995) Biosynthesis of fatty acid and polyketide metabolites. *Nat Prod Rep* 12: 1–32

2 Hutchinson CR (1999) Microbial polyketide synthases: more and more prolific. *Proc Natl Acad Sci USA* 96(7): 3336–8

3 Carreras CW, Santi DV (1998) Engineering of modular polyketide synthases to produce novel polyketides. *Curr Opin Biotechnol* 9: 403–11

4 Hopwood DA (1997) Genetic contributions to the understanding of polyketide synthesis. *Chem Rev* 97(7): 2465–97

5 Hutchinson CR (1998) Combinatorial biosynthesis for new drug discovery. *Curr Opin Microbiol* 1(3): 319–29

6 Carreras CW, Pieper R, Khosla C (1997) The chemistry and biology of fatty acid, polyketide, and nonribosomal peptide biosynthesis. *Topics Curr Chem* 188: 85–126

7 Magnuson K, Jackowski S, Rock CO, Cronan JEJ (1993) Regulation of fatty acid biosynthesis in *Escherichia coli*. *Microbiol Rev* 57(3): 522–42

8 Wakil S (1989) Fatty Acid Synthase, A proficient multifunctional enzyme. *Biochemistry* 28: 4523–30
9 Smith S, The animal fatty acid synthase: one gene, one polypeptide, seven enzymes. *FASEB J* 8(15): 1248–59
10 Fernández-Moreno MA, Martínez E, Boto L, Hopwood DA, Malpartida F (1992) Nucleotide sequence and deduced functions of a set of cotranscribed genes of *Streptomyces coelicolor* A3(2) including the polyketide synthase for the antibiotic actinorhodin. *J Biol Chem* 267(27): 19 278–90
11 Sherman DH, Malpartida F, Bibb MJ, Kieser HM, Bibb MJ, Hopwood DA (1989) Structure and deduced function of the granaticin-producing polyketide synthase gene cluster of *Streptomyces violaceoruber* Tü22. *EMBO J* 8(9): 2717–25
12 Bibb MJ, Biró S, Motamedi H, Collins JF, Hutchinson CR (1989) Analysis of the nucleotide sequence of the *Streptomyces glaucescens tcmI* genes provides key information about the enzymology of polyketide biosynthesis. *EMBO J* 8(9): 2727–36
13 McDaniel R, Ebert-Khosla S, Hopwood DA, Khosla C (1993) Engineered biosynthesis of novel polyketides. *Science* 262: 1546–1550
14 McDaniel R, Ebert-Khosla S, Hopwood DA, Khosla C (1995) Rational design of aromatic polyketide natural products by recombinant assembly of enzymatic subunits. *Nature* 375: 549–54
15 Cortes J, Haydock SF, Roberts GA, Bevitt DJ, Leadlay PF (1990) An unusually large multifunctional polypeptide in the erythromycin-producing polyketide synthase of *Saccharopolyspora erythraea*. *Nature* 348: 176–8
16 Donadio S, Staver MJ, JB M, Swanson SJ, Katz L (1991) Modular organization of genes required for complex polyketide biosynthesis. *Science* 252: 675–9
17 Macneil DJ, Gewain KM, Ruby CL, Dezeny G, Gibbons PH, Macneil T (1992) Analysis of *Streptomyces avermitilis* genes required for avermectin biosynthesis utilizing a novel integration vector. *Gene* 111: 61–8
18 Schwecke T, Aparacio JF, Molnar I, König A, Khaw L, Haydock SF, Oliynyk M, Caffrey P, Cortes J, Lester JB, Böhm GAet al (1995) The biosynthetic gene cluster for the polyketide immunosuppressant rapamycin. *Proc Natl Acad Sci USA* 92: 7839–43
19 Kao CM, Pieper R, Cane DE, Khosla C (1996) Evidence for two catalytically independent clusters of active sites in a functional modular polyketide synthase. *Biochemistry* 35(38): 12 363–8
20 Staunton J, Caffrey P, Aparicio JF, Roberts GA, Bethell SS, Leadlay PF (1996) Evidence for a double-helical structure for modular polyketide synthases. *Nat Struct Biol* 3(2): 188–92
21 Kao CM, Katz L, Khosla C (1994) Engineered biosynthesis of a complete macrolactone in a heterologous host. *Science* 265: 509–12
22 Hutchinson CR, Fujii I (1995) Polyketide synthase gene manipulation: A structure-function approach in engineering novel antibiotics. *Annu Rev Microbiol* 49: 201–38
23 Kao CM, Luo G, Katz L, Cane DE, Khosla C (1994) Engineered biosynthesis of a triketide lactone from an incomplete modular polyketide synthase. *J Amer Chem Soc* 116(25): 11 612–11 613
24 Cortes J, Wiesmann KE, Roberts GA, Brown MJ, Staunton J, Leadlay PF (1995) Repositioning of a domain in a modular polyketide synthase to promote specific chain cleavage. *Science* 268(5216): 1487–9
25 Kao CM, Luo G, Katz L, Cane DE, Khosla C (1995) Manipulation of macrolide ring size by directed mutagenesis of a modular polyketide synthase. *J Amer Chem Soc* 117: 9105–6
26 Kao CM, Luo G, Katz L, Cane DE, Khosla C (1996) Engineered biosynthesis of structurally diverse tetraketides by a trimodular polyketide synthase. *J Amer Chem Soc* 118: 9184–5
27 Oliynyk M, Brown MJ, Cortes J, Staunton J, Leadlay PF (1996) A hybrid modular polyketide synthase obtained by domain swapping. *Chem Biol* 3(10): 833–9
28 Liu L, Thamchaipenet A, Fu H, Betlach M, Ashley G (1997) Biosynthesis of 2-nor-6-deoxyerythronolide B by rationally designed domain substitutions. *J Amer Chem Soc* 119: 10 553–10 554
29 Ruan X, Pereda A, Stassi D, Zeidner D, Summers R, Jackson M, Shivakumar A, Kakavas S, Staver M, Donadio S, Katz L (1997) Acyltransferase domain substitutions in erythromycin polyketide synthase yield novel erythromycin derivatives. *J Bacteriol* 179: 6416–6425
30 Stassi DL, Kakavas SJ, Renolds KA, Gunawardana G, Swanson S, Zeidner D, Jackson M, Liu H, Buko A, Katz L (1998) Ethyl-substituted erythromycin derivatives produced by metabolic engineering. *Proc Natl Acad Sci USA* 95: 7305–9
31 Donadio S, McAlpine JB, Sheldon PJ, Jackson M, Katz L (1993) An erythromycin analog produced by reprogramming of polyketide synthesis. *Proc Natl Acad Sci USA* 90: 7119–23
32 McDaniel RN, Kao C, Fu H, Hevezi P, Gustafsson C, Betlach M, Ashley G, Dane D, Khosla C

(1997) Gain-of-function mutagenesis of a modular polyketide synthase. *J Amer Chem Soc* 119: 4309–4310
33 Kao C, McPherson M, McDaniel R, Fu H, Cane D, Khosla C (1997) Gain of function mutagenesis of the erythromycin polyketide synthase. 2. Engineered biosynthesis of an eight-membered ring tetraketide lactone. *J Amer Chem Soc* 119: 11 339–11 340
34 Kao CM, McPherson M, McDaniel RN, Fu H, Cane DE, Khosla C (1998) Alcohol stereochemistry in polyketide backbones is controlled by the β-ketoreductase domains of modular polyketide synthases. *J Amer Chem Soc* 120(10): 2478–9
35 Marsden AF, Wilkinson B, Cortes J, Dunster NJ, Staunton J, PF L (1998) Engineering broader specificity into an antibiotic-producing polyketide synthase. *Science* 279(5348): 199–20
36 Dutton CJ, Gibson SP, Goudie AC, Holdom KS, Pacey MS, Ruddock JC, Bu'Lock JD, Richards MK (1991) Novel avermectins produced by mutational biosynthesis. *J Antibiot* 44: 357–65
37 Jacobsen JR, Hutchinson CR, Cane DE, Khosla C (1997) Precursor-directed biosynthesis of erythromycin analogs by an engineered polyketide synthase. *Science* 277(5324): 367–9
38 Tsukamoto N, Chuck JA, Luo G, Kao CM, Khosla C, Cane DE (1996) 6-deoxyerythronolide B synthase 1 is specifically acylated by a diketide intermediate at the β-ketoacyl-acyl carrier protein synthase domain of module 2. *Biochemistry* 35(48): 15 244–8
39 Weber JM, Wierman CK, Hutchinson CR (1985) Genetic analysis of erythromycin production in *Streptomyces erythreus*. *J Bacteriol* 164: 425–33
40 McDaniel RM, Thamchaipenet A, Gustafsson C, Fu H, Betlach M, Ashley G (1999) Multiple genetic modifications of the erythromycin polyketide synthase to produce a library of novel "unnatural" natural products. *Proc Natl Acad Sci USA* 96(5): 1846–51
41 Xue Q, Hutchinson CR, Santi DV (1999) A Multi-plasmid Approach to Preparing Large Libraries of Polyketides. *Proc Natl Acad Sci USA* 96(21): 11 740–5
42 Gokhale RS, Tsuji SY, Cane DE, Khosla C (1999) Dissecting and exploiting intermodular communication in polyketide synthases. *Science* 284(5413): 482–5
43 Omura S, Tanaka H (1984) Production and antimicrobial activity of macrolides. *In*: S Omura (ed.): *Macrolide antibiotics: chemistry, biology, and practice*. Academic Press, Orlando, FL. p. 3–36
44 Langtry HD, Brigden RN (1997) Clarithromycin. A review of its efficacy in the treatment of respiratory tract infections in immunocompetent patients. *Drugs* 53(6): 973–1004
45 Agouridas C, Denis A, Auger J-M, Benedetti Y, Bonnefoy A, Bretin F, Chantot J-F, Dussarat A, Fromentin C, D'Ambrieres SG et al (1998) Synthesis and antibacterial activity of ketolides (6-O-methyl-3-oxoerythromycin derivatives): a new class of antibacterials highly potent against macrolide-resistant and -susceptible respiratory pathogens. *J Med Chem* 41: 4080–100
46 Mrozik H (1993) Advances in research and development of avermectins. *In*: PA Hedin (ed.): *Natural and Engineered Pest Management Agents*. 54–73
47 Shoop WL, Mrozik H, Fisher MH (1995) Structure and activity of avermectins and milbemycins in animal health. *Vet Parasitol* 59: 139–56
48 Prichard R (1994) Antithelmintic resistance. *Vet Parasitol* 54: 259–68
49 Lacey E, Gill JH, Power ML, Rickards RW, O'Shea MG, Rothschild JM (1995) Bafilolides, potent inhibitors of the motility and development of the free-living stages of parasitic nematodes. *Int J Parasitol* 25: 349–57
50 Thang TT, Lukacs G, Omura S, Bartner P, Boxler DL, Brambilla R, Mallams AK, Morton JB, Reichert R et al (1978) Megalomicins. 6. Tertiary glycosidic macrolide antibiotics. A structural revision by carbon-13 nuclear magnetic resonance and x-ray crystallography. *J Amer Chem Soc* 100: 663–6
51 Bonay P, Duran-Chica I, Fresno M, Alarcon B, Alcina A (1998) Antiparasitic effects of the intra-Golgi transport inhibitor megalomicin. *Antimcrob Agents Chemother* 42: 2668–73
52 Hoefle G, Bedorf N, Gerth K, Irschick, Reichenbach H (1996) Epothilones A and B: Antifungal and Cytotoxic Compounds from *Sorangium cellulosum* (Myxobacteria). Production, Physicochemical and Biological Properties. *J Antibiot* 49: 560–3
53 Bolag DM, McQueney PA, Zhu J, Hensens O, Koupal L, Liesch J, Goetz M, Lazarides E, Woods CM (1995) Epothilones, a new class of microtubule-stabilizing agents with a taxol-like mechanism of action. *Cancer Res* 55(11): 2325–33
54 Liu J, Farmer JD, Friedman JJ, Weissman I, Schreiber SL (1991) Calcineurin is a common target of cyclophilin-cyclosporin A and FKBP-FK506 complexes. *Cell* 66: 807–15
55 Sabatini DM, Erdjument-Bromage H, Lui M, Tempst P, Snyder SH (1994) RAFT1: a mammalian protein that binds to FKBP12 in a rapamycin-dependent fashion and is homologous to yeast

TORs. *Cell* 78: 35–43
56 Hamilton GS, Steiner JP (1998) Immunophilins: Beyond Immunosuppression. *J Med Chem* 41(26): 5119–43
57 Lyons WE, George EB, Dawson TM, Steiner JP, Snyder SH (1994) Immunosuppressant FK506 promotes neurite outgrowth in cultures of PC-12 cells and sensory ganglia. *Proc Natl Acad Sci USA* 91: 3191–5
58 Steiner JP, Connolly MA, Valentine HL, Hamilton GS, Dawson TM, Hester L, Snyder SH (1997) Neurotrophic actions of nonimmunosuppressive analogues of immunosuppressive drugs FK506, rapamycin and cyclosporin A. *Nat Med* 3: 421–8
59 Fehr T, Sanglier J-J, Schuler W, Gschwind L, Ponelle M, Schiling W, Wioland C (1996) Antascomicins A, B, C, D and E. Novel FKBP12 binding compounds from a *Micromonospora* strain. *J Antibiot* 49: 230–3
60 Kudoh S (1998) Erythromycin treatment in diffuse panbronchiolitis. *Curr Opin Pulm Med* 4: 116–21
61 Takizawa H, Desaki M, Ohtoshi T, Kawasaki S, Kohyama T, Sato M, Tanaka M, Kasama T, Kobayashi K, Nakajima J, Ito K (1997) Erythromycin modulates IL-8 expression in normal and inflamed human bronchial epithelial cells. *Amer J Respir Crit Care Med* 156–71: 266–
62 Sunazuka T, Takizawa H, Desaki M, Suzuki K, Obata R, Otoguro K, Omura S (1999) Effects of erythromycin and its derivatives on interleukin-8 release by human bronchial epithelial cell line BEAS-2B cells. *J Antibiot* 52: 71–4
63 Kohyama T, Takizawa H, Kawasaki S, Akiyama N, Sato M, Ito K (1999) Fourteen-member macrolides inhibit interleukin-8 release by human eosinophils from atopic donors. *Antimicrob Agents Chemother* 43(4): 907–11
64 Mitsuyama T, Tanaka T, Hidaka K, M A, Hara N (1995) Inhibition by erythromycin of superoxide anion production by human polymorphonuclear leukocytes through the action of cyclic AMP-dependent protein kinase. *Respiration* 62: 269–73
65 Mitsuyama T, Hidaka K, Furono T, Hara N (1997) Neutrophil-induced endothelial cell damage: inhibition by a 14-membered ring macrolide through the action of nitric oxide. *Int Arch Allergy Immunol* 114–5: 111
66 Mitsuyama T, Hidaka K, Furono T, Hara N (1997) Release of nitric oxide and expression of constitutive nitric oxide synthase of human endothelial cells: enhancement by a 14-membered ring macrolide. *Mol Cell Biochem* 181: 157–61
67 Okudaira H, Mori A, Mikami T, Kaminuma O, Ohmura T, Hoshino A, Suko M (1997) Selective suppression of IL-5 synthesis by OM-1 – Pinpoint treatment of atopic diseases by IL-5 gene transcription inhibitor. *Int Arch Allergy Immunol* 113: 331–4
68 Peters TL (1993) Erythromycin and other macrolides as prokinetic agents. *Gastroenterology* 105: 1886–99
69 Peeters T (1999) Motilide Web Site. *http://www.med.kuleuven.ac.be/med/gih/motilid.htm*
70 Lartey PA, Nellans HN, Faghih R, Petersen A, Edwards CM, Freiberg L, Quigley S, Marsh K, Klein LL, Plattner JJ (1995) Synthesis of 4"-deoxy motilides: Identification of a potent and orally active prokinetic drug candidate. *J Med Chem* 38: 1793-8.

New Approaches to Drug Development
ed. by P. Jollès

Physiological functions of protein kinase inhibitors

John J. Turchi and Mengfei Ho

Department of Biochemistry and Molecular Biology, Wright State University School of Medicine, Dayton, Ohio 45435, USA

Summary. Protein kinases are key regulatory enzymes involved in a multitude of biochemical pathways. This chapter will describe the current research on targeting specific protein kinases with inhibitors in attempts to disrupt flux through specific pathways. Targeting specific kinases presents a distinct challenge as there are hundreds of individual kinase enzymes that use ATP as a substrate to phosphorylate specific target molecules. The challenge clearly lies in obtaining specificity for a given kinase, thus allowing inhibition or activation of a specific pathway. This chapter will focus on two areas of kinase inhibitors, those that target the MAP kinase pathway and those directed against the phosphatidylinositol-3 kinase (PI-3K) related kinase family. The cellular and physiological effects of inhibition of the various pathways controlled by these kinases will be reviewed.

Mitogen-activated protein kinase

Three-kinase module assembly

Mitogen-activated protein kinases (MAP kinases or MAPKs) are protein kinases that become active in response to extracellular stimuli, including cytokines, growth factors, neurotransmitters, hormones and various environmental stimuli. Each active MAPK phosphorylates a subset of transcriptional activators to regulate genes responsible for proliferation, apoptosis and morphogenesis. The common theme of MAPK signaling pathways is the assembly of three protein kinase modules. MAPK is activated by MAP kinase kinase or MKK, which itself needs to be activated by MAP kinase kinase kinase or MKKK. In mammalian cells, there are at least 14 MKKKs, 7 MKKs, and 12 MAPKs [1]. The MAP kinase family can be categorized into five subfamilies, the ERK1/2, ERK3/4, p38, JNK and ERK5. The dual phosphorylation sites in the activation loop of these MAPK subfamilies are T-E-Y, S-E-G, T-G-Y, T-P-Y and T-E-Y, respectively. However, the recognition between each MAPK and its specific MKK is primarily through the tertiary structure, and the sequence of the phosphorylation site alone is not adequate for specific MKK/MAPK recognition [2, 3]. As shown in Figure 1, an intricate and specific MKK/MAPK activation scheme can operate in a given cell.

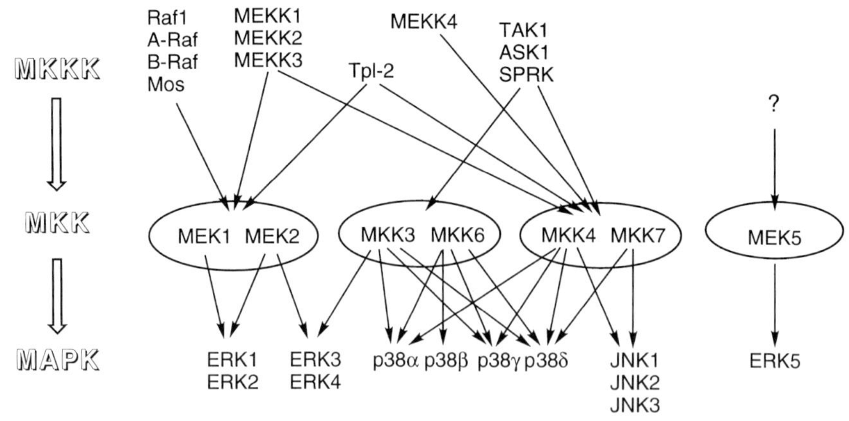

Figure 1. MAP kinase signaling pathways.

Raf/MEK/ERK signaling pathway

The Raf/MEK/ERK pathway is the first identified three-sequential protein kinase module. Raf1 is activated by protein kinase C (PKC) at S449, a tyrosine kinase such as Src at Y340, Y341, or in combination with binding the small G-protein Ras [4]. These Raf1-activating signals are initiated from cytokine receptors, tyrosine kinase receptors, G-protein coupled receptors or cellular adhesion. In addition to there being numerous entry points into the Raf/MEK/ERK signaling pathway, Raf1 is not the only MKKK that activates the MEK/ERK pathway. Other known ERK1,2 activators include A-Raf, B-Raf, Mos, Tpl-2, MEKK 1, 2 and 3. This diverse entry to activating ERK1,2 allows the outcome of ERK1,2 activation to be as diverse as the cell types and tissues. Activation of ERK1,2 is in general correlated with cellular proliferation, but the strength and the duration of the signal determine whether the cell proliferates, differentiates or undergoes other changes [1].

Inhibitors specific for Raf/MEK/ERK signaling pathway

Currently there are no Raf inhibitors that act on Raf directly, but inhibitors are available that decrease Raf activity indirectly. CGP 69846A is a 20-mer phosphorothioate oligodeoxynucleotide directed against human C-raf kinase gene expression [5]. This antisense nucleotide, in combination with anti-cancer drugs, has shown additive effects against human small-cell lung cancers [6]. Raf1 exists in complex with HSP90 and other proteins in cells. Natural products such as herbimycin A [7] and geldanamycin [8] bind HSP90 and disrupt the Raf1-HSP90 multimolecular complex, leading to destabilization of Raf-1.

Radicicol causes selective depletion of Raf kinase and disrupts the Ras- activated signaling pathway, possibly through a similar mechanism [9].

PD098059 blocks the activation of MEK by Raf [10]. This inhibitor appears to bind the unactivated MEK and prevents phosphorylation at the activation loop. The constitutively active mutant MEK1(S217E,S221E) also binds the inhibitor, but the active, doubly phosphorylated MEK1 does not. PD098059 is about 10 times more potent against MEK1 than MEK2 at IC_{50} = 2–7 μM and 50 μM, respectively. U0126 is an inhibitor specific for MEK1/2 with an affinity 100 times higher than PD098059 [11]. Both PD098059 and U0126 appear to be specific tools for shutting off the Raf1/MEK1,2/ERK1,2 signaling pathway.

p38 pathway inhibitor

SK&F 86002 is a pyridinylimidazole compound initially known as an anti-inflammatory agent that inhibits interleukin (IL)-1 synthesis in human monocytes and later identified as a p38 inhibitor (IC_{50} = 260 nM) [12]. SB203580 and SB202190 are two potent pyridinylimidazole inhibitors for p38 with IC_{50} = 42 nM and 50 nM, respectively [13]. They bind both the inactivated and activated forms of p38. Analogs of SB203580, such as RWJ 68354, have been reported to be about 7 times more potent against p38, and about 5 times against p38-dependent tumor necrosis factor α (TNF-α) release [14]. These inhibitors were active only against p38α, p38β but not p38γ [15] or p38δ [16].

JNK specific inhibitor

CEP1347/KT7515 is a derivative of a broad-spectrum indolocarbazole inhibitor K-252a. CEP1347 is a less potent kinase inhibitor than K-252a but it blocks JNK1 activation and can reverse apoptosis in neurons [17]. This compound is specific for the JNK1 pathway but not the p38- or ERK-activated pathways. JNK1 does not appear to be the direct target, and it is not clear which upstream activator is the real target for CEP1347 [17]. The p38 inhibitor SB203580 is known to inhibit JNK2 at 10 μM, but not JNK1 at this concentration [18, 19].

Crosstalk between different MAPK pathways

As outlined in Figure 1, there are a limited number of MAPKs, and the recognition between MKK/MAPK is relatively specific. However, each MKK/MAPK pathway can be activated by multiple MKKKs, and each MKKK can be activated by multiple upstream signals. The MEK1,2/ERK1,2 pathway can be activated by the Raf subfamily and by the MEKK1,2,3 subfamily. Raf activation could be the result of a broad spectrum of upstream signals including those from growth factor receptors, cytokine receptors, G-protein coupled receptors and through mediators such as JAK/STAT, Ras, Src, PI-3K and PKC [1]. Activation of MEKK1 has been shown to affect multiple down-stream pathways, including ERK, JNK, NFκB. MEKK1-4 do not activate the p38 pathway in some cells; however, the JNK-specific MKK4,7 are capable of activating the p38δ isoform [16]. In T-cells all three MAPK pathways, ERK, p38 and JNK, are critical for regulation of TNFα expression [20].

NF-κB pathway

NF-κB is a transcription factor which activates groups of genes in response to extracellular stimuli that promote inflammation and apoptosis [21]. The mechanism for the NF-κB activation has some similarity to the MAPK pathways, particularly the sequential phosphorylation activation and cross-talk with the MAPK pathways. The cytoplasmic form of NF-κB is the p50/p65 heterodimer sequestered in a complex with IκB inhibitor proteins. Phosphorylation of Ser32 and Ser36 in IκB-α, or Ser19 and Ser23 in IκB-β, by IκB kinase (IKK) results in dissociation of the NF-κB complex, and subsequent ubiquitination and degradation of IκB-α and nuclear translocation of the NF-κB heterodimer. The inactive form of IKK is a heterocomplex of IKK-α, IKK-β and non-kinase unit IKK-γ. Activation of IKK is mediated through phosphorylation on Ser176,180 in IKK-α or Ser177,181 in IKK-β. NF-κB inducing kinase (NIK) and MEKK1,2,3 are the known activators for IKK [22].

NF-κB inhibitors

Several anti-inflammatory sesquiterpene lactones are known to inhibit transcriptional activation by NF-κB. Parthenolide was reported to prevent degradation of IκB-α and IκB-β, [23] while Helenalin selectively alkylates the p65 subunit of NF-κB without affecting the nuclear translocation of NF-κB [24]. The anti-inflammatory drug BAY 11-7082 inhibits activation of NF-κB and it was found to inhibit cytokine-inducible IκB-α phosphorylation at concentrations of 10 μM [25]. Aspirin also inhibits NF-κB, in addition to its more well-known role as a cyclooxygenase inhibitor. Aspirin has been shown to inhibit IκB-β but not IκB-α at concentrations of 50 μM [26]. These two inhibitors are important leads for designing specific and potent anti-arthritis agents.

PI-3 kinases; lipids, protein or both

Members of the PI-3K superfamily are also involved in signal transduction pathways in response to external stimuli. PI-3K family members catalyze the phosphorylation of specific lipids, specific proteins and in some cases both. Lipid phosphorylation occurs at the D-3 position of the inositol ring of PI, PI(4)P and PI(4,5)P_2, generating PI(3)P, PI(3,4)P_2 and PI(3,4,5)P_3, respectively. The hydrolysis of these phospholipids results in the release of the various inositol-phosphates that serve as important second messengers in a variety of signal transduction pathways. The mammalian p110α/p85 PI-3K has also been shown to catalyze protein phosphorylation, specifically autophosphorylation of the p85 regulatory subunit which results in a decrease in kinase activity [27].

PI-3K-related kinases

Other PI-3K family members exhibit only protein kinase activity and these have been classified as PI-3K-related kinases. This class consists of large molecular weight proteins that all share homology in a C-terminal lipid kinase domain similar to the mammalian p110α PI-3K [28]. The protein kinase activity attributed to various members of the PI-3K-related kinases are specific for serine and threonine and often includes the ability to autophosphorylate the catalytic subunit and in some cases the regulatory subunits. PI-3K-related kinases play important roles in a variety of processes, including cell cycle regulation and the cellular response to DNA damage [29]. PI-3K-related kinase family members include DNA-PK (DNA-dependent protein kinase), FRAP (FKBP12/rapamycin-binding protein), ATM (ataxia telangiectasia mutated) and the ATR (ataxia telangiectasia and Rad3-like) kinase. While all four of these protein kinases share sequence homology with the mammalian lipid kinase p110α PI-3 kinase, each has no detectable lipid kinase activity. FRAP kinase is involved in cell cycle regulation in response to extracellular signals. DNA-PK is required for double-strand break (DSB) repair in a non-homologous end joining (NHEJ) pathway [30]. Similarly, ATM is involved in the cellular response to DNA damage and these cells are hypersensitive to ionizing radiation (IR) and exhibit defects in cell cycle checkpoint regulation as measured by radioresistant DNA synthesis [31]. ATR has also been implicated in DNA damage-response signaling pathways as alterations are observed in ATR-deficient cells in response to both UV- and IR-induced DNA damage [32]. Specific and general inhibitors of the PI-3K-related kinases have been used to elucidate the pathways controlled by these kinases as well as to achieve a beneficial physiological effect.

FRAP kinase inhibition by rapamycin

FRAP (also called mTOR and RAFT 1) is involved in cell cycle progression from the G_1 to S phase of the cell cycle. FRAP-dependent signaling pathways are sensitive to the immunosuppressant drug rapamycin which results in a cell cycle arrest in G_1 phase [33]. Rapamycin exerts its inhibitory effect on FRAP via a direct interaction with the 12 kDa cellular protein FKBP12 (FK506 binding protein). A ternary complex between FKBP12-rapamycin and FRAP results in a decrease in FRAP phosphorylation activity. The rapamycin-FKBP12 binding domain of FRAP is required for kinase activity and progression through the G_1 phase of the cell cycle and has been mapped to be N-terminal in relation to the kinase domain [34].

Events downstream of FRAP include the regulation of cyclin-dependent kinase (cdk) inhibitor expression and effects on the efficiency of protein synthesis. The first pathway has been studied in a variety of cells where cell cycle arrest is observed upon rapamycin treatment. A series of particularly relevant studies using IL-2-activated T cells demonstrated that rapamycin-induced G_1 arrest occurs via maintaining the expression of the cdk inhibitor, $p27_{kip}$ [35]. Interestingly, IL-2 alone was shown to induce expression of p21, another cdk inhibitor which was inhibited by rapamycin. Therefore, it was suggested that $p27_{kip}$ is responsible for regulating the exit from G_o into G_1 phase of the cell cycle while p21 regulates progression through G_1 in actively cycling cells. Interestingly, the immunosuppressant activity of rapamycin can be observed independent of IL-2, as immunosuppression is also observed in IL-2 knockout mice upon treatment with rapamycin. This raises the possibility that other cytokines can be regulated by rapamycin to mediate the immunosuppressive effect.

FRAP effects on protein synthesis can be attributed to two components, the ribosomal p70 S6 kinase ($p70_{S6K}$) and the eukaryotic initiation factor 4E (eIF4E) binding factor (4E-BP1/PHAS-1). FRAP has been shown to regulate $p70_{S6K}$ *in vivo*, but it is still unclear if there are intermediate steps in the pathway [36]. FRAP has been demonstrated to directly phosphorylate the eIF4E binding factor 4E-BF1 *in vitro* and potentially *in vivo*. The role of 4E-BF1 is to bind eIF4E and inhibit the ability of the protein to bind capped mRNAs. The binding of eIF4E is a limiting step in translation for mRNAs with 5' untranslated regions that can form secondary structures. Phosphorylation of 4E-BF1 results in dissociation of the complex and releases eIF4E to function in initiation of translation. Rapamycin-dependent inhibition of FRAP results in an increase in unphosphorylated 4E-BP1 and hence the inactivation of eIF4E and a subsequent decrease in protein synthesis [37]. The immunosuppressive effect of rapamycin is mediated by FRAP to induce cell cycle arrest by maintaining expression of $p27_{kip}$ and by decreasing the expression of a subset of proteins at the level of translation.

DNA-PK, ATM and ATR pathways

The human DNA-PK is a heterotrimer consisting of a dimeric DNA-binding subunit termed Ku and a 460 kDa catalytic subunit termed DNA-PKcs. The C-terminal of DNA-PKcs contains the lipid kinase domain homologous to PI-3K.

A number of *in vitro* targets of DNA-PK have been identified including p53, the 32 kDa subunit of replication protein A (RPA) and MDM2 [38, 39]. Definitive determination of the *in vivo* targets of DNA-PK remains. This is partly a result of the considerable overlap of DNA-PK with ATM and ATR in terms of target proteins and pathways regulated. One unique feature of DNA-PK is the requirement for DNA to stimulate protein kinase activity, and specifically DNA termini [40, 41]. A variety of different DNA structures have been assessed and the overriding feature is that all substrates which support phosphorylation by DNA-PK have discontinuities, usually either single-strand breaks or DSBs. This result is consistent with the Ku DNA-binding activity which preferentially binds at DNA ends. The DNA-end binding activity of Ku is also consistent with the physiological role of DNA-PK which is in the processing of DSBs in DNA. The variable-diversity joining (V(D)J) recombination pathway proceeds via a DSB intermediate and DNA-PK is required for this process. Likewise, the NHEJ pathway for DSB repair requires DNA-PK. Considering these activities, in retrospect it was not surprising that mutations in any of the three DNA-PK subunits result in hypersensitivity to IR.

Similar to DNA-PK, mutations in ATM result in hypersensitivity to IR. In addition, AT cells also exhibit defects in cell cycle checkpoint arrest following DNA damage, which is characterized by radioresistant DNA synthesis [42, 43]. Following a DNA damage event, induction of cell cycle arrest is initiated, presumably to allow time for repair of the damage. The defect in AT results in continued cell cycle progression following IR-induced DNA damage, such that cells are capable of entry into S-phase. This provides a convenient end-point and determination of DNA synthesis levels following IR treatment gives a measure of checkpoint abrogation. AT is also characterized by neurological defects and an increased rate of cancers including lymphoma and leukaemia [44]. Consistent with this finding, ATM has recently been shown to have tumour suppressor functions in a lymphoma cell line and ATM mutations were observed in B-cell chronic lymphocytic leukaemia [45, 46]. Clearly the similarities suggest that there may be some redundancy in the roles of DNA-PK, and ATM in the cell's response to DNA damage, and specifically DSBs. Both ATM [47, 48] and DNA-PK can phosphorylate p53 in response to DNA damage and more recently this was demonstrated for another PI-3K-related kinase ATR [49]. ATR also possesses protein kinase activity [50] and defects in cell cycle checkpoint arrest in response to DNA damage are observed in cells containing mutant ATR [32, 51]. A role in chromosomal structure and alignment has also been proposed as both ATM and ATR have recently been shown to be associated with meiotic chromosomes, ATM after chromosome pairing and ATR prior to chromosome pairing [50]. Despite these similarities, phenotypic differences are observed in cells harboring AT or DNA-PK mutations, which suggests there are unique pathways controlled by each protein.

Inhibition of DNA-PK and ATM by wortmannin

Wortmannin is a fungal metabolite that was originally found to inhibit the mammalian PI-3K [52, 53]. Inactivation of PI-3K proceeds via the covalent modification of an active site lysine and is irreversible [54]. The specificity of this compound was originally thought to be fairly high and in fact other classes of kinases are relatively insensitive to wortmannin. Wortmannin inhibition of p110α PI-3K has an IC_{50} in the 1–5 nM range and the next most sensitive class of kinases was found to be the myosin light chain kinase family with inhibition observed in the 300 nM range [55]. However, when DNA-PK and ATM were found to have homology with PI-3K, a study of the effect of wortmannin also showed inhibition with IC_{50} values of 16 nM for DNA-PK and 150 nM for ATM [56]. Wortmannin has also been reported to inhibit FRAP at sub µM concentration [57]. ATR is relatively insensitive to wortmannin with IC_{50} values in the low µM range [56]. In light of the diverse pathways that are regulated by the PI-3K family and the family of related kinases and similar IC_{50} values, a reevaluation of which kinases targeted by wortmannin regulate which pathways is needed.

The physiological effects of wortmannin have been studied in a variety of cell types and can be divided into cell cycle effects, where the target is likely to be p110α PI-3K and possibly FRAP, and DNA damage recognition, where the target is likely to be DNA-PK ATR or ATM. Wortmannin has been found to be an effective sensitizer to IR, rendering treated cells more sensitive to IR [58, 59]. This effect is especially relevant to cancer chemotherapy, where sensitization of tumours to IR can be an effective means to treat a variety of cancers, including head and neck cancer. This sensitization effect of wortmannin is correlated with the ability to repair DSBs and not to the amount of damaged induced by the IR treatment [59]. It has also been demonstrated that the repair of single-strand breaks is not affected by wortmannin. Consistent with this finding, wortmannin does not sensitize cells to mitoycin C or UV irradiation while the sensitization effect is observed in cells treated with bleomycin, a drug that induces double strand DNA breaks [60]. The IR-sensitization activity of wortmannin was concluded to be the result of inhibition of DNA-PK, based on similarities in the inhibition curve for DNA-PK kinase activity and the IR-sensitization profile, and the finding that AT cells were sensitized to IR by wortmannin [61]. Conflicting data, however, were obtained using a different AT cell line where wortmannin was unable to sensitize to IR [60]. Considering the overlapping target proteins and the similar phenotypes of ATM –/– and DNA-PK –/– cells, it is clearly possible that wortmannin effects on both ATM and DNA-PK could be responsible for the sensitization effect.

Other evidence also points to wortmannin-dependent sensitization to IR resulting from specific inhibition of DNA-PK. We isolated and purified the human DNA-PK based on its ability to bind duplex DNA damaged with the chemotherapeutic drug cisplatin [62]. Cisplatin forms a series of intra- and inter-strand DNA adducts that inhibit transcription and replication, induce cell cycle arrest and ultimately induce apoptosis. We demonstrated that the Ku subunits of DNA-PK bind cisplatin- damaged DNA with an affinity similar to that for undamaged DNA. Interestingly, when Ku is bound to cisplatin-damaged DNA, DNA-PKcs is unable to catalyze phosphorylation of target proteins [62, 63]. The conclusion is that cisplatin-damaged DNA is an inhibitor of DNA-PK. If the hypothesis that wortmannin-dependent inhibition of DNA-PK results in radiosensitization, then cisplatin should also serve to sensitize cells to IR. In fact cisplatin can be effective as a clinical radiosensitizer and numerous trials are underway to better characterize the dose and scheduling to optimize the radiosensitization effect. Initial hypotheses suggested that the sensitization activity of cisplatin was the result of an increase in DSBs in treated cells. However, more recently, the effect of cisplatin in radiosensitization was shown to be via a decreased ability to repair DSBs, and not the generation of a greater number of DSBs following exposure to IR [64]. Collectively, these results argue that the cisplatin sensitization to IR results from the inhibition of DNA-PK.

Other effects of wortmannin inhibition of PI-3K

In addition to sensitization to IR, wortmannin has been reported to induce a variety of other physiological effects. These include antitumor activity that was observed *in vitro* using cells in culture; however, the antitumor was not observed in an *in vivo* mouse model. Wortmannin was hypothesized to target PI-3K in this study; however, IC_{50} values in the low μM range were reported, suggesting that the effect may be the result of other PI-3K or related kinases [65]. Signaling from a variety of cellular receptors is affected by wortmannin. Erythropoietin-induced differentiation is inhibited by wortmannin and decreases in both haemoglobin and glycophorin A expression was observed [66]. Wortmannin has also been shown to inhibit cellular receptor trafficking in two different systems. An endocytic pathway was inhibited by wortmannin where alterations in the transport of the platelet-derived growth factor (PDGF) receptor were observed [67]. Effects of insulin transport were observed in response to wortmannin. The effect was mediated by the glucose transporter 4 (GLUT 4) translocation which was inhibited by wortmannin in a cell culture model [68]. Subsequently, using a perfused rat heart model, insulin-mediated translocation of GLUT 4 was demonstrated to be inhibited by wortmannin while translocation induced by ischaemia was not altered [69].

Concluding remarks

While significant advances have been achieved in the recent past concerning the identification of protein kinase inhibitors and their physiological effects, clearly there is much work to be done. The study of natural products has resulted in the identification of a number of interesting compounds that exhibit varying degrees of specificity in terms of targeting specific kinases and the pathways they control. The challenge will now largely be to modify these potential lead compounds to achieve a greater degree of specificity for the individual kinases. Inhibitors specific for a given class or family of kinases are available in some cases; however, the more difficult task will be to target the specific kinases within a given family. Considering that kinases are placed in the same family based largely on a significant degree of homology in the active site of the kinase, rational design of inhibitors will be required. This approach relies on extensive structural and kinetic characterization of the individual kinases which is an area of active investigation.

Acknowledgments
Research in the authors' lab has been supported by the National Institutes of Health (JJT) and the American Heart Association (MH). We thank the members of our labs for their help in preparing this review.

References

1 Widmann C, Gibson S, Jarpe MB, Johnson GL (1999) Mitogen-activated protein kinase: conservation of a three-kinase module from yeast to human. *Physiol Rev* 79: 143–80
2 Jiang Y, Li Z, Schwarz EM, Lin A, Guan K, Ulevitch RJ, Han J (1997) Structure-function studies of p38 mitogen-activated protein kinase. Loop 12 influences substrate specificity and autophosphorylation, but not upstream kinase selection. *J Biol Chem* 272: 11 096–102
3 Robinson MJ, Cheng M, Khokhlatchev A, Ebert D, Ahn N, Guan KL, Stein B, Goldsmith E, Cobb MH (1996) Contributions of the mitogen-activated protein (MAP) kinase backbone and phosphorylation loop to MEK specificity. *J Biol Chem* 271: 29 734–29 739
4 Morrison DK, Cutler RE (1997) The complexity of Raf-1 regulation. *Curr Opin Cell Biol* 9: 174–179
5 Monia BP, Sasmor H, Johnston JF, Freier SM, Lesnik EA, Muller M, Geiger T, Altmann KH, Moser H, Fabbro D (1996) Sequence-specific antitumor activity of a phosphorothioate oligodeoxyribonucleotide targeted to human C-raf kinase supports an antisense mechanism of action *in vivo*. *Proc Natl Acad Sci USA* 93: 15 481–15 484
6 Geiger T, Muller M, Monia BP, Fabbro D (1997) Antitumor activity of a C-raf antisense oligonucleotide in combination with standard chemotherapeutic agents against various human tumors transplanted subcutaneously into nude mice. *Clin Cancer Res* 3: 1179–85
7 Schneider C, Sepp-Lorenzino L, Nimmesgern E, Ouerfelli O, Danishefsky S, Rosen N, Hartl FU (1996) Pharmacologic shifting of a balance between protein refolding and degradation mediated by Hsp90. *Proc Natl Acad Sci USA* 93: 14 536–14 541
8 Schulte TW, Blagosklonny MV, Romanova L, Mushinski JF, Monia BP, Johnston JF, Nguyen P, Trepel J, Neckers LM (1996) Destabilization of Raf-1 by geldanamycin leads to disruption of the Raf- 1-MEK-mitogen-activated protein kinase signalling pathway. *Mol Cell Biol* 16: 5839–5845
9 Soga S, Kozawa T, Narumi H, Akinaga S, Irie K, Matsumoto K, Sharma SV, Nakano H, Mizukami T, Hara M (1998) Radicicol leads to selective depletion of Raf kinase and disrupts K-Ras- activated aberrant signaling pathway. *J Biol Chem* 273: 822–828
10 Alessi DR, Cuenda A, Cohen P, Dudley DT, Saltiel AR (1995) PD 098059 is a specific inhibitor of the activation of mitogen-activated protein kinase kinase *in vitro* and *in vivo*. *J Biol Chem* 270: 27 489–27 494
11 Favata MF, Horiuchi KY, Manos EJ, Daulerio AJ, Stradley DA, Feeser WS, Van Dyk DE, Pitts WJ, Earl RA, Hobbs F et al (1998) Identification of a novel inhibitor of mitogen-activated protein kinase kinase. *J Biol Chem* 273: 18 623–18 632
12 Gallagher TF, Seibel GL, Kassis S, Laydon JT, Blumenthal MJ, Lee JC, Lee D, Boehm JC,

Fier-Thompson SM, Abt JW et al (1997) Regulation of stress-induced cytokine production by pyridinylimidazoles; inhibition of CSBP kinase. *Bioorgan Med Chem* 5: 49–64
13 Young PR, McLaughlin MM, Kumar S, Kassis S, Doyle ML, McNulty D, Gallagher TF, Fisher S, McDonnell PC, Carr SA (1997) Pyridinyl imidazole inhibitors of p38 mitogen-activated protein kinase bind in the ATP site. *J Biol Chem* 272: 12 116–12 121
14 Henry JR, Rupert KC, Dodd JH, Turchi IJ, Wadsworth SA, Cavender DE, Fahmy B, Olini GC, Davis JE, Pellegrino-Gensey JL et al (1998) 6-Amino-2-(4-fluorophenyl)-4-methoxy-3-(4-pyridyl)-1H-pyrrolo[2, 3-b]pyridine (RWJ 68354): a potent and selective p38 kinase inhibitor. *J Med Chem* 41: 4196–4198
15 Enslen H, Raingeaud J, Davis RJ (1998) Selective activation of p38 mitogen-activated protein (MAP) kinase isoforms by the MAP kinase kinases MKK3 and MKK6. *J Biol Chem* 273: 1741–1748
16 Hu MC, Wang YP, Mikhail A, Qiu WR, Tan TH (1999) Murine p38-delta mitogen-activated protein kinase, a developmentally regulated protein kinase that is activated by stress and proinflammatory cytokines. *J Biol Chem* 274: 7095–7102
17 Maroney AC, Glicksman MA, Basma AN, Walton KM, Knight E Jr, Murphy CA, Bartlett BA, Finn JP, Angeles T, Matsuda Y et al (1998) Motoneuron apoptosis is blocked by CEP-1347 (KT 7515), a novel inhibitor of the JNK signaling pathway. *J Neurosci* 18: 104–111
18 Clerk A, Sugden PH (1998) The p38-MAPK inhibitor, SB203580, inhibits cardiac stress-activated protein kinases/c-Jun N-terminal kinases (SAPKs/JNKs). *FEBS Lett* 426: 93–96
19 Harada J, Sugimoto M (1999) An inhibitor of p38 and JNK MAP kinases prevents activation of caspase and apoptosis of cultured cerebellar granule neurons. *Jpn J Pharmacol* 79: 369–378
20 Hoffmeyer A, Grosse-Wilde A, Flory E, Neufeld B, Kunz M, Rapp UR, Ludwig S (1999) Different mitogen-activated protein kinase signaling pathways cooperate to regulate tumor necrosis factor alpha gene expression in T lymphocytes. *J Biol Chem* 274: 4319–4327
21 May MJ, Ghosh S (1999) IkappaB kinases: kinsmen with different crafts. *Science* 284: 271–273
22 Zhao Q, Lee FS (1999) Mitogen-activated protein kinase/ERK kinase kinases 2 and 3 activate nuclear factor-kappaB through IkappaB kinase-alpha and IkappaB kinase-beta. *J Biol Chem* 274: 8355–8358
23 Hehner SP, Heinrich M, Bork PM, Vogt M, Ratter F, Lehmann V, Schulze-Osthoff K, Droge W, Schmitz ML (1998) Sesquiterpene lactones specifically inhibit activation of NF-kappa B by preventing the degradation of I kappa B-alpha and I kappa B-beta. *J Biol Chem* 273: 1288–1297
24 Lyss G, Knorre A, Schmidt TJ, Pahl HL, Merfort I (1998) The anti-inflammatory sesquiterpene lactone helenalin inhibits the transcription factor NF-kappaB by directly targeting p65. *J Biol Chem* 273: 33 508–33 516
25 Pierce JW, Schoenleber R, Jesmok G, Best J, Moore SA, Collins T, Gerritsen ME (1997) Novel inhibitors of cytokine-induced IkappaBalpha phosphorylation and endothelial cell adhesion molecule expression show anti-inflammatory effects *in vivo*. *J Biol Chem* 272: 21 096–21 103
26 Yin MJ, Yamamoto Y, Gaynor RB (1998) The anti-inflammatory agents aspirin and salicylate inhibit the activity of I(kappa)B kinase-beta. *Nature* 396: 77–80
27 Fruman D, Meyers R, Cantley L (1998) Phosphoinositide kinases. *Annu Rev Biochem* 67: 481–507
28 Keith C, Schreiber S (1995) PIK-related kinases: DNA repair, recombination, and cell cycle checkpoints. *Science* 270: 50–51
29 Abraham R (1996) Phosphatidylinositol 3-kinase related kinases. *Curr Opin Immunol* 8: 412–418
30 Critchlow SE, Jackson SP (1998) DNA-end-joining: from yeast to man. *Trends Biochem Sci* 23: 394–398
31 Lavin MF, Shiloh Y (1997) The genetic defect in ataxia-telangiectasia. *Annu Rev Immunol* 15: 177–202
32 Wright J, Keegan K, Herendeen D, Bentley N, Carr A, Hoekstra M, Concannon P (1998) Protein kinase mutants of human ATR increase sensitivity to UV and ionizing radiation and abrogate cell cycle checkpoint control. *Proc Natl Acad Sci USA* 95: 7445–7450
33 Brown E, Albers M, Shin T, Ichikawa K, Keith C, Lane W, Schreiber S (1994) A mammalian protein targeted by G1-arresting rapamycin-receptor complex. *Nature* 369: 756–758
34 Vilella-Bach M, Nuzzi P, Fang YM, Chen J (1999) The FKBP12-rapamycin-binding domain is required for FKBP12-rapamycin-associated protein kinase activity and G(1) progression. *J Biol Chem* 274: 4266–4272
35 Nourse J, Firpo E, Flanagan W, Coats S, Polyak K, Lee M, Massague J, Crabtree G, Roberts J

(1994) Interleukin-2-mediated elimination of the p27Kip1 cyclin-dependent kinase inhibitor prevented by rapamycin. *Nature* 372: 570–573
36 Brown E, Beal P, Keith C, Chen J, Shin T, Schreiber S (1995) Control of p70 s6 kinase by kinase activity of FRAP *in vivo*. *Nature* 377: 441–446
37 Beretta L, Gingras A, Svitkin Y, Hall M, Sonenberg N (1996) Rapamycin blocks the phosphorylation of 4E-BP1 and inhibits cap-dependent initiation of translation. *EMBO J* 15: 658–664
38 Anderson CW (1993) DNA damage and the DNA-activated protein kinase. *Trends Biochem Sci* 18: 433–437
39 Mayo LD, Turchi JJ, Berberich SJ (1997) Mdm-2 phosphorylation by DNA-dependent protein kinase prevents interaction with p53. *Cancer Res* 57: 5013–5016
40 Gottlieb TM, Jackson SP (1993) The DNA-dependent protein kinase: requirement for DNA ends and association with Ku antigen. *Cell* 72: 131–142
41 Morozov VE, Falzon M, Anderson CW, Kuff EL (1994) DNA-dependent protein kinase is activated by nicks and larger single-stranded gaps. *J Biol Chem* 269: 16 684–16 688
42 Xu Y, Ashley T, Brainerd E, Bronson R, Meyn M, Baltimore D (1996) Targeted disruption of ATM leads to growth retardation, chromosomal fragmentation during meiosis, immune defects, and thymic lymphoma. *Gene Develop* 10: 2411–2422
43 Xu Y, Baltimore D (1996) Dual roles of ATM in the cellular response to radiation and in cell growth control. *Gene Develop* 10: 2401–2410
44 Canman C, Lim D (1998) The role of ATM in DNA damage responses and cancer. *Oncogene* 17: 3301–3308
45 Stankovic T, Weber P, Stewart G, Bedenham T, Murray J, Byrd P, Moss P, Taylor A (1999) Inactivation of ataxia telangiectasia mutated gene in B-cell chronic lymphocytic leukaemia. *Lancet* 353: 26–29
46 Bullrich F, Rasio D, Kitada S, Starostik P, Kipps T, Keating M, Albitar M, Reed J, Croce C (1999) ATM mutations in B-cell chronic lymphocytic leukemia. *Cancer Res* 59: 24–27
47 Canman C, Lim D, Cimprich K, Taya Y, Tamai K, Sakaguchi K, Appella E, Kastan M, Siliciano J (1998) Activation of the ATM kinase by ionizing radiation and phosphorylation of p53. *Science* 281: 1677–1679
48 Banin S, Moyal L, Shieh S, Taya Y, Anderson C, Chessa L, Smorodinsky N, Prives C, Reiss Y, Shiloh Y et al (1998) Enhanced phosphorylation of p53 by ATM in response to DNA damage. *Science* 281: 1674–1677
49 Tibbetts R, Brumbaugh K, Williams J, Sarkaria J, Cliby W, Shieh S, Taya Y, Prives C, Abraham R (1999) A role for ATR in the DNA damage-induced phosphorylation of p53. *Gene Develop* 13: 152–157
50 Canman C, Kastan M (1997) Role of p53 in apoptosis. *Adv Pharm* 41: 429–460
51 Cliby W, Roberts C, Cimprich K, Stringer C, Lamb J, Schreiber S, Friend S (1998) Overexpression of a kinase-inactive ATR protein causes sensitivity to DNA-damaging agents and defects in cell cycle checkpoints. *EMBO J* 17: 159–169
52 Powis G, Bonjouklian R, Berggren M, Gallegos A, Abraham R, Ashendel C, Zalkow L, Matter W, Dodge J, Grindey G et al (1994) Wortmannin, a potent and selective inhibitor of phosphatidylinositol-3-kinase. *Cancer Res* 54: 2419–2423
53 Arcaro A, Wymann M (1993) Wortmannin is a potent phosphatidylinositol 3-kinase inhibitor: the role of phosphatidylinositol 3,4,5-trisphosphate in neutrophil responses. *Biochem J* 296: 297–301
54 Wymann M, Bulgarelli-Leva G, Zvelebil M, Pirola L, Vanhaesebroeck B, Waterfield M, Panayotou G (1996) Wortmannin inactivates phosphoinositide 3-kinase by covalent modification of Lys-802, a residue involved in the phosphate transfer reaction. *Mol Cell Biol* 16: 1722–1733
55 Nakanishi S, Kakita S, Takahashi I, Kawahara K, Tsukuda E, Sano T, Yamada K, Yoshida M, Kase H, Matsuda Y et al (1992) Wortmannin, a microbial product inhibitor of myosin light chain kinase. *J Biol Chem* 267: 2157–2163
56 Sarkaria J, Tibbetts R, Busby E, Kennedy A, Hill D, Abraham R (1998) Inhibition of phosphoinositide 3-kinase related kinases by the radiosensitizing agent wortmannin. *Cancer Res* 58: 4375–4382
57 Brunn G, Williams J, Sabers C, Wiederrecht G, Lawrence JJ, Abraham R (1996) Direct inhibition of the signaling functions of the mammalian target of rapamycin by the phosphoinositide 3-kinase inhibitors, wortmannin and LY294002. *EMBO J* 15: 5256 5267
58 Price B, Youmell M (1996) The phosphatidylinositol 3-kinase inhibitor wortmannin sensitizes murine fibroblasts and human tumor cells to radiation and blocks induction of p53 following DNA

damage. *Cancer Res* 56: 246–250
59 Boulton S, Kyle S, Yalcintepe L, Durkacz B (1996) Wortmannin is a potent inhibitor of DNA double strand break but not single strand break repair in Chinese hamster ovary cells. *Carcinogenesis* 17: 2285–2290
60 Hosoi Y, Miyachi H, Matsumoto Y, Ikehata H, Komura J, Ishii K, Zhao H, Yoshida M, Takai Y, Yamada S et al (1998) A phosphatidylinositol 3-kinase inhibitor wortmannin induces radioresistant DNA synthesis and sensitizes cells to bleomycin and ionizing radiation. *Int J Cancer* 78: 642–647
61 Chernikova S, Wells R, Elkind M (1999) Wortmannin sensitizes mammalian cells to radiation by inhibiting the DNA-dependent protein kinase-mediated rejoining of double-strand breaks. *Radiat Res* 151: 159–166
62 Turchi JJ, Henkels K (1996) Human Ku autoantigen binds cisplatin-damaged DNA but fails to stimulate human DNA-activated protein kinase. *J Biol Chem* 271: 13 861–13 867
63 Turchi JJ, Patrick SM, Henkels KM (1997) Mechanism of DNA-dependent protein kinase inhibition by cis-diamminedichloroplatinum(II)-damaged DNA. *Biochemistry* 36: 7586–7593
64 Dolling JA, Boreham DR, Brown DL, Mitchel REJ, Raaphorst GP (1998) Modulation of radiation-induced strand break repair by cisplatin in mammalian cells. *Int J Radiat Biol* 74: 61–66
65 Schultz RM, Merriman RL, Andis SL, Bonjouklian R, Grindey GB, Rutherford PG, Gallegos A, Massey K, Powis G (1995) *In vitro* and *in vivo* antitumor activity of the phosphatidylinositol-3-kinase inhibitor, wortmannin. *Anticancer Res* 15: 1135–1139
66 Kubota Y, Tanaka T, Yamaoka G, Yamaguchi M, Ohnishi H, Kawanishi K, Takahara J, Irino S (1996) Wortmannin, a specific inhibitor of phosphatidylinositol-3-kinase, inhibits erythropoietin-induced erythroid differentiation of K562 cells. *Leukemia* 10: 720–726
67 Shpetner H, Joly M, Hartley D, Corvera S (1996) Potential sites of PI-3 kinase function in the endocytic pathway revealed by the PI-3 kinase inhibitor, wortmannin. *J Cell Biol* 132: 595–605
68 Evans J, Honer C, Womelsdorf B, Kaplan E, Bell P (1995) The effects of wortmannin, a potent inhibitor of phosphatidylinositol 3-kinase, on insulin-stimulated glucose transport, GLUT4 translocation, antilipolysis, and DNA synthesis. *Cell Signal* 7: 365–376
69 Egert S, Nguyen N, Brosius F, Schwaiger M (1997) Effects of wortmannin on insulin- and ischemia-induced stimulation of GLUT4 translocation and FDG uptake in perfused rat hearts. *Cardiovasc Res* 35: 283–293

New Approaches to Drug Development
ed. by P. Jollès

Molecular approaches to receptors as targets for drug discovery

Dean B. Evans, Peter Traxler and Carlos García-Echeverría

Oncology Research, Novartis Pharma AG, CH-4002 Basel, Switzerland

Summary. Many receptors have been selected as viable drug discovery targets. One particular class of receptors that have received much interest and so far relatively good success are the receptor protein tyrosine kinases (RPTKs). Typically, RPTKs are activated following the binding of the peptide growth factor ligand to its receptor. The RPTKs play crucial roles in signal transduction pathways that regulate a number of cellular functions, such as cell differentiation and proliferation, both under normal physiological conditions as well as in a variety of pathological disorders. A variety of different tumour types have been shown to have dysfunctional RPTKs, either as a result of excess production of the growth factor, the receptor or both, or via mutations in the RPTKs structure. Irrespective of the cause, this leads to the over-activity of the particular RPTK system and in turn to the aberrant and inappropriate cellular signalling within the tumour cell. RPTKs are attractive targets in the search for therapeutic agents, not only against cancers but also against many other disease indications. Although an ever-increasing number of RPTKs have been selected as viable molecular targets for drug discovery programmes, four examples will be covered in this article. These are the epidermal growth factor receptor (EGF-R), platelet-derived growth factor receptor (PDGF-R), fibroblast growth factor receptor (FGR-R) and vascular endothelial growth factor receptor (VEGF-R), with the main emphasis of interest being on their role in oncology.

Introduction

With the onset of modern molecular, biochemical and cell biology technologies combined with advances made in synthetic chemistry, computer-assisted molecular modelling and crystallography, the number of receptors that are being utilized as molecular targets in drug discovery has greatly increased. Based on the particular biological processes that are regulated following the initiation of receptor signal transduction, the types of molecular approaches that have been considered in order to alter receptor function have predominantly included the design of either agonist or antagonist ligands to repress or activate the particular receptor signal transduction pathways. These agonist or antagonist ligands may represent molecules that either exert partial or full modulatory effects on the receptor. In addition to small molecular weight compounds, use has been made of other chemical entities, such as peptides, proteins and peptido-mimetic ligands. Whilst many approaches have utilized direct interactions via the ligand-binding domain of the receptor, a number of alternative approaches have been utilized which affect distinct sites on the receptor structure. For example, in cases where receptors need to dimerize to initiate downstream signal transduction pathways, methods aimed at the disruption of receptor dimerization have been approached. Additionally, the disruption of the interaction between the receptor with signal transducers has been explored, such as in the case of interactions of Grb2-SH2 with tyrosine kinase receptors [1–6]. The use of antibodies, antisense oligonucleotides, binding-proteins and soluble receptor fragments has also been designed.

Whilst there are many possibilities that have been utilized to target numerous classes of receptors, this review will focus on one particular family of receptors which have proven to be very attractive therapeutic targets for molecular disruption, namely, the receptor protein tyrosine kinases (RPTKs). Typically, RPTKs are activated following the binding of the peptide growth factor ligand to its receptor. The RPTKs play crucial roles in signal transduction pathways that regulate a number of cellular functions, such as cell differentiation and proliferation, both under normal physiological conditions as well as in a variety of pathological disorders. RPTKs are therefore attractive targets in the search for therapeutic agents, not only against cancers but also against many other disease indications. A variety of different tumour types have been shown to have dysfunctional RPTKs, either as a result of excess production of the growth factor, the receptor or both, or via mutations in the RPTKs structure. Irrespective of the cause, this leads to the over-activity of the particular RPTK system and in turn to the aberrant and inappropriate cellular signalling within the tumour cell.

About ten years ago, when research in the signal transduction field was initiated in many pharmaceutical companies, the epidermal growth factor receptor (EGF-R) was chosen as a "proto-type" drug discovery RPTK target. Meanwhile, many other RPTKs have been detected and found to be attractive targets for research programs in both oncological and non-oncological indications. Novel members of the family of the EGF-R PTK, such as the c-erbB2/HER-2/neu RPTK, have been identified and are used as targets for medicinal chemistry programs. In the case of the c-erbB2/HER-2/neu RPTK, the use of a recombinant humanized anti-HER2 antibody (Herceptin) has proven clinically successful [7]. In a phase-III clinical trial there is evidence that systemic administration of anti-HER2 antibodies (Herceptin) in patients with c-erbB2/HER-2/neu over-expressing primary tumours, can increase the time to recurrence of and overall response rates in metastatic breast cancer. In recent years a switch from the EGF-R target to other RTPKs (e.g., the PDGF-R, VEGF-R, FGF-R, etc.) has been observed. In the last decade, hundreds, or even thousands, of small molecular weight tyrosine kinases inhibitors have been described in the scientific or patent literature. The majority of these compounds have only served as experimental tools to establish *in vitro* or cellular assay systems and to prove their functionality. Only a few examples have demonstrated *in vivo* efficacy in relevant tumour models in nude mice. The first compounds, in most cases inhibitors of the EGF-R tyrosine kinase, that have entered preclinical development failed to fulfil many of the stringent pharmacodynamic (efficacy, selectivity), pharmacokinetic (resorption, metabolism), toxicological and technical (synthesis, formulation) requirements and therefore have never entered clinical trials. Despite some of these set-backs, the majority of these hurdles have been successfully overcome with successive generations of compounds. As a consequence, there are already several low molecular weight compounds that have been described as being more or less selective RPTK inhibitors that have reached early phases of clinical trials [8–10] (see Fig. 1 for details). In addition, several candidates are in a late stage of preclinical development and are close to entering Phase-I trials [8–10].

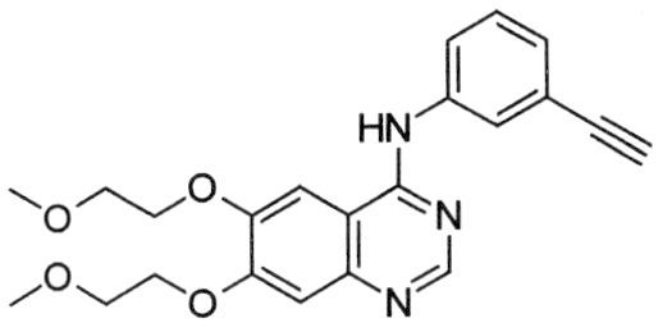

1

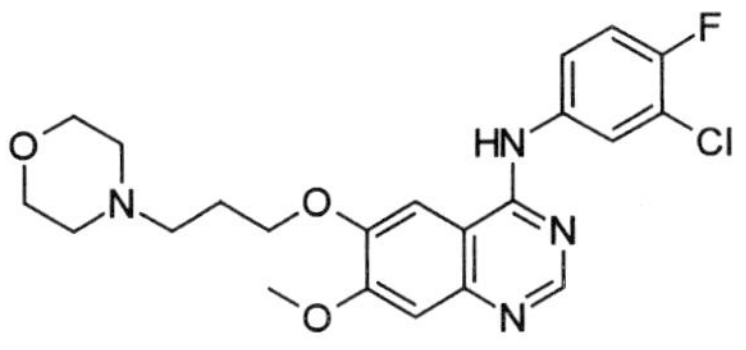

2

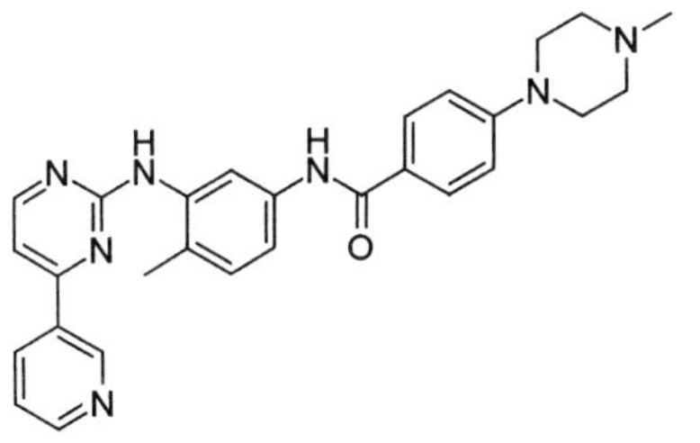

3

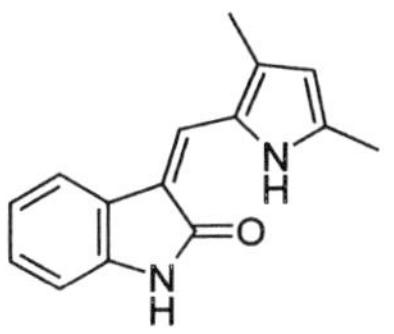

4

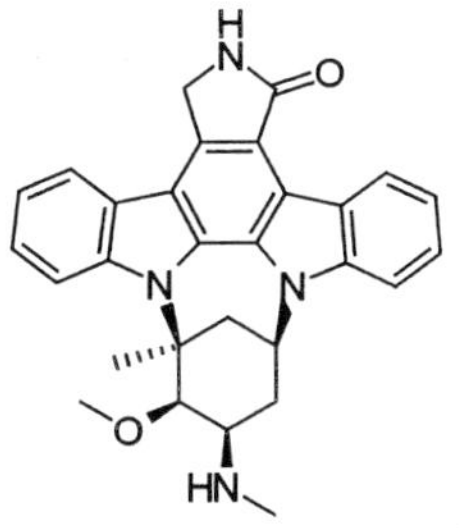

5

Figure 1. Examples of ATP site-directed receptor tyrosine kinase inhibitors in clinical trials: **1** (CP 358774; OSI Pharmaceuticals/Pfizer, EGF-R kinase), **2** (ZD 1839; Zeneca, EGF-R kinase), **3** (SU 5416; Sugen, VEGF-R kinase) and **4** (SU 5416; Sugen, VEGF-R kinase) and 5 (staurosporine).

The reader is directed to a very useful web site called “The Protein Kinase Resource” (http://www.sdsc.edu/kinases) which covers a very comprehensive list of protein kinases including the RPTKs [11]. “The Protein Kinase Resource” is a web-accessible resource of information on protein kinases which includes tools for structural and computational analyses and in addition provides links to other related web resources. Another useful general source of reference is “The Protein Kinase Factsbook: Protein-Tyrosine Kinases” [12].

When research activities in the pharmaceutical industries were initiated to aim at the disruption of signal transduction pathways using small molecular weight inhibitors, it was considered to be a real challenge, especially when based on aspects of selectivity. RPTKs use the gamma phosphate of adenosine triphosphate (ATP) to produce monoesters via protein phenolic groups on tyrosine residues as phosphate group acceptor sites. Common to all the RPTKs is a region known as the catalytic or kinase domain which serves several functions. These include the correct binding and orientation of the ATP phosphate donor in a complex with Mn^{2+} or Mg^{2+} divalent cations, binding and location of the RPTK substrate(s) and transfer of the gamma phosphate of ATP to the substrate, resulting in protein phosphorylation and initiation of the subsequent signal transduction cascades. Therefore, following receptor activation, a plethora of functional responses are modulated, based on the particular interactions of the receptor with different signal transducers.

RPTKs contain a general organization composed of an extracellular ligand-binding domain, a hydrophobic single transmembrane spanning region and an intracellular domain which not only contains the kinase domain but also key docking-sites required for interactions of signal transducers with the RPTK. In addition to receptors in which a ligand has been identified, there are some RPTKs which so far remain classified as orphan receptors and are still waiting to find a physiological ligand. In terms of primary structure, the amino acid sequences of the kinase domain show high conservation and are conveniently divided into sub-domains I–XI [13]. Generally, association of the growth factor with its receptor leads to receptor activation via receptor dimerization or oligomerization. The juxta-positioning of the individual RPTK leads to changes in intra/intermolecular conformations and autophosphorylation of the tyrosine residues located within the cytoplasmic domain of the receptor. In turn this leads to either substrate phosphorylation or creation of phospho-tyrosine docking sites for Src homology 2 (SH2)-containing signal transducers.

Although an ever-increasing number of RPTKs have been selected as viable molecular targets for drug discovery programmes, four examples will be now be covered. These are the EGF-R, platelet-derived growth factor receptor (PDGF-R), fibroblast growth factor receptor (FGR-R) and vascular endothelial growth factor receptor (VEGF-R) with the main emphasis of interest being on their role in oncology.

Epidermal growth factor receptor (EGF-R)

Refer to Voldborg et al., 1997 [14] for a recent review. One of the first growth factors to be identified was epidermal growth factor (EGF) which was later shown to bind to the EGF-R. The EGF-R is a growth factor receptor that induces cell proliferation and cell differentiation upon binding and activation by one of a number of its now known ligands. The EGF-R consists of a single polypeptide chain of 1186 amino acids that is expressed on the cell membrane of many different cell types. The extracellular N-terminal region has been divided up into four main domains (I–IV) with ligand interaction occurring with the domain III. In addition to EGF, several other ligands have been shown to bind to the EGF-R, including transforming growth factor α (TGF-α), betacellulin (BTC) and Heparin-binding EGF-like growth factor (HB-EGF).

Following ligand binding, the EGF-R receptor dimerizes potentially either as a homodimer or as a heterodimer with other members of the EGF-R receptor family. Receptor dimerization results in allosteric alterations in the intracellular portion of the receptor, leading to the activation of the intracellular tyrosine kinase activity and auto-phosphorylation of five cytoplasmic-located tyrosine residues on the EGF-R (Y992, Y1068, Y1086, Y1148 and Y1173). Once phosphorylated, these phospho-tyrosine residues then represent recognition sites for interactions of the SH2-containing substrates required for further transduction of the signal. If these autophosphorylation sites are mutated, then the interactions of these proteins with the EGF-R are severely impaired along with further signalling. SH2-containing proteins known to interact with the EGF-R include PLCγ1, GAP, *syp*, p85 PI3K, Shc, Grb-2 and Nck. Down-regulation of the EGF-R involves the internalization of the active receptor and targeting to lysosomes for degradation. Receptor desensitization, resulting in a reduced capacity for signalling, also occurs through phosphorylation of intracellular threonine and serine residues. Following ligand binding and receptor activation, the EGF-Rs cluster in regions known as coated pits. These are endocytosed into vesicles that eventually fuse with the lysosomes, resulting in both the EGF-R and the bound ligand being degraded. The tyrosine kinase activity of the EGF-R is required for the ligand-induced endocytosis, since kinase-inactive mutants are not shunted into this degradative process.

Much evidence has accumulated over recent years to implicate the EGF-R and its family members in the development and progression of numerous human tumours. In a large proportion of tumours, expression of the EGF-R can be changed as a result of over-expression or via mutations. In addition, the increased expression of the EGF-R ligands or co-expression of both receptor and ligand(s) within the tumour cells has the potential for autocrine regulation leading to continued EGF-R activation and an over-active receptor pathway. Consequently, aberrant EGF-R activity can lead to uncontrolled cellular processes, including cell proliferation leading to the development of a malignant condition. Gene amplifications have also been detected in a number of major tumour types and EGF-R over-expression has been documented in tumours such as breast, ovarian, bladder, lung, glioblastomas and squamous carcinomas. The EGF-R is used as a tumour prognostic marker because its over-expression has been correlated with a poor prognosis in several tumour types.

Since tyrosine phosphorylation is a critical requisite for the whole process of EGF-R activation, inactivation of the EGF-R via the kinase domain would demobilize the receptor and block signal transduction. Based on the clinical epidemiology and feasibility, the EGF-R has been viewed as an ideal and viable target in drug discovery programmes. The progress in synthetic chemistry to achieve this aim, via the use of small molecular weight ATP-competitive inhibitors of the kinase domain, is discussed later in this article.

Vascular endothelial growth factor receptor (VEGF-R)

Refer to Neufeld et al., 1999 [15] for a recent review. Angiogenesis represents an essential event in a variety of physiological and pathological processes that involve the formation of new blood vessels from an existing vascular network. Under normal physiological conditions, angiogenesis is restricted to processes associated with embryogenesis, ovulation and wound healing. However, under pathological conditions, angiogenesis is also associated with tumour growth and the formation of tumour metastases, inflammation, rheumatoid arthritis, ocular neovascularization and in psoriasis. Many different cytokines and growth factors have been shown to exert angiogenic activities. One of the main proteins involved is called vascular endothelial growth factor (VEGF), which appears to be a key factor in pathological situations involving neovascularization as well as with enhanced vascular permeability. The family of VEGF (VEGF-Rs) include VEGF-R1 (Flt-1, Fms-like tyrosine kinase) and VEGF-R2 (KDR, kinase insert domain-containing receptor) which belong to a larger family including the related PDGF-R, c-Kit, c-Fms, Flt-3 and Flt-4 receptors. Expression of VEGF-Rs is usually low in normal tissues and only increases during the development of pathological states associated with neovascularization. The VEGF-Rs are characterized by the presence of seven immunoglobulin-like domains in their extracellular domain, a single transmembrane-spanning region and an intracellular split tyrosine kinase domain. VEGF-R1 and VEGF-R2 are both glycosylated, with the latter receptor only capable of undergoing autophosphorylation in response to VEGF when it is in its final glycosylated form. The ligands that bind to VEGF-R2 are VEGF-A, VEGF-C and VEGF-D, whereas VEGF-R1 binds VEGF-A, VEGF-B and also the related placenta growth factors (PIGF). The VEGF-Rs form dimers that undergo autophosphorylation on cytoplamsic tyrosine residues. The VEGF-R homo- or heterodimer is held together by the interaction of the individual monomers via a shared dimerized VEGF ligand. This interaction is further stabilized via the dimerization of immunoglobulin-like domain 4 of the VEGF-R. Whilst activation of the VEGF-R2 by VEGF leads to a mitogenic response, VEGF-R1 following autophosphorylation does not promote cell proliferation. However, both VEGF-Rs induce cell migration. These observations suggest that these two receptors are capable of differential signal transduction in response to VEGF. However, there is limited information so far that accounts for this. VEGF is known to promote the binding and phosphorylation of the adaptor proteins Shc and Nyc, binding of Grb2 and the protein tyrosine phosphatases SHP-1 and SHP-2 to the VEGF-R2 and to activation of the downstream MAP kinase activity. Subsequent effects associated with VEGF-Rs functionality in the angiogenesis process include the production of proteases needed for the breakdown of blood vessel basement membranes, expression of certain integrins associated wth angiogenesis and stimulation of cell migration and proliferation.

The expression of VEGF is increased under hypoxic conditions around solid tumours, thereby potentiating the tumour-associated VEGF-induced angiogenesis. Inhibition of VEGF activity using neutralizing antibodies or antisense approaches directed to the VEGF-R leads to an inhibition of tumour growth and regression. Since the expression of VEGF receptors is practically restricted to endothelial cells, the tumour cells would not be targeted directly by any VEGF-R-based therapeutics. Rather it is the host vasculature cells which are targeted and this in turn affects tumour growth. Recently, the successful crystallization of the VEGF-R2 kinase domain has been reported [16], which should aid further medicinal chemistry activities.

Fibroblast growth factor receptor (FGF-R)

Refer to Klint and Claesson-Welsh 1999 [17] for a recent review. The FGF family of proteins is currently composed of about twenty different factors that show 30–70% identity in their primary amino acid sequences. FGFs are potent mitogens for a wide variety of different cell types and bind to four structurally-related FGF-Rs. The biological effects of FGFs are manifested as a result of binding to and activation of a family of FGF-Rs, resulting in transmission of intracellular signal cascades. Different FGF family members are known to activate the FGF-R sub-types to varying degrees and this appears to be related to the relative ligand-binding affinities with each receptor type. Dimerization of FGF-R monomers occurs upon ligand binding and this is required for the activation of the tyrosine kinase domain and receptor trans-phosphorylation. Receptor activation leads to autophosphorylation of cytoplasmic tyrosine residues that serve as high-affinity docking sites for a variety of SH2 domain-containing signal transduction molecules. These molecules transduce the signal cascades from the receptor and eventually result in the biological responses.

The four FGF-Rs have a similar overall structural organization. Alternative splicing of the extracellular domain creates variants composed of two or three immunoglobulin (Ig)-like domains. A distinct feature, that is critical for all of the FGF-Rs function, is a row of eight continuous acidic residues, referred to as the "acidic box". This acidic box is located between the first and second Ig-like domains. The extracellular portion of the FGF-Rs is connected via a single transmembrane-spanning domain to the cytoplasmic region composed of a juxtamembrane, kinase and C-terminal domains. The juxtamembrane domain of the FGF-Rs is much longer than those found in other tyrosine kinase receptors. Within the juxtamembrane region of the FGF-R1 and FGF-R2, a single tyrosine phosphorylation site exists. Howcvcr, thc FGF-R3 and FGF-R 4 lack such tyrosine residues in this domain. Like the VEGF-Rs, the tyrosine kinase domain of the FGF-Rs is divided into two parts separated via a short non-catalytic insert of approximately 15 amino acids. In this insert region of the FGF-R1 and FGF-R2, two phosphorylatable tyrosine residues are found. In the FGF-R3 a single tyrosine phosphorylation site is found whilst there are none in FGF-R-4. The latter observation is likely to account for the lower mitogenic potential of the FGF-R-4 than is found with the other FGF-Rs. The C-terminal tail region of the FGF-Rs contains a number of tyrosine residues, of which some are located at identical positions in this receptor class.

Following the binding of FGF, this leads to either homo- or heterodimerization of FGF-Rs. The mechanisms for FGF receptor dimerization are not yet as well defined as compared with other growth factor/receptor interaction models. The FGF ligands are generally thought to be monomeric growth factors. As for many other RPTKs, dimerization followed by conformational alterations of the FGF-R appears to be an important requirement for activation of the tyrosine kinase activity. The locations of the tyrosine phosphorylation sites in FGF-R1 have been studied in more detail compared to the other FGF-Rs. Excluding the kinase domain, there are five cytoplasmic tyrosine residues located in the receptor. Whereas Y463, Y583, Y585 and Y766 are potential phosphorylation and SH2-docking sites, Y776 does not appear to be phosphorylated. Once phosphorylated, only Y766 has been shown to interact with SH2-containing proteins, in this case binding phospholipase Cγ. However, tyrosine phosphorylation of Crk and the downstream signal transduction depends on the Y463. Within the FGF-R1 kinase domain, three tyrosine phosphorylation sites have been identified, namely, Y653, Y654 and Y730. Y653 and Y654 appear to be involved in the regulation of the kinase activity, since their mutation results in the loss of kinase function. The FGF-R1 also contains a number of serine residues, some, but not all, of which are phosphorylated following activation of the receptor.

FGFs are known to promote *in vitro* endothelial cell migration, proliferation and differentiation. In a similar way to VEGF, FGFs appear to play a major role *in vivo* in the regulation of angiogenesis. FGFs also have a direct effect on the tumour cells and a number of FGF family members were identified as oncogenes. One mechanism of action includes the over-production of the FGFs as a consequence of activating viral integrations, leading to persistent activation of the receptors. Another potential molecular mechanism would involve FGF-R mutations or gene rearrangements, creating constitutively active receptors leading to cell transformation. Recently, several manuscripts describe the successful crystallization of the FGF-R1 kinase domain in the presence or absence of ATP or synthetic compounds [18–20].

Platelet-derived growth factor receptor (PDGF-R)

Refer to Heldin et al. [21] for a recent review. Platelet-derived growth factor (PDGF) is an effective and potent stimulator of growth and motility of several types of cells. Structurally, PDGF is a dimeric protein composed of disulphide bond-linked A or B polypetide chains which may combine to form homodimers (AA or BB) or heterodimers (AB). These peptide ligands bind to two structurally similar PDGF-Rs termed α and β. The PDGF-Rs initiate similar although not fully identical cell effects. Activities regulated by the PGDF-Rs include stimulation of cell proliferation, actin filament rearrangements (stress fibres, cell margin veil-like structures and circular ruffles), chemotaxis, Ca^{2+} mobilization and protection against apoptosis. The PDGF ligand isoforms exist as dimers and in a similar way to VEGF have two sites for interaction with the PDGF-Rs. Therefore, one PDGF peptide is capable of binding to two PDGF-Rs at one time. Due to the existence of two different ligands and receptors, a number of permutations exist for interactions: α/α PDGF-R interacts with A/A, A/B and B/B, β/β PDGF-R interacts with A/B and B/B whilst β/β PDGF-R only interacts with B/B ligands. The α and β PDGF-Rs both contain five Ig domains in their extracellular region, a single transmembrane-spanning domain and a cytoplasmic region containing a split-insert kinase domain. The PDGF ligands bind to the three N-terminal Ig regions of the receptor and induce receptor dimerization, which is considered to occur via the fourth Ig domain. Alignment of the two receptors then allows for the autophosphorylation of receptor tyrosine residues and regulation of the kinase activity. Y857 represents the only autophosphorylation site within the kinase domain of the β PDGF-R and this residue is conserved in the α PDGF-R (Y849) and is important for receptor kinase activity. The other eleven autophosphorylation sites in the β PDGF-R are found outside the kinase domain and constitute interaction sites for SH2-containing proteins. A number of different SH2-containing proteins have been shown to associate with these different autophosphorylation sites (PI3K, PLC-γ, Src, GAP, Grb2, Grb7, Grb10, Shc, Nck, Crk, SHP-2 and Stat1, 3 and 5) which mediate different cellular responses. Alterations in the PDGF-R signalling can therefore lead to the aberrant regulation of different signal transduction pathways implicated in tumourigenesis.

Examples of medicinal chemistry approaches to inhibit receptor tyrosine kinases and integrated responses

Nowadays, the identification and development of therapeutic agents for disease states linked to the overexpression or abnormal activation of RPTKs [8] [22–24] is an important and a major field of research in both academic institutions and pharmaceutical companies. Several medicinal chemistry approaches have been pursued to inhibit deregulated RPTKs and currently, several ATP site-directed RPTK inhibitors are in clinical trials [8–10] (1–4, Fig. 1). Here we briefly review and discuss some of the strategies and advances that have been made over the past years in the development of novel therapeutic agents for RPTK drug targets.

ATP site-directed inhibitors of protein kinases

Initially, inhibition of RPTKs by ATP site-directed compounds was considered less advantageous than substrate-based inhibitors due, among other factors, to the assumption that the catalytic domain among protein kinases is highly conserved and selectivity with ATP-competitive inhibitors would be very difficult to achieve. The ATP site-directed inhibitors in early phases of clinical trials (Fig. 1) or near clinical development [8–9] and the X-ray structures of protein kinase domains [25], in most cases complexed with this class of inhibitors, have confirmed that, at least for some specific targets, the initial concerns for this strategy were maybe overcome.

Over the past years, numerous reviews have covered the design, synthesis and biological evaluation of ATP site-directed inhibitors of RPTKs (for recent reviews, see refs. [25–27]). The requirements for a clinical candidate have been pursued with some success by optimising lead structures (for further details on these compounds and structures refer to [26] e.g., quinazolines (**1**, **2**) [28–29], pyrido[d]- [30–31] and pyrimido[d]-pyrimidines [32], dianilinophthalimides [33–34], pyrazolo[d]pyrimidine [35], pyrrolo[d]pyrimidines [36], phenylamino-pyrimidines (**4**) [37–38], 1-oxo-3-aryl-1H-indene-2-carboxylic acid derivatives [39] and susbtituted indolin-2-ones (**3**) [40–41] or natural products (e.g., staurosporine, **5**)[1] using state-of-the-art medicinal chemistry technologies. ATP site-directed inhibitors that discriminate between closely related protein kinases have been obtained by exploiting subtle, sequence-related structural features of the ATP-binding domain of the target enzyme. In the absence of the X-ray structure of the kinase (e.g., EGF-R protein tyrosine kinase), pharmacophore models [43–46] (Fig. 2) have been constructed and further refined with structure-activity relationships (SAR) studies. These models are now generally applicable and have paved the way to exploit structure-based drug design approaches to increase potency and/or selectivity within a lead series.

As mentioned above, several ATP site-directed inhibitors for the treatment of cancers are currently in different phases of clinical trials [8–9] (Fig. 1). The utility of this class of compounds as therapeutic agents, as well as the validity of the ATP-competitive mechanism approach, will be clearer when the results from the clinical trials become public knowledge within the next couple of years. Positive results will then validate the ATP-competitive mechanism approach and represent a real "proof of concept" of the utility of receptor tyrosine kinase inhibitors in the clinic.

Peptide-based active site-directed inhibitors and non-peptidic competitive inhibitors of protein substrates

As for any other class of enzymes, selectivity for RPTKs is dictated at the molecular level by the sequence of the amino acids flanking the tyrosine

[1] In general, natural products have exhibited little specificity for individual receptor tyrosine kinases and they merely serve as generic inhibitors that can be used as internal standards in high-throughput screening assays or for co-crystallisation purposes. However, staurosporine derivatives have been reported as rather selective and potent TrkA inhibitors (e.g., CEP-751) [42].

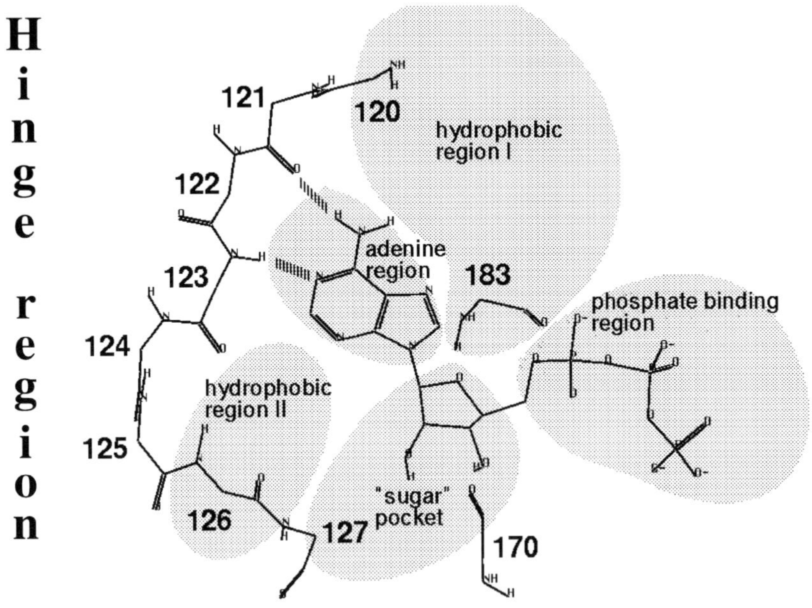

Figure 2. Protein kinase pharmacophore model as constructed by Dr Pascal Furet (Novartis Pharma AG, Basel, Switzerland) for the EGF-R. High binding affinity and improved selectivity for a specific target have been obtained by exploiting the sequence related structural features present in the five regions depicted in this figure. This figure is reproduced with the kind permission of Ashley Publications Ltd, London, UK, from the publication Traxler P. (1998) Tyrosine kinase inhibitors in cancer treatment (part II). Expert Opinion on Therapeutic Patents 8: 1599–1625.

residue to be phosphorylated and by the secondary and/or tertiary structures of the domain engulfing the target site. The majority of the reported compounds that compete with the protein substrate are peptides in which the phosphorylatable tyrosine residue of the native substrate sequence has been replaced with a non-phosphorylatable α-amino acid [47]. The modified peptides have shown very poor inhibitory activity with K_i values in the low mM range [48–50]. The poor performance of these inhibitors, together with the known hurdles associated with peptide-based compounds (e.g., bioavailability, cell penetration), justified the lack of efforts and progress so far in this approach.

The identification of the natural product erbstatin [51] (**6**, Fig. 3) as a substrate-competitive inhibitor of the EGF-R [52][2] fueled the search for inhibitors directed to the tyrosine kinase-substrate binding site. A wide variety of tyrphostins – *tyr*osine *phos*phorylation *in*hibitor*s* constructed around the benzylidenemalononitrile template (**7**, Fig. 3) – with mixed and unpredictable inhibition patterns against ATP and peptide-substrate have been reported in the last four years [53–55]. However none of them have emerged as a viable clinical

[2] It was found later that erbstatin is a competitive inhibitor with respect to both the substrate and the ATP-binding sites [53].

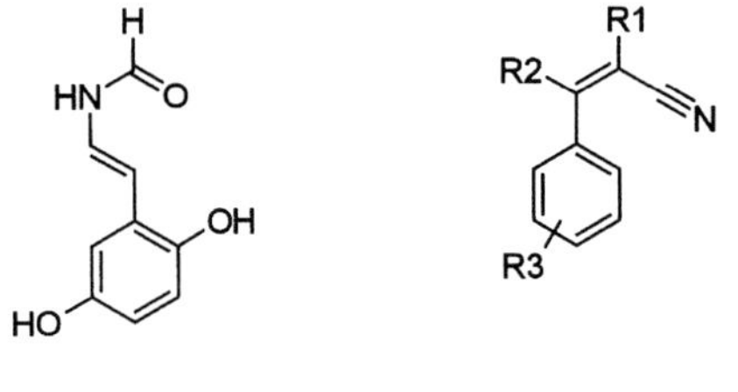

6 7

Figure 3. Mixed substrate and ATP site-directed receptor tyrosine kinase inhibitors.

drug candidate. Potency, selectivity and non-specific effects in cellular assays can be considered as important limitations for this class of compounds [47], [56–57].

Inhibition of sub-cellular protein-protein interactions initiated by receptor tyrosine kinases – Antagonists of Src homology 2 (SH2) domains

A growing number of the elements identified in intracellular signalling pathways are proteins that physically interact with each other via domains that are specifically recognised by amino acid sequences [58–59]. The phosphotyrosine residues of activated RPTKs and the amino acids located immediately *C*-terminal to it serve as recognition motifs for SH2 domains. Following binding to the receptor and intracellular translocation of other elements, the SH2 proteins trigger signal transduction cascades that affect cell growth and transformation [60]. Agents that specifically disrupt these protein-protein interactions could potentially shut down deregulated signal transduction pathways and represent an intervention point for blocking human malignancies.

Degenerate phosphotyrosyl peptide libraries [61–62] have shown that the sequence specificities of SH2 domains for phosphotyrosyl peptides reside in the portion of the peptide immediately *C*-terminal of the phosphotyrosine residue and synthetic [63] and phage display libraries [64] approaches have

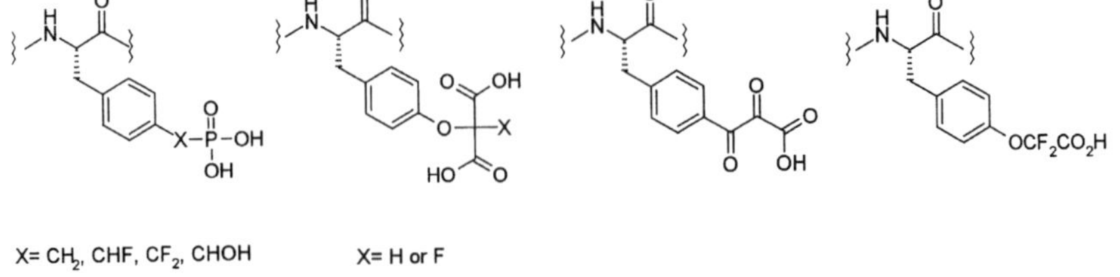

Figure 4. Examples of phosphotyrosine isosteres used in the synthesis of antagonists of SH2 domains.

been used to optimise these residues. Starting with the minimal recognition motif of the target domain, structural information has been exploited to optimise the potency of and confer selectivity on short peptide sequences [1–6], [65]. To confer activity on intracellular systems, major medicinal chemistry efforts have been focused on the identification of phosphotyrosine mimetics [66–72] (Fig. 4) that are stable to phosphatases and, when incorporated into optimised sequences, are able to cross the cell membrane [73]. Highly potent and selective antagonists of SH2 domains have been reported (for a recent review, see ref [65]; e.g., 8–11, Fig. 5). However, the development of clinical candidates for the above signal transduction targets still remains unclear.

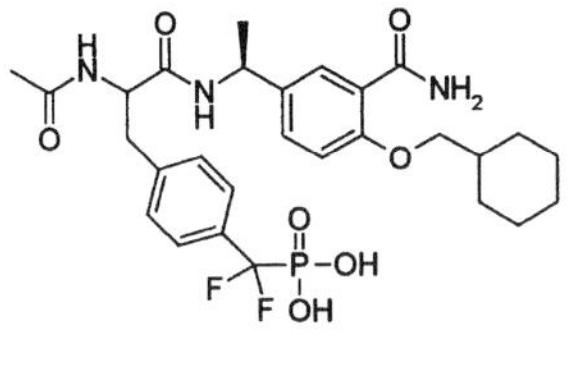

8

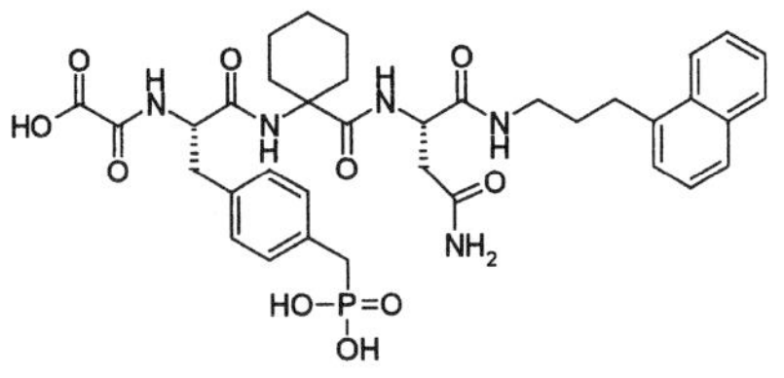

9

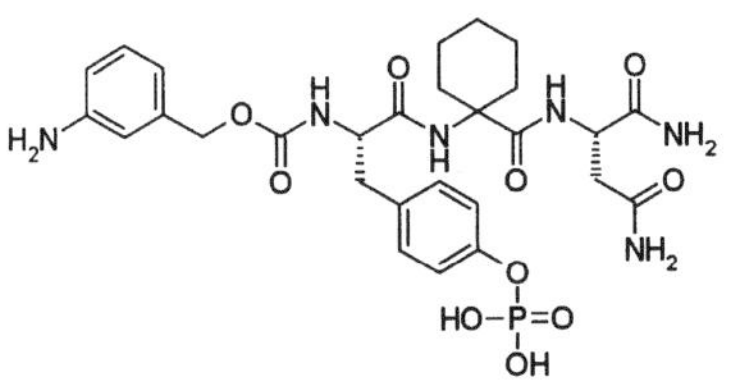

10

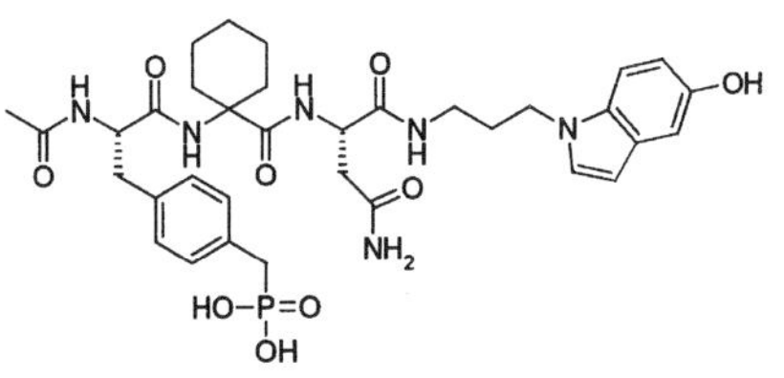

11

Figure 5. Chemical structures of "*de novo*" design antagonists of SH2 domains: **8** (Src SH2) and **9–11** (Grb2-SH2).

Outlook

Based on the close interface and combination of both biological and chemical technologies, many drug candidates have been generated which target several RPTKs, especially at the ATP-binding site of the kinase domain. A number of these are currently being evaluated in clinical trials and many others will likely progress from preclinical testing to studies in cancer patients. Whilst these compounds have preclinically validated the approach of blocking RPTK activity leading to anti-tumour effects in animal models, their clinical utility is currently under investigation. Success here will underscore the rational design of novel therapeutics based on strong clinical epidemiology and harnessing together of multi-disciples in the modern drug discovery processes.

Acknowledgement
The authors are grateful to Dr Pascal Furet (Novartis Pharma AG, Basel, Switzerland) for construction and allowing the use of the CAMM pharmacophore model presented in Figure 2. This figure is reproduced with the kind permission of Ashley Publications Ltd, London, UK.

References

1 Schoepfer J, Fretz H, Gay B, Furet P, García-Echeverría C, End N, Caravatti G (1999) Highly potent inhibitors of the Grb2-SH2 domain. *Bioorg Med Chem Lett* 9: 221–226
2 García-Echeverría C, Furet P, Gay B, Fretz H, Rahuel J, Schoepfer J, Caravatti G (1998) Potent antagonists of the SH2 domain of Grb2: optimization of the X+1 position of 3-amino-Z-Tyr(PO_3H_2)-X^{+1}-Asn-NH_2. *J Med Chem* 41: 1741–1744
3 Furet P, Gay B, Caravatti G, García-Echeverría C, Rahuel J, Schoepfer J, Fretz H (1998) Structure-based design and synthesis of high affinity tripeptide ligands of the Grb2-SH2 domain. *J Med Chem* 41: 3442–3449
4 Rahuel J, García-Echeverría C, Furet P, Strauss A, Caravatti G, Fretz H, Schoepfer J, Gay B (1998) Structural basis for the high affinity of amino-aromatic SH2 phosphopeptide ligands. *J Mol Biol* 279: 1013–1022
5 Schoepfer J, Gay B, Caravatti G, García-Echeverría C, Fretz H, Rahuel J, Furet P (1998) Structure-based design of peptidomimetic ligands of the Grb2-SH2 domain. *Bioorg Med Chem Lett* 8: 2865–2870
6 Furet P, Gay B, García-Echeverría C, Rahuel J, Fretz H, Schoepfer J, Caravatti G (1997) Discovery of 3-aminobenzyloxycarbonyl as an N-terminal group conferring high affinity to the minimal phosphopeptide sequence recognized by the Grb2-SH2 domain. *J Med Chem* 40: 3551–3556
7 Ross JE, Fletcher J (1998) The HER-2/neu onogene in breast cancer: prognostic factor, predictive factor, and target for therapy. *Stem Cells* 16: 413–428
8 Strawn LM, Shawver LK (1998) Tyrosine kinases in disease: overview of kinase inhibitors as therapeutic agents and current drugs in clinical trials. *Expert Opin Invest Drugs* 7: 553–573
9 Bridges AJ (1998) Current progress towards the development of tyrosine kinase inhibitors as anticancer agents. *Emerg Drugs* 3: 279–292
10 Lavelle F (1997) American Association for *Cancer Res* 1997: progress and new hope in the fight against cancer, April 12–16, 1997, San Diego, California. *Expert Opin Invest Drugs* 6: 771–775
11 Smith CM, Shindayalov IN, Veretnik S, Gribskov M, Taylor SS, Ten Eyck LF, Bourne PE (1997) The protein kinase resource. *Trends Biochem Sci* 22: 444–446
12 Hardie G, Hanks S (eds) (1995) *The protein kinase facts book: protein-tyrosine kinases*. Academic Press, London
13 Hanks S, Quinn AM (1991) Protein kinase catalytic domain sequence database: Identification of conserved features of primary structure and classification of family members. *Meth Enzymol* 200: 38–62
14 Voldborg BR, Damstrup L, Spang-Thomsen M, Poulsen HS (1997) Epidermal growth factor receptor (EGF-R) and EGF-R mutations, function and possible role in clinical trials. *Ann Oncol* 12: 1197–1206
15 Neufeld G, Cohen T, Gengrinovitch S, Poltorak Z (1999) Vascular endothelial growth factor (VEGF) and its receptor. *FASEB J* 13: 9–22
16 McTigue MA, Wickersham JA, Pinko C, Showalter RE, Parast CV, Tempczyk-Russell A, Gehring MR, Mroczkowski B, Kan C-C, Villafranca JE et al (1999) Crystal structure of the kinase domain of human vascular endothelial growth factor receptor 2: a key enzyme in angiogenesis. *Structure* 7: 319–330
17 Klint P, Claesson-Welsh L (1999) Signal transduction by fibroblast growth factor receptors. *Front Biosci* 4: D165–D177
18 Mohammadi M, Froum S, Hamby JM, Schroeder MC, Panek RL, Lu GH, Eliseenkova AV, Green D, Schlessinger J, Hubbard SR (1998) Crystal structure of an angiogenesis inhibitor bound to the FGF receptor tyrosine kinase domain. *EMBO J* 17: 5896–5904
19 Mohammadi M, McMahon G, Sun L, Tang C, Hirth P, Yeh BK, Hubbard SR, Schlessinger J (1997) Structure of the tyrosine kinase domain of fibroblast growth factor receptor in complex

with inhibitors. *Science* 276: 955–960
20 Mohammadi M, Schlessinger J, Hubbard SR (1996) Structure of the FGF receptor tyrosine kinase domain reveals a novel autoinhibitory mechanism. *Cell* 86: 577–587
21 Heldin CH, Östman A, Rönnstrand L (1998) Signal transduction via platelet-derived growth factor receptors. *Biochim Biophys Acta* 1378: F79–F113
22 Aaronson AA (1991) Growth factors and cancer. *Science* 254: 1146–1153
23 Boutin JA (1994) Tyrosine protein kinase inhibition and cancer. *Int J Biochem* 26: 1203–1226
24 Levitzki A (1994) Signal-transduction therapy. *Eur J Biochem* 226: 1–13
25 McMahon G, Sun L, Liang C, Tang C (1998) Protein kinase inhibitors: structural determinants for target specificity. *Curr Opin Drug Discover Develop* 1: 131–146
26 Traxler P (1998) Tyrosine kinase inhibitors in cancer treatment (part II). *Expert Opin Ther Patents* 8: 1599–1625
27 Hollis Showalter HD, Kraker AJ (1997) Small molecule inhibitors of the platelet-derived growth factor receptor, the fibroblast growth factor receptor, and src family tyrosine kinases. *Pharmacol Ther* 76: 55–71
28 Moyer JD, Barbacci EG, Iwata KK, Arnold L, Boman B, Cunningham A, DiOrio C, Doty J, Morin MJ, Moyer MP et al (1997) Induction of apoptosis and cell cycle arrest by CP-358,774, an inhibitor of epidermal growth factor receptor tyrosine kinase. *Cancer Res* 57: 4838–4848
29 Woodburn JR, Barker AJ, Gibson KH, Ashton SE, Wakeling AE, Curry BJ, Scarlett L, Henthorn LR (1997) ZD 1839, an epidermal growth factor tyrosine kinase inhibitors selected for clinical development. *Proc Amer Assoc Cancer Res* 38: 6333
30 Rewcastle GW, Murray DK, Elliott WL, Fry DW, Howard CT, Nelson JM, Roberts BJ, Vincent PW, Showalter HD, Winters RT et al (1998) Tyrosine kinase inhibitors. 14. structure-activity relationships for methyl-amino-substituted derivatives of 4-[(3-bromophenyl)amino]-6-(methylamino)-pyrido[3, 4-d]pyrimidine (PD 158780), a potent and specific inhibitor of the tyrosine kinase activity of receptors for the EGF family of growth factors. *J Med Chem* 41: 742–751
31 Rewcastle GF, Palmer BD, Thompson AM, Bridges AJ, Cody DR, Zhou HM, Fry DW, McMichael A, Denny WA (1996) Tyrosine kinase inhibitors. 10. Isomeric 4-[(3-bromophenyl)-amino]pyrido[*d*]pyrimidines are potent ATP binding site inhibitors of the tyrosine kinase function of the epidermal growth factor receptor. *J Med Chem* 39: 1823–1835
32 Rewcastle GF, Bridges AJ, Fry DW, Rubin JR, Denny WA (1997) Tyrosine kinase inhibitors. 12. Synthesis and structure-activity relationships for 6-substituted 4-(phenylamino)pyrimido[5, 4-*d*]pyrimidies designed as inhibitors of the epidermal growth factor receptor. *J Med Chem* 40: 1820–1826
33 Buchdunger E, Trinks U, Mett H, Regenass U, Muller M, Meyer T, Beilstein P, Wirz B, Schneider P, Traxler P et al (1995) 4,5-Bis(4-fluoroanilino)phthalimide: a selective inhibitor of the EGF receptor signal transduction pathway with potent *in vivo* antitumor activity. *Clin Cancer Res* 1: 813–821
34 Buchdunger E, Trinks U, Mett H, Regenass U, Muller M, Meyer T, McGlynn E, Pinna LA, Traxler P, Lydon NB (1994) 4,5-Dianilinophthalimide: a protein-tyrosine kinase inhibitor with selectivity for the epidermal growth factor receptor signal transduction pathway and potent *in vivo* antitumor activity. *Proc Natl Acad Sci USA* 91: 2334–2338
35 Traxler P, Bold G, Frei J, Lang M, Lydon N, Mett H, Buchdunger E, Meyer T, Mueller M, Furet P (1997) Use of a pharmacophore model for the design of EGF-R tyrosine kinase inhibitors: 4-(phenylamino)pyrazolo[3, 4-d]pyrimidines. *J Med Chem* 40: 3601–3616
36 Traxler PM, Furet P, Mett H, Buchdunger E, Meyer T, Lydon N (1996) 4-(Phenylamino)pyrrolopyrimidines: potent and selective, ATP site directed inhibitors of the EGF-receptor protein tyrosine kinase. *J Med Chem* 39: 2285–2292
37 Zimmermann J, Buchdunger E, Mett H, Meyer T, Lydon NB (1997) Potent and selective inhibitors of the ABL-kinase: phenylaminopyridine (PAP) derivatives. *Bioorg Med Chem Lett* 7: 187–192
38 Zimmermann J, Caravatti G, Mett H, Meyer T, Mueller M, Lydon NB, Fabbro D (1996) Phenylamino-pyrimidine (PAP) derivatives: a new class of potent and selective inhibitors of protein kinase C (PKC). *Arch Pharm (Weinheim, Ger.)* 329: 371–376
39 Barvian MR, Panek RL, Lu GH, Kraker AJ, Amar A, Hartl BG, Hamby JM, Showalter HDH (1997) 1-Oxo-3-aryl-1*H*-indene-2-carboxylic acid derivatives as selective inhibitors of fibroblast growth factor receptor-1 tyrosine kinase. *Bioorg Med Chem Lett* 7: 2903–2908
40 Fong TAT, Shawver LK, Sun L, Tang C, App H, Powell TJ, Kim YH, Schreck R, Wang X, Risau W et al (1999) SU5416 is a potent and selective inhibitor of the vascular endothelial growth fac-

tor receptor (Flk-1/KDR) that inhibits tyrosine kinase catalysis, tumor vascularization, and growth of multiple tumor types. *Cancer Res* 59: 99–106
41 Sun L, Tran F, App H, Hirth P, McMahon G, Tang C (1998) Synthesis and biological evaluations of 3-substituted indolin-2-ones: a novel class of tyrosine kinase inhibitors that exhibit selectivity toward particular RTKs. *J Med Chem* 41: 2588–2603
42 Dionne CA, Camoratto AM, Jani JP, Emerson E, Neff NT, Vaught JL, Murakata C, Djakiew D, Lamb J, Bova S et al (1998) Cell cycle-independent death of prostate adenocarcinoma is induced by the trk tyrosine kinase inhibitor CEP-751 (KT6587). *Clin Cancer Res* 4: 1887–1898
43 Furet P, Caravatti G, Lydon N, Priestle JP, Sowadski JM, Trinks U, Traxler P (1995) Modelling study of protein kinase inhibitors: binding mode of staurosporine and origin of the selectivity of CGP52411. *J Computer-Aided Molec Design* 9: 465–472
44 Trumpp-Kallmeyer S, Showalter HD (1998) Use of a pharmacophore model for the design of EGF-R tyrosine kinase inhibitors: 4-(phenylamino)pyrazolo[3,4-d]pyrimidines. *Chemtracts* 11: 550–560
45 Trumpp-Kallmeyer S, Rubin JR, Humblet C, Hamby JM, Showalter HDH (1998) Development of a binding model to protein tyrosine kinases for substituted pyrido[2,3-*d*]pyrimidine inhibitors. *J Med Chem* 41: 1752–1763
46 Traxler P, Green J, Mett H, Séquin U, Furet P (1999) Use of a pharmacophore model for the design of EGF-R tyrosine kinase inhibitors: isoflavones and 3-phenyl-4(1*H*)-quinolones. *J Med Chem* 42: 1018–1026
47 Lawrence DS, Niu J (1998) Protein kinase inhibitors: the tyrosine-specific protein kinases. *Pharmacol Ther* 77: 81–114
48 Fry DW, McMichael A, Singh J, Dobrusin EM, McNamara DJ (1994) Design of a potent peptide inhibitor of the epidermal growth factor receptor tyrosine kinase utilizing sequences based on the natural phosphorylation sites of phospholipase C-gamma 1. *Peptides* 15: 951–957
49 Yuan CJ, Jakes S, Elliott S, Graves DJ (1990) A rationale for the design of an inhibitor of tyrosyl kinase. *J Biol Chem* 265: 16 205–16 209
50 Shoelson SE, White MF, Kahn CR (1989) Nonphosphorytable substrate analogs selectively block autophosphorylation and activation of the insulin receptor, epidermal growth factor receptor, and pp60v-src kinases. *J Biol Chem* 264: 7831–7836
51 Umezawa H, Imoto M, Sawa T, Issihiki K, Matsuda N, Uchida T, Finuma H, Hamada M, Takeuchi T (1986) Studies on a new epidermal growth factor kinase inhibitors, erbstatin, produce by MH435-hF3. *J Antibiot* 39: 170–173
52 Imoto M, Umezawa K, Isshiki K, Kunimoto S, Sawa T, Takeuchi T, Umezawa H (1987) Kinetic studies of tyrosine kinase inhibition by erbstatin. *J Antibiot* 40: 1471–1473
53 Posner I, Engel M, Gazit A, Levitzki A (1994) Kinetics of inhibition by tyrphostins of the tyrosine kinase activity of the epidermal growth factor receptor and analysis by a new computer program. *Mol Pharmacol* 45: 673–683
54 Levitzki A, Gazit A (1991) Tyrphostins as molecular tools and potential antiproliferative drugs. *Trends Pharmacol Sci* 12: 171–174
55 Levitzki A, Gazit A (1995) Tyrosine kinase inhibition: an approach to drug development. *Science* 267: 1782–1788
56 Agbotounou WK, Mousset S, Piperno S, Pierre M, Jacquemin-Sablon A, Pierre J (1994) Activation of the mitogen-activated protein kinase cascade by tyrphostin. *Biochem Pharmacol* 48: 505–515
57 Young SW, Poole RC, Hudson AT, Halestrap AP, Denton RM, Tavaré JM (1993) Effects of tyrosine kinase inhibitors on protein kinase-independent systems. *FEBS Lett* 316: 278–282
58 Wishart MJ, Dixon JE (1998) Gathering STYX: phosphatase-like form predicts functions for unique protein-interaction domains. *Trends Biochem Sci* 23: 301–306
59 Mayer BJ, Gupta R (1998) Functions of SH2 and SH3 domains. *Curr Topic Microbiol Immunol* 228: 1–22
60 Zhou SY, Cantley LC (1995) Recognition and specificity in protein tyrosine kinase-mediated signalling. *Trends Biochem Sci* 20: 470–475
61 Songyang Z, Shoelson SE, McGlade J, Olivier P, Pawson T, Bustelo XR, Barbacid M, Sabe H, Hanafusa H, Yi T et al (1994) Specific motifs recognized by the SH2 domain of Csk, 3BP2, fps/fes, Grb-2, HCP, SHC, syk, and Vav. *Mol Cell Biol* 14: 2777–2785
62 Songyang Z, Shoelson SE, Chaudhuri M, Gish G, Pawson T, Haser WG, King F, Roberts T, Ratnofsky S, Lechleider RJ et al (1993) SH2 domains recognize specific phosphopeptide

sequences. *Cell* 72: 767–778
63 Müller K, Gombert FO, Manning U, Grossmüller F, Graff P, Zaegel H, Zuber JF, Freuler F, Tschopp C, Baumann G (1996) Rapid identification of phosphopeptide ligands for SH2 domains. *J Mol Biol* 28: 16 500–16 505
64 Gram H, Schmitz R, Zuber JF, Baumann G (1997) Identification of phosphopeptide ligands for the Src-homology 2 (SH2) domain of Grb2 by phage display. *Eur J Biochem* 246: 633–637
65 Sawyer TK (1998) Src homology-2 domains: structure, mechanisms, and drug discovery. *Biopolymers (Peptide Science)* 47: 243–261
66 Fretz H (1998) O-(Carboxydifluoromethyl)-L-tyrosine: design and synthesis of a novel non phosphorous-containing phosphotyrosine isostere. *Tetrahedron* 54: 4849–4858
67 Fretz H (1996) Solid phase synthesis of peptides containing novel L-phenylalanine substituted with vicinal tricarbonyl moieties. *Tetrahedron Lett* 37: 8479–8482
68 Fretz H (1996) Synthesis of novel L-phenylalanine derivatives substituted with a keto ylide as stable precursor of a vicinal tricarbonyl moiety. *Tetrahedron Lett* 37: 8475–8478
69 Burke TR Jr, Ye B, Akamatsu M, Ford H, Yan X, Kole HK, Wolf G, Shoelson SE, Roller PP (1996) 4'-O-[2-(2-fluoromalonyl)]-L-tyrosine: a phosphotyrosyl mimic for the preparation of signal transduction inhibitory peptides. *J Med Chem* 39: 1021–1027
70 Ye B, Burke TR Jr (1995) L-O-(2-Malonyl)tyrosine (L-OMT), a new phosphotyrosyl mimic suitably protected for solid-phase synthesis of signal transduction inhibitory peptides. *Tetrahedron Lett* 36: 4733–4736
71 Burke TR Jr, Smyth MS, Nomizu M, Otaka A, Roller PR (1993) Preparation of fluoro- and hydroxy-4-(phosphonomethyl)-D,L-phenylalanine suitably protected for solid-phase synthesis of peptides containing hydrolytically stable analogs of O-phosphotyrosine. *J Org Chem* 58: 1336–1340
72 Shoelson SE, Chatterjee S, Chaudhuri M, Burke TR Jr (1991) Solid-phase synthesis of nonhydrolyzable phosphotyrosyl peptide analogues with N^{α}-Fmoc-(O,O-di-*t*-butyl)phosphono-*p*-methylphenylalanine. *Tetrahedron Lett* 32: 6061–6064
73 Yao Z-J, King CR, Cao T, Kelley J, Milne GWA, Voigt JH, Burke TR Jr (1999) Potent inhibition of Grb2 SH2 domain binding by non-phosphate-containing ligands. *J Med Chem* 42: 25–35

New Approaches to Drug Development
ed. by P. Jollès

Peroxisome proliferator activated receptor agonists

Sander Kersten and Walter Wahli

Institut de Biologie Animale, Bâtiment de Biologie, Université de Lausanne, CH-1015 Lausanne, Switzerland

Summary. The peroxisome proliferator activated receptors (PPARs) represent a group of ligand-activated transcription factors that mediate the biological effects of various drugs, such as fibrates and thiazolidinediones. Three PPAR subtypes can be distinguished, α, β and γ, each of which has an unique pattern of expression among vertebrate tissues. The PPARα receptor is activated by hypolipidemic drugs of the fibrate class, and regulates the expression of numerous genes involved in fatty acid catabolism. The PPARγ receptor is activated by hypoglycaemic drugs of the thiazolidinedione class, and is an important determinant of adipocyte differentiation. Little is currently known about PPARβ. A heavy research effort is currently directed towards the identification of novel high-affinity, high-specificity agonists and antagonists that may be used in the treatment of hyperglycaemia, hyperlipidemia, and other diseases of metabolic origin.

Peroxisome proliferator-activated receptor agonists

The peroxisome proliferator-activated receptors (PPARs) belong to the superfamily of nuclear hormone receptors (NHRs). This superfamily encompasses a large collection of transcription factors grouped together based on both structural and functional characteristics [1]. NHRs regulate transcription in response to small lipophilic compounds such as retinoic acid (RAR, RXR), vitamin D (VDR), thyroid hormone (T3R), pregnanes (PXR), or cholesterol (LXR).

All NHRs share a modular structure consisting of several domains, each with a distinct function, that are conserved to varying degrees between the receptors [1]. The best-studied domains are the DNA-binding domain (DBD) and the ligand-binding domain (LBD). The DBD contains the structural motif that allows the receptor to bind to a specific nucleotide sequence on the DNA called the hormone response element (HRE). The LBD is, as its name implies, responsible for binding of the ligand. It also mediates protein-protein interactions between two molecules of the receptor itself, or between different NHRs. This latter property is particularly important for receptors such as PPARs, which can only bind to DNA as a heterodimer with the common binding partner RXR [2].

As alluded to above, NHRs recognize a specific sequence of nucleotides in the promoter of target genes. For PPARs and many other NHRs, these HREs consist of a direct repeat of the consensus hexanucleotide AGGTCA (half-site), with one base pair in between the two hexanucleotides (called DR-1) [3, 4]. Besides this core element, binding to DNA by PPARs is also determined by a 5' extension consisting of several nucleotides [5]. PPARs bind to these elements only as a heterodimer with RXR, with PPAR occupying the 5' half-site [6]. According to the general model for NHRs, upon binding of the ligand, co-repressors are dislodged from the receptor, while co-activators are recruited. The transcription of the gene is stimulated via a complex sequence of events that involves remodelling of chromatin via histone acetylation and de-acetylation [7].

Three distinct PPAR subtypes have been identified, α, β, and γ, each of which has an unique expression pattern among vertebrate tissues. In rats PPARα is most highly expressed in brown adipose tissue, followed by liver, kidney, heart and skeletal muscle [8] (Escher and Wahli, unpublished data). PPARγ is mostly highly expressed in brown and white adipose tissue, and, to a lesser extent, in other organs such as the colon, while PPARβ is found at high levels in almost any tissue examined [8] (Escher and Wahli, unpublished data). As a group, the PPARs can bind both endogenous ligands (such as eicosanoids and fatty acids) and synthetic ligands, resulting in stimulation of transcription of target genes [9–12]. The list of currently known target genes encompasses a relatively homogenous group of genes that participate in various aspects of lipid metabolism, such as uptake of fatty acids through membranes, fatty acid binding in cells, fatty acid oxidation (both peroxisomal and mitochondrial), and lipoprotein assembly and transport (Tabs 1 and 2). It includes such genes as lipoprotein lipase [13], fatty acid binding protein [14], acyl CoA oxidase [3, 4], apo AII [15], carnitine palmitoyl transferase [16, 17] and many others.

Currently the best-studied PPAR-subtype is PPARα. It has been shown, both *in vitro* and *in vivo*, that this receptor plays an important role in the oxi-

Table 1. Genes for which a PPAR response element (PPRE) has been identified

Liver
acyl-CoA oxidase
enoyl-coA hydratase-3-hydroxyacyl-CoA dehydrogenase
acyl-CoA synthetase
cyp4A6 (fatty acid ω-hydroxylase)
hydroxymethylglutaryl-CoA synthase
Rev-erbα
apo A-I (human)
apo A-II (human)
liver fatty acid binding protein
NADP malate dehydrogenase
medium chain acyl-CoA dehydrogenase
White adipose tissue
phospho-enol pyruvate carboxykinase
adipocyte fatty acid binding protein
lipoprotein lipase
fatty acid transport protein
Muscle
carnitine palmitoyl transferase I

Table 2. Genes that are upregulated by WY14643 or down-regulated in PPARα-null mice

Liver
fatty acid translocase
cyp4A1, cyp4A2, cyp4A3, cyp4A10
very long chain acyl-CoA dehydrogenase
long chain acyl-CoA dehydrogenase
short chain acyl-CoA dehydrogenase
trifunctional protein (long chain specific hydratase, 3-hydroxyacyl-CoA dehydogenase, and 3-ketoacyl-CoA thiolase)
short chain 3-hydroxyacyl-CoA dehydrogenase
short chain specific 3-ketoacyl-CoA thiolase
acetoacyl-CoA thiolase
very long chain acyl-CoA synthetase
carnitine palmitoyl transferase I and II
mitochondrial thioesterase I
peroxisomal hydratse and 3-hydroxyacyl-CoA dehydrogenase
D-type peroxisomal hydratse and 3-hydroxyacyl-CoA dehydrogenase
peroxisomal thiolase
uncoupling protein 2

dation of fatty acids in the liver by activating the expression of many genes involved in this metabolic pathway [18, 19]. By stimulating fatty acid oxidation, PPARα is essential for the adaptive response to nutritional challenges that demand an increased hepatic fatty acid oxidation, such as fasting. PPARα also mediates the effect of peroxisome proliferators on hepatic cell proliferation. Indeed, mice that lack PPARα are refractory to the effects of peroxisome proliferators [20].

While PPARα functions in lipid catabolism, the role of PPARγ seems to be geared towards anabolic pathways, particularly in adipose tissue. Activation of PPARγ in pre-adipocytes is sufficient to direct the cell towards the adipocyte differentiation programme, leading to a dramatic build-up of lipid inside the cell [21]. Other tissues where PPARγ may play a role include the colon, where a stimulating effect of PPARγ activators on colon polyp formation could be demonstrated [22, 23], and monocytes/macrophages where PPARγ may play a role in cytokine production [24–27].

Very little is currently known about PPARβ. Its ubiquitous expression suggests that it may play a role in some basic aspect of lipid metabolism but no hard data are available to support this notion.

PPAR agonists

Nuclear receptor agonists can be defined as compounds that *activate* a certain nuclear receptor, while antagonists are compounds that *de-activate* the receptor. Antagonists usually do not actively suppress transactivation but inhibit agonist-mediated transcriptional activation by competing with the agonist for the ligand-binding site. The antagonist itself is unable to elicit the transcriptional activation function of the receptor. Currently, not many nuclear receptor antagonists are known. This is especially true for the PPARs, for which to our knowledge no good antagonists are available as yet.

Nuclear receptor agonists can be both natural (endogenous) or synthetic. The overall approach towards the identification of either one is similar, but there are also important differences. For a complete description of how to identify natural agonists for NHRs see Kersten and Wahli [28].

The identification of synthetic agonists now commonly starts with a chemical-based library. A chemical-based library consists of a huge collection of compounds that can be tested for potential activity in various drug-screening assays. In the case of NHRs and PPARs, the whole library or parts of the library are screened in a transactivation assay to identify compounds that display agonistic activity. Once a compound has been identified as an activator, the direct binding of this compound to the receptor can be tested by various ligand-binding assays [28].

Natural PPAR agonists

A unique feature of the crystallographic structure of the ligand-binding domain of PPAR receptors is that the ligand-binding cavity is unusually large [29–31]. Because of this, the cavity is able to accommodate a large number of compounds, which explains the relatively low ligand-specificity of PPARs [31]. Most of these ligands are either fatty acids or compounds derived from fatty acids. This appears to be true for all three PPARs. Nevertheless, there are clear differences between the ligand-specificity of the three PPAR subtypes.

The ligand-specificity of PPARα has been the best characterized. PPARα seems to have a clear preference for long-chain unsaturated fatty acids over saturated fatty acids [10, 11]. A recent equilibrium binding study indicates that the dissociation constants for binding of unsaturated fatty acids to mouse PPARα are on the order of several nM, while binding of saturated fatty acids was too weak to be measured [32]. Although the dissociation constants obtained in this study were several orders of magnitude lower than those observed in previous studies using non-equilibrium binding methods, the qualitative information derived from these studies is very similar. This study also confirmed that LTB4, a fatty acid derivative, binds to PPARα with high affinity, although not quite as high as unsaturated fatty acids [32, 33]. Other fatty acids or fatty acid derivatives that have shown high-affinity binding to PPARα include phytanic acid and 8S-HETE [9, 10, 34].

Only two reports have looked at the binding of natural agonists to the PPARβ subtype. Although these studies used PPARβ from different species, the main conclusion is shared in the sense that poly-unsaturated fatty acids, such as 18:2, 18:3 and 20:4, are the most potent PPARβ ligands [10, 11]. No fatty acid derivative or prostaglandin has yet been described as a high-affinity PPARβ agonist.

In the case of PPARγ, the most potent agonist described thus far is 15-deoxy-$D_{12,14}$-prostaglandin J_2 [35, 36]. No quantitative analysis of the binding affinity of 15-deoxy-$D_{12,14}$-prostaglandin J_2 to PPARγ has yet been carried out, but in qualitative analysis it scores higher than any other endogenous ligands, including polyunsaturated fatty acids or other eicosanoids [11].

Synthetic PPAR agonists

PPARs have been postulated to play a role in a variety of diseases, including chronic ailments such as diabetes, obesity, cardiovascular disease and cancer. Synthetic receptor agonists that selectively activate a certain PPAR subtype (such as fibrates for PPARα and thiazolidinediones for PPARγ), are currently employed in the treatment of some of these conditions. Most of these agonists were developed independently of PPARs, and were later demonstrated to be potent PPAR ligands. Unfortunately, some of these drugs, such as the PPARγ agonist troglitazone, carry along a low risk for severe side-effects [45], and accordingly there is a continuing demand for low-toxicity, high-specificity PPAR agonists.

PPARα agonists

Synthetic PPARα agonists can be divided into three major groups, based on their chemical class, their biological effects in humans and rodents, and their affinity and specificity for PPARα. The first group are the fibrate hypolipidemic drugs, most of which are relatively specific agonists for PPARα, and whose biological effect in rodents and humans is almost certainly mediated via PPARα. The second group of PPARα agonists are the prostaglandin I2 and leukotriene B4 analogs. The third group encompasses a diverse array of compounds such as the arachidonic acid analog and the cyclo-oxygenase inhibitor ETYA, and various herbicides, pesticides, phtalate esters and other man-made xenobiotics. Since none of these compounds currently have any medical application, they fall beyond the scope of this chapter.

Fibrates
Fibrates are potent hypolipidemic drugs that are extensively used in the clinic to treat hyperlipidemia. They include the drugs bezafibrate, clofibrate, ciprofibrate, fenofibrate, and nafenopin. Most fibrates bind to PPARα with moderate affinity except for WY14643, for which recent studies have measured a dissociation constant in the nanomolar range [32]. Most fibrates are relatively specific for PPARα except for bezafibrate, which, at least in *Xenopus*, is a potent agonist for PPARβ, and ciprofibrate, which also binds well to *Xenopus* PPARγ [11].

In humans, fibrate treatment has two major lipid-lowering effects that are both beneficial to the management of hyperlipidemic disorders: first, they decrease the concentration of plasma triglycerides, and second, they increase the plasma HDL concentration [37]. The former effect is believed to be achieved by stimulation of hepatic fatty acid oxidation via activation of PPARα, thereby directing fatty acids away from esterification and subsequent secretion as VLDL. In addition, by activating PPARα, fibrates suppress the hepatic production of apoCIII, which is a potent inhibitor of lipoprotein lipase [37].

In contrast to humans, in rodents the plasma HDL concentration is decreased by fibrate treatment, which renders rodents relatively unsuitable as a model to study the biological effects of fibrates *in vivo* [18]. This is due to the opposite effect of fibrates in humans and rodents on the expression of apoA-I, via a mechanism that involves the nuclear receptor Rev-erbα [38].

Inasmuch as the hypolipidemic action of fibrates is advantageous to the treatment of atherosclerosis, it is believed that this mechanism is not exclusively responsible for the atherosclerosis-reducing effect of fibrate treatment. Recent evidence has shown that PPARα plays an important role in the inflammatory response, by altering cytokine production and/or redox status [32, 39, 40]. Future research efforts should be directed towards more precisely establishing the role of PPARα and PPARα -agonists in inflammation.

Prostaglandin I2 and Leukotriene B4 analogs

Prostaglandin I2 analogs such as iloprost are used for their vasodilatory effect in the treatment of a variety of clinical disorders, including scleroderma, pulmonary hypertension, and Raynaud's phenomenon. Although their primary effect is by mimicking prostacyclins, part of their action may also be due to activation of PPARα [41]. A recent study has shown that the EC_{50} for binding of carbaprostacyclin to PPARα is in the same order of magnitude as that for WY14643 [10].

Recently, leukotriene B4 was identified as a ligand for PPARα [33]. Subsequent studies have shown that two LTB4 analogs are also able to bind and activate this receptor (Devchand and Wahli, unpublished data).

PPARγ agonists

PPARγ has been implicated in several important disorders, such as atherosclerosis, diabetes, obesity and cancer. As a result, a massive effort is currently directed towards the development of specific PPARγ agonists and antagonists. PPARγ agonists can be divided into two main groups. The first group of PPARγ agonists consists of the thiazolidinediones (TZDs) [42], which includes troglitazone, pioglitazone, ciglitazone, and risoglitazone (also named BRL49653). Troglitazone is currently the only TZD approved for use on patients (Rezulin). The second group of PPARγ agonists are the non-steroidal anti-inflammatory drugs (NSAIDs) [43], which are relatively low-affinity agonists for both PPARα and PPARγ, and whose primary biological action is not mediated via PPARs.

Thiazolidinediones

TZDs quickly rose to the medical spotlight when they appeared to be a relatively safe and efficacious alternative to the historical treatment of type-II diabetes with sulphonylureas or biguanides. Instead of stimulating insulin production in the pancreatic islets (sulphonylureas), or suppressing glucose output in the liver (biguanides), TZDs act by improving peripheral insulin resistance. Besides correcting hyperglycaemia and hyperinsulinaemia, TZDs also lowers plasma triglyceride levels [44]. It is believed that the hypoglycaemic effect of TZDs (troglitazone) is entirely achieved by an increase in insulin-mediated peripheral glucose disposal in muscle [45, 46]

TZDs bind to PPARγ with an affinity ranging from 40 nM (rosiglitazone) [42], to several μM (troglitazone and others). They are potent inducers of adipogenesis in pre-adipocytes expressing PPARγ [34, 35], and are able to stimulate expression of PPARγ target genes such as lipoprotein lipase (LPL) and fatty acid translocase (FAT) in adipose tissue [13, 47]. The stimulatory effect of TZDs on the expression of LPL and FAT probably explains their hypotriglyceridemic effect. What is less understood is how TZDs reduce insulin resistance in skeletal muscle. This is because skeletal muscle, which accounts entirely for the TZD-mediated increase in glucose disposal [46], expresses only trace amounts of PPARγ. To explain this paradox, a mechanism has been proposed by which TZDs divert fatty acids away from skeletal muscle by increasing fatty acid uptake in adipose tissue, and as a result reduce the deleterious effects of fatty acids on muscle insulin action. However, mice that lack adipose tissue are still able to benefit from the action of TZDs [48], suggesting that adipose tissue is dispensable for mediating the hypoglycaemic effects of TZDs.

Non-steroidal anti-inflammatory drugs

The NSAIDs encompass a chemically diverse group of compounds that are used to treat a variety of disorders of both inflammatory and non-inflammatory origin, including ordinary headaches, fever and thrombosis. NSAIDs include many common household names such as aspirin (acetyl-salicylic acid), and ibuprofen. Most of their pharmacological effects can be attributed to inhibition of the activity of cyclo-oxygenase 1 and 2 (COX-1 and COX-2). However, certain effects, such as suppression of intestinal polyp formation, require much higher doses of NSAIDs, suggesting that NSAIDs have other molecular targets. One of these targets might be PPARs. Lehmann et al. [43] demonstrated that certain NSAIDs such as indomethacin and fenoprofen can activate PPARα and PPARγ at relatively high concentrations, and indomethacin was able to compete with rosiglitazone for binding to PPARγ. However, the binding seemed weak, casting some doubt on the medical relevance of these findings.

PPARβ agonists

Recently, Berger et al. [49] reported the characterization of an array of non-TZD agonists for PPARγ and PPARβ. One of these compounds, L-165041, seems to represent a relatively specific, high-affinity agonist for PPARβ. Animal studies showed that this compound does not lower plasma glucose or triglyceride levels in db/db mice, but does affect the concentration of plasma cholesterol [49]. Another PPARβ-specific agonist that has been reported is GW2443 [50]. No clinical data relating to this compound are available as yet.

Conclusion

The PPARs represent a fascinating group of transcription factors that have received a huge amount of attention, partly for their postulated link with various chronic diseases such as diabetes, atherosclerosis, obesity and cancer. At the moment, agonists for PPARα are widely used in the treatment of hyperlipidemia. There is evidence suggesting that the anti-atherosclerotic action of fibrates is also partly due to modulation of the inflammatory response, perhaps linking PPARα agonists to other diseases of inflammatory origin. PPARγ agonists that belong to the class of TZDs have become extremely popular in recent years for their potent hypoglycaemic action in patients that suffer from insulin resistance. The recent association of PPARγ with other diseases such as cancer and atherosclerosis suggests that PPARγ agonists and antagonists may find a wider range of medical applications in the future.

References

1 Gronemeyer H, Laudet V (1995) Transcription factors 3: nuclear receptors. *Protein Profile* 2: 1173–308

2 Kliewer SA, Umesono K, Noonan DJ, Heyman RA, Evans RM (1992) Convergence of 9-cis retinoic acid and peroxisome proliferator signalling pathways through heterodimer formation of their receptors. *Nature* 358: 771–4

3 Dreyer C, Krey G, Keller H, Givel F, Helftenbein G, Wahli W (1992) Control of the peroxisomal beta-oxidation pathway by a novel family of nuclear hormone receptors. *Cell* 68: 879–87

4 Tugwood JD, Issemann I, Anderson RG, Bundell KR, McPheat WL, Green S (1992) The mouse peroxisome proliferator activated receptor recognizes a response element in the 5' flanking sequence of the rat acyl CoA oxidase gene. *EMBO J* 11: 433–9

5 Juge-Aubry C, Pernin A, Favez T, Burger AG, Wahli W, Meier CA, Desvergne B (1997) DNA binding properties of peroxisome proliferator-activated receptor subtypes on various natural peroxisome proliferator response elements. Importance of the 5'-flanking region. *J Biol Chem* 272: 25 252–9

6 IJpenberg A, Jeannin E, Wahli W, Desvergne B (1997) Polarity and specific sequence requirements of peroxisome proliferator-activated receptor (PPAR)/retinoid X receptor heterodimer binding to DNA. A functional analysis of the malic enzyme gene PPAR response element. *J Biol Chem* 272: 20 108–17

7 Jenster G (1998) Coactivators and corepressors as mediators of nuclear receptor function: an update. *Mol Cell Endocrinol* 143: 1–7

8 Braissant O, Foufelle F, Scotto C, Dauca M, Wahli W (1996) Differential expression of peroxisome proliferator-activated receptors (PPARs): tissue distribution of PPAR-alpha, beta, and-gamma in the adult rat. *Endocrinology* 137: 354–66

9 Yu K, Bayona W, Kallen CB, Harding HP, Ravera CP, McMahon G, Brown M, Lazar MA (1995) Differential activation of peroxisome proliferator-activated receptors by eicosanoids. *J Biol Chem* 270: 23 975–83

10 Forman BM, Chen J, Evans RM (1997) Hypolipidemic drugs, polyunsaturated fatty acids, and eicosanoids are ligands for peroxisome proliferator-activated receptors alpha and delta. *Proc Natl Acad Sci USA* 94: 4312–7

11 Krey G, Braissant O, L'Horset F, Kalkhoven E, Perroud M, Parker MG, Wahli W (1997) Fatty acids, eicosanoids, and hypolipidemic agents identified as ligands of peroxisome proliferator-activated receptors by coactivator-dependent receptor ligand assay. *Mol Endocrinol* 11: 779–91

12 Kliewer SA, Sundseth SS, Jones SA, Brown PJ, Wisely GB, Koble CS, Devchand P, Wahli W, Willson TM, Lenhard JM, Lehmann JM (1997) Fatty acids and eicosanoids regulate gene expression through direct interactions with peroxisome proliferator-activated receptors alpha and gamma. *Proc Natl Acad Sci USA* 94: 4318–23

13 Schoonjans K, Peinado-Onsurbe J, Lefebvre AM, Heyman RA, Briggs M, Deeb S, Staels B, Auwerx J (1996) PPARalpha and PPARgamma activators direct a distinct tissue-specific transcriptional response via a PPRE in the lipoprotein lipase gene. *EMBO J* 15: 5336–48

14 Issemann I, Prince R, Tugwood J, Green S (1992) A role for fatty acids and liver fatty acid binding protein in peroxisome proliferation? *Biochem Soc Trans* 20: 824–7

15 Vu-Dac N, Schoonjans K, Kosykh V, Dallongeville J, Fruchart JC, Staels B, Auwerx J (1995) Fibrates increase human apolipoprotein A-II expression through activation of the peroxisome proliferator-activated receptor. *J Clin Invest* 96: 741–50

16 Mascaro C, Acosta E, Ortiz JA, Marrero PF, Hegardt FG, Haro D (1998) Control of human muscle-type carnitine palmitoyltransferase I gene transcription by peroxisome proliferator-activated receptor. *J Biol Chem* 273: 8560–3

17 Brandt JM, Djouadi F, Kelly DP (1998) Fatty acids activate transcription of the muscle carnitine palmitoyltransferase I gene in cardiac myocytes via the peroxisome proliferator-activated receptor alpha. *J Biol Chem* 273: 23 786–92

18 Peters JM, Hennuyer N, Staels B, Fruchart JC, Fievet C, Gonzalez FJ, Auwerx J (1997) Alterations in lipoprotein metabolism in peroxisome proliferator-activated receptor alpha-deficient mice. *J Biol Chem* 272: 27 307–12

19 Aoyama T, Peters JM, Iritani N, Nakajima T, Furihata K, Hashimoto T, Gonzalez FJ (1998) Altered constitutive expression of fatty acid-metabolizing enzymes in mice lacking the peroxisome proliferator-activated receptor alpha (PPARalpha). *J Biol Chem* 273: 5678–84

20 Lee SS, Pineau T, Drago J, Lee EJ, Owens JW, Kroetz DL, Fernandez-Salguero PM, Westphal H, Gonzalez FJ (1995) Targeted disruption of the alpha isoform of the peroxisome proliferator-activated receptor gene in mice results in abolishment of the pleiotropic effects of peroxisome proliferators. *Mol Cell Biol* 15: 3012–22

21 Tontonoz P, Hu E, Spiegelman BM (1994) Stimulation of adipogenesis in fibroblasts by PPAR gamma 2, a lipid-activated transcription factor. *Cell* 79: 1147–56

22 Saez E, Tontonoz P, Nelson MC, Alvarez JG, Ming UT, Baird SM, Thomazy VA, Evans RM (1998) Activators of the nuclear receptor PPARgamma enhance colon polyp formation. *Nat Med* 4: 1058–61

23 Lefebvre AM, Chen I, Desreumaux P, Najib J, Fruchart JC, Geboes K, Briggs M, Heyman R, Auwerx J (1998) Activation of the peroxisome proliferator-activated receptor gamma promotes the development of colon tumors in C57BL/6J-APCMin/+ mice. *Nat Med* 4: 1053–7

24 Jiang C, Ting AT, Seed B (1998) PPAR-gamma agonists inhibit production of monocyte inflammatory cytokines. *Nature* 391: 79–82

25 Ricote M, Li AC, Willson TM, Kelly CJ, Glass CK (1998) The peroxisome proliferator-activated receptor-gamma is a negative regulator of macrophage activation. *Cell* 93: 241–252

26 Tontonoz P, Nagy L, Alvarez JG, Thomazy VA, Evans RM (1998) PPARgamma promotes monocyte/macrophage differentiation and uptake of oxidized LDL. *Cell* 93: 241–252

27 Nagy L, Tontonoz P, Alvarez JG, Chen H, Evans RM (1998) Oxidized LDL regulates macrophage gene expression through ligand activation of PPARgamma. *Cell* 93: 229–240

28 Kersten S, Wahli W (1999) Adopting orphans: finding ligands. *In*: D Picard (ed.): *Steroid/nuclear receptorsuperfamily: a practical approach*. Practical Approach Series, Oxford University Press,

Oxford, *in press*
29 Nolte RT, Wisely GB, Westin S, Cobb JE, Lambert MH, Kurokawa R, Rosenfeld MG, Willson TM, Glass CK et al (1998) Ligand binding and co-activator assembly of the peroxisome proliferator-activated receptor-gamma. *Nature* 395: 137–43
30 Uppenberg J, Svensson C, Jaki M, Bertilsson G, Jendeberg L, Berkenstam A (1998) Crystal structure of the ligand binding domain of the human nuclear receptor PPARgamma. *J Biol Chem* 273: 31 108–12
31 Xu EM, Lambert MH, Montana VG, Parks DJ, Blanchard SG, Brown PJ, Sternbach DD, Lehmann JM, Wisely GB, Willson TM et al (1999) Molecular recognition of fatty acids by peroxisome proliferator activated receptors. *Mol Cell 3: 397–403*
32 Lin Q, Ruuska SE, Shaw NS, Dong D, Noy N (1999) Ligand selectivity of the peroxisome proliferator-activated receptor alpha. *Biochemistry* 38: 185–90
33 Devchand PR, Keller H, Peters JM, Vazquez M, Gonzalez FJ, Wahli W (1996) The PPARalpha-leukotriene B4 pathway to inflammation control. *Nature* 384: 39–43
34 Ellinghaus P, Wolfrum C, Assmann G, Spener F, Seedorf U (1999) Phytanic acid activates the peroxisome proliferator-activated receptor alpha (PPARalpha) in sterol carrier protein 2-/ sterol carrier protein x-deficient mice. *J Biol Chem* 274: 2766–72
35 Forman BM, Tontonoz P, Chen J, Brun RP, Spiegelman BM, Evans RM (1995) 15-Deoxy-delta 12, 14-prostaglandin J2 is a ligand for the adipocyte determination factor PPAR gamma. *Cell* 83: 803–12
36 Kliewer SA, Lenhard JM, Willson TM, Patel I, Morris DC, Lehmann JM (1995) A prostaglandin J2 metabolite binds peroxisome proliferator-activated receptor gamma and promotes adipocyte differentiation. *Cell* 83: 813–9
37 Staels B, Dallongeville J, Auwerx J, Schoonjans K, Leitersdorf E, Fruchart JC (1998) Mechanism of action of fibrates on lipid and lipoprotein metabolism. *Circulation* 98: 2088–93
38 Vu-Dac N, Chopin-Delannoy S, Gervois P, Bonnelye E, Martin G, Fruchart JC, Laudet V, Staels B (1998) The nuclear receptors peroxisome proliferator-activated receptor alpha and Rev-erbalpha mediate the species-specific regulation of apolipoprotein A-I expression by fibrates. *J Biol Chem* 273: 25 713–20
39 Staels B, Koenig W, Habib A, Merval R, Lebret M, Torra IP, Delerive P, Fadel A, Chinetti G, Fruchart JC et al (1998) Activation of human aortic smooth-muscle cells is inhibited by PPARalpha but not by PPARgamma activators. *Nature* 393: 790–3
40 Poynter ME, Daynes RA (1998) Peroxisome proliferator-activated receptor alpha activation modulates cellular redox status, represses nuclear factor-kappaB signaling, and reduces inflammatory cytokine production in aging. *J Biol Chem* 273: 32 833–41
41 Hertz R, Berman I, Keppler D, Bar-Tana J (1996) Activation of gene transcription by prostacyclin analogues is mediated by the peroxisome-proliferators-activated receptor (PPAR). *Eur J Biochem* 235: 242–7
42 Lehmann JM, Moore LB, Smith-Oliver TA, Wilkison WO, Willson TM, Kliewer SA (1995) An antidiabetic thiazolidinedione is a high affinity ligand for peroxisome proliferator-activated receptor gamma (PPAR gamma). *J Biol Chem* 270: 12 953–6
43 Lehmann JM, Lenhard JM, Oliver BB, Ringold GM, Kliewer SA (1997) Peroxisome proliferator-activated receptors alpha and gamma are activated by indomethacin and other non-steroidal anti-inflammatory drugs. *J Biol Chem* 272: 3406–10
44 Fujiwara T, Yoshioka S, Yoshioka T, Ushiyama I, Horikoshi H (1988) Characterization of new oral antidiabetic agent CS-045. Studies in KK and ob/ob mice and Zucker fatty rats. *Diabetes* 37: 1549–58
45 Imura H (1998) A novel antidiabetic drug, troglitazone – reason for hope and concern. *N Engl J Med* 338: 908–9
46 Inzucchi SE, Maggs DG, Spollett GR, Page SL, Rife FS, Walton V, Shulman GI (1998) Efficacy and metabolic effects of metformin and troglitazone in type II diabetes mellitus. *N Engl J Med* 338: 867–72
47 Motojima K, Passilly P, Peters JM, Gonzalez FJ, Latruffe N (1998) Expression of putative fatty acid transporter genes are regulated by peroxisome proliferator-activated receptor alpha and gamma activators in a tissue- and inducer-specific manner. *J Biol Chem* 273: 16 710–4
48 Burant CF, Sreenan S, Hirano K, Tai TA, Lohmiller J, Lukens J, Davidson NO, Ross S, Graves RA (1997) Troglitazone action is independent of adipose tissue. *J Clin Invest* 100: 2900–8
49 Berger J, Leibowitz MD, Doebber TW, Elbrecht A, Zhang B, Zhou G, Biswas C, Cullinan CA,

Hayes NS, Li Y et al (1999) Novel peroxisome proliferator-activated receptor (PPAR) gamma and PPARdelta ligands produce distinct biological effects. *J Biol Chem* 274: 6718–6725
50 Brown PJ, Smith-Oliver TA, Charifson PS, Tomkinson NC, Fivush AM, Sternbach DD, Wade LE, Orband-Miller L, Parks DJ, Blanchard SG et al (1998) Identification of peroxisome proliferator-activated receptor ligands from a biased chemical library. *Chem Biol* 4: 909–18

New Approaches to Drug Development
ed. by P. Jollès

Tomorrow's anticancer agents: inhibitors of Ras farnesylation

Dominique Perrin and Bridget T. Hill

Division de Cancérologie Expérimentale, Centre de Recherche Pierre Fabre, F-81106 Castres, France

Summary. The purpose of this chapter is to concentrate on what can be considered as definite milestones on the way from examples of inhibitors of farnesyl-protein transferase (FPTase) to candidate drugs actually being considered for or already being evaluated in clinical trials. Emphasis will be placed on results obtained using experimental tumour models *in vivo*, with a detailed discussion of these results and of the questions which remain to be studied or are still unanswered. The data discussed here are almost exclusively based on published reports, with only brief reference, in the chapter "use of the FPTase inhibitors in the clinic", to some of the newer compounds reported on during recent meetings, details of which have not yet appeared in the peer-reviewed literature. For those requiring a more extensive review of the catalogue of FPTase inhibitors now discovered, some excellent reviews have been committed to this purpose [1–3].

Introduction

Ras proteins

Ras proteins are small (21 kD) GTPases which form part of several signalling pathways controlling notably gene expression and proliferation. In recent years, an increasing number of elements upstream and downstream of Ras, implicated in several interconnected pathways, (Fig. 1) have been described (cf. reviews [4–6]). Four distinct forms of Ras have so far been identified in mammalian cells, namely Harvey (H)-Ras, neuroblastoma (N)-Ras, Kirsten (K)-Ras A and K-Ras B, corresponding to 3 distinct genes, with K-Ras A and B being produced by alternate splicing. The precise function of each isoform is still unclear. However, mice null for both N- and H-*ras* were viable, whilst embryos null for K-*ras* died before birth [7, 8]. Furthermore, preferential activation of phosphatidylinositol 3-kinase (PI3K) or mitogen-activated protein kinase (MAPK) pathways associated with H- and K-Ras, respectively, has been reported [9]. These results are indicative of a potential specificity of the individual "wiring" of these Ras isoforms. The constitutive activation of Ras proteins in tumours appears to contribute to their malignant growth properties and members of this gene family are mutationally activated in a large number of malignancies. Ras mutations are present in around 25% of all cancers, notably 30% of lung cancer, 50% of colonic and 90% of pancreatic tumours [10, 11]. A single amino acid substitution, most often in positions 12, 13, 59

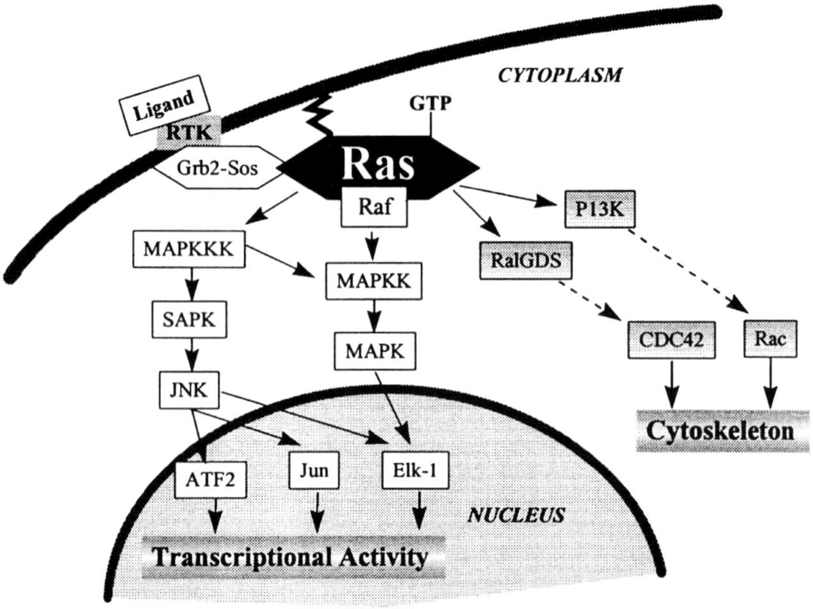

Figure 1. Ras signalling pathways. Upon binding of its ligand, a tyrosine kinase receptor autophosphorylates and recruits the cytosolic complex Grb2-Sos to the plasma membrane. In turn, Sos binds Ras localised to the plasma membrane, and activates it by promoting the exchange of GDP for GTP. Activated (GTP-bound) Ras recruits Raf to the plasma membrane and activates it, which in turn induces the phosphorylation and activation of mitogen activated protein kinase (MAPK) kinase (MAPKK), which phosphorylates and activates MAPK. Phosphorylated MAPK activates the transcription factor Elk-1. Independently of Raf, Ras activates the MAPKK kinase (MAPKKK) pathway, involved in stress-response, leading to transcriptional activity via a cascade involving stress-activated protein kinase (SAPK), Jun kinase (JNK) and transcription factors like ATF2, Jun and Elk-1. MAPKKK can actually bypass Raf and activate the MAPKK cascade. Moreover, Ras controls the dynamics of the actin cytoskeleton by interacting with phosphatidylinositol 3-kinase (PI3K) and RalGDS via complex pathways (4–6).

or 61, is sufficient to abolish the GTPase activity of Ras proteins and therefore to lock them into their GTP-bound form, constitutively activating them (for a review see [10]). Whilst in myeloid and lymphoid diseases, as well as in melanoma and liver carcinoma, notably N-*ras* mutations are prevalent, K-*ras* mutations predominate in lung, pancreas and colonic carcinomas [11].

Constitutive activation of Ras in tumours appears to contribute to malignant growth. Inactivation of mutated K-*ras* in human colon tumour cell lines induced either a total abrogation of their tumourigenicity [12] *in vitro* and *in vivo* or at least a slowing of their growth rates [13]. Moreover, transgenic mice expressing either the v-H- or v-N-*ras* oncogene under the control of the mouse mammary tumour virus (MMTV) developed mammary and salivary or mammary and lymphoid tumours respectively, in a stochastic and time-dependent manner. These results emphasise the role of Ras in malignant growth, most probably in combination with other randomly occurring mutations.

The activation of normal Ras proteins is recognised as an essential step in the signalling cascade of mitogenesis and normal Ras is considered necessary for the growth activity of numerous autocrine loops in tumours [14]. Moreover, the expression of constitutively-activated Ras has been shown to induce the expression of angiogenic factor VEGF (vascular endothelial growth factor) in rat intestine epithelial and in human colon tumour cell lines [15]. These data suggest an involvement of Ras in angiogenesis and, therefore, in facilitating tumour growth. Furthermore, it has been suggested that beside its promoting effects on cell growth, oncogenic Ras could play a role in tumourigenesis by inhibition of the apoptosis phenomenon induced by the escape of epithelial cells from monolayer growth [16]. In tumour cells, constitutively active (GTP-bound) Ras could alter the delicate balance between proliferation and apoptosis by inducing growth signal independent cell proliferation and/or promoting escape from death via apoptosis. However, it seems that the anti-apoptotic function of Ras is dependent on the genetic background of cells. For instance, expression of activated H-*ras* in embryonic fibroblasts derived from $p53^{+/+}$ mice sensitised the cells to various anticancer agents via increased apoptosis, although this did not occur in the same cell type derived from $p53^{-/-}$mice [17]. Whilst a role for mutated/activated Ras in carcinogenesis is widely accepted, an often underappreciated concept is the importance of the overexpression of normal Ras in this phenomenon, since a deregulation of wild-type Ras isoforms is often observed in tumours (see [18] as a review).

Therefore, blocking Ras activity should result not only in an inhibition of the growth of tumours expressing mutationally-activated Ras, but also those which depend on the activation of molecules upstream of Ras, like growth factor receptors, and could interfere with antiapoptotic signalling and angiogenesis. Moreover, inhibitors of Ras should be expected to induce few or no side effects on the growth of normal cells which do not rely on a constitutively-activated Ras protein for their proliferation.

Ras localisation to the plasma membrane is essential for its signal-transducing abilities, a localisation which depends on complex post-translational processing of the Ras protein. The first step in the processing of Ras proteins consists of the addition of a farnesyl (a 15 carbon prenyl) group to a cysteine of the C-terminal tetrapeptide CAAX (where A is an aliphatic amino acid and X preferentially either methionine or serine), catalysed by the enzyme FPTase [19]. Further processing involves proteolysis of the AAX terminal amino acids [20], methylation (Fig. 2) and, in the case of H- and N-Ras, a palmitoylation step. However, these latter steps have been reported as not being essential for Ras-transforming activity [19].

Strategies for blocking Ras function

One of the possible approachs to blocking Ras function is to target its mRNA. Anti-*ras* ribozymes have been successfully delivered to cells before implanta-

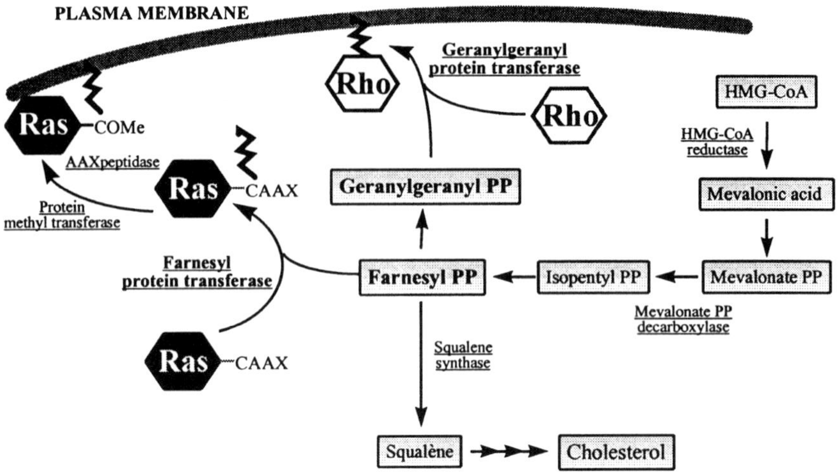

Figure 2. Ras-processing and cholesterol biosynthesis pathway. Ras is sequentially processed by farnesyl-protein transferase (FPTase) which transfers a farnesyl moiety to the cysteine residue of its CAAX carboxyterminus, an AAX peptidase which removes the last three C-terminal amino acids, and a protein methyl transferase which methylates the prenylated cysteine.

tion into nude mice via either transfection [21] or a retroviral vector [22]. In both cases, the presence of the ribozyme reverted the transformed phenotype of the cells, the amounts of *ras* mRNA and protein were downregulated and growth in nude mice was inhibited markedly. However, a clear demonstration of delivery and effectiveness of a ribozyme against an established tumour *in vivo* is still awaited.

On the other hand, the H-*ras* antisense phosphorothioate-based oligonucleotide ISIS-2503 has been shown to downregulate the amount of H-*ras* mRNA and to inhibit potently the growth of a panel of human cell lines (namely MDA MB231, A549, H69, SW620 and T24) xenografted onto nude mice) at 25 mg/kg daily, administered by the intravenous (i.v.) or intraperitoneal (i.p.) routes [23]. On the basis of these impressive *in vivo* results, clinical trials of ISIS-2503 were initiated at the end of 1997, and results are eagerly awaited.

An alternative strategy is to ***target downstream effectors*** of the Ras protein, in order to block the signalling pathway; putative targets could be Raf, MAPK, or PI3K which is an oncogene overexpressed in 40% of ovarian cancers [24].

An ***immunological approach*** has been attempted too, whose basis together with a summary of preclinical findings can be found in an excellent review by Abrams et al. [25]. In the clinical setting, antigen-presenting cells were loaded *ex vivo* with a synthetic Ras peptide corresponding to the mutation present in the individual patient's tumour and then reintroduced into that same patient. An immune response could be seen in 2 out of 5 patients in the absence of side-effects [26].

Another strategy is to inhibit the processing of Ras protein by ***depleting the intracellular store of farnesylpyrophosphate*** (FPP) via an interference with the mevalonate biosynthesis pathway, as exemplified by compounds like lovastatin or phenylacetate. In a phase-I clinical study [27] of lovastatin involving 88 patients with solid tumours, one minor response (45% reduction in size) was observed in one patient with a recurrent glioma out of 24 patients with brain tumours. The dose-limiting toxicity encountered was myopathy. During a phase-I/II clinical trial of lovastatin, doses of up to 30 mg/kg per day with or without radiation were administered to two groups of nine glioma patients. Without any detectable toxicity other than joint pain in 2 patients, 4 minor or partial responses were recorded in the lovastatin-plus-radiation arm, against only 2 in the lovastatin-alone arm [28]. Similarly, phenylacetate, which blocks the enzyme mevalonate-pyrophosphate decarboxylase, has demonstrated anti-tumour activity in tissue culture [29] and experimental animal models [30]. Phenylacetate can suppress the growth of various *ras*-transformed cells through interference with Ras isoprenylation [31]. In the clinic, limited efficacy against refractory malignant glioma (1/7 patients) and a 50% decrease in prostate-specific antigen level (1/9 patients) has been described. Dose-limiting toxicity was a reversible central nervous system depression [32].

Certain compounds, described in more detail elsewhere [33], like limonene and its derivatives, with a yet ***uncharacterised mode of action***, have been reported as potential inhibitors of the Ras signalling pathway. Interestingly, limonene [34] and its metabolite perillyl alcohol [35], which have demonstrated anti-proliferative activity especially in *ras*-transformed cells, are currently being evaluated as anti-cancer agents in clinical trials. The maximal tolerated dose (MTD) for limonene was defined as 8 g/m^2 per day and one partial response in a breast cancer patient was maintained for 11 months at this dose in a phase-I study; however, no responses were detected in a subsequent limited phase-II evaluation in breast cancer [34].

However, the lion's share of published literature in this field relates to ***inhibitors of Ras processing***, and especially of FPTase, with only a few reports being published on inhibitors of prenylated protein methyl transferase, or of peptidase (see review [33]). This is based on the fact that farnesylation of Ras is considered as being the key step in Ras processing. Moreover, the synthesis of inhibitors of enzymes with well-defined substrates, namely CAAX peptides and FPP, is a task familiar to the pharmaceutical industry, whilst targeting protein/protein interactions is as yet a more demanding challenge. Moreover, farnesylation affects only a few proteins in the cell, thus effectively limiting the risks for potential side-effects. The rationale behind the development of specific inhibitors of FPTase was that these compounds would affect primarily cells relying on an activated Ras protein for growth, and so fewer side-effects would be expected than from "classical" anticancer drugs which are predominantly DNA-damaging agents. However, there was major initial apprehension that novel side-effects, linked for instance to potential interference with visual signal transduction (see section below) might "rear their ugly heads", before the first *in vivo* data became available.

Inhibitors of farnesyl protein transferase

The target: the FPTase enzyme

FPTase is a heterodimeric metalloenzyme composed of a 48 kD α and a 46 kD β subunit, whose activity depends strictly on the presence of both zinc and magnesium ions. FPTase catalyses the addition of a farnesyl (a 15 carbon prenyl) group to a cysteine residue of proteins carrying the carboxy terminal tetrapeptide CAAX (where A is an aliphatic amino acid and X preferentially either methionine or serine). A great deal of work (reviewed in [1]) has been devoted to studying this protein, notably involving X-ray crystallography, nuclear magnetic resonance, kinetic and cross-linking studies. In short, the enzyme preferentially binds both substrates via its β subunit, with the α subunit contributing to the stability of the β subunit and to the catalytic activity. The enzyme preferentially binds FPP, then the substrate protein and releases the farnesylated protein upon binding of a new FPP substrate. FPTase not only farnesylates proteins of the Ras family (H-, N- and K-Ras), but also Rho B, a protein controlling the dynamics of the actin cytoskeleton, the nuclear lamins, controlling the stability of the perinuclear membrane, as well as transducin and rhodopsin kinase, two proteins involved in the visual signal-transduction pathway [19, 20].

Interestingly, it has been shown recently that the FPTase enzyme itself might have transforming abilities. NIH3T3 cells transfected with the genes coding for the α and β subunits of human FPTase resulted in a 3- to 13-fold increase in protein expression and a parallel 1.5- to 3-fold increase in enzyme activity [36]. Platelet-derived growth factor (PDGF) and basic fibroblast growth factor (bFGF) induced enhanced MAPK activity, DNA synthesis and anchorage-dependent growth, as well as enhanced colony formation in soft agar. Development of tumours in nude mice was also noted, with tumourigenicity being correlated to FPTase enzyme activities in the various clones isolated [36]. Moreover, in a study involving ovarian carcinoma biopsies, all K-Ras mutated tumours displayed an increase in the level of FPTase mRNA [37]. Similarly, transformation of the thyroid cell line FRTL-5 with activated K-*ras* and H-*ras* was associated with an increase in FPTase activity, and K-*ras* transformation specifically induced an increased prenylation of small G proteins [38]. Overall, these results point to a mechanism of regulation of FPTase expression and activity by Ras proteins.

Squalene synthase (SSase) is another enzyme which uses FPP as a substrate and it is an important enzyme in the cholesterol biosynthesis pathway, so that a specificity for inhibitors towards FPTase as opposed to SSase has been considered essential to avoid perturbation of cholesterol synthesis.

Geranylgeranyl-protein transferase I (GGPTase I) is an enzyme related to FPTase, both being heterodimeric proteins which share the same α-subunit [39]. GGPTase I catalyses transfer of a 20 carbon prenyl group to a cysteine of a somewhat different protein CAAX C-terminus (where A is an aliphatic amino acid and X either leucine or proline), and processes important proteins such as the γ-subunit of trimeric GTP-binding proteins, Rap 1A, involved in growth control, RhoA which controls the actin cytoskeleton, and CDC42, which controls the cell cycle [19]. Overall, in cells, geranylgeranylation is 5 to 10 times more widespread than farnesylation. Still another protein using geranylgeranylpyrophosphate as a substrate is geranylgeranyl-protein transferase II, which processes proteins of the Rab family, involved in vesicle traffic and exo-endocytosis [40].

Therefore, to date most of the chemistry has focused on the search for potent and specific inhibitors of FPTase, relative to GGPTase I, so as to minimise any interference with geranylgeranylation occurring in the cell.

Screening procedures for FPTase inhibitors

From a methodological point of view, the search for inhibitors classically involves a three-step process. Firstly, enzymatic assays based upon farnesylation of H-Ras or a CAAX-based peptide by a mammalian FPTase are performed, with the efficacy of the inhibitor towards the enzyme being indicated by its IC_{50} value (i.e., the concentration decreasing the activity of the enzyme by 50%). Specificity for FPTase *versus* GGPTase is assessed using a mammalian GGPTase I and RhoA or a CAAX-based peptide as substrate. Then *in vitro* assays are usually carried out with *ras*-transformed NIH3T3 or Rat1 cells. Processing of H-Ras and Rap 1A, used as surrogates for farnesylation and geranylation respectively, is then generally assessed, together with evaluations of cytotoxicity under conditions of anchorage-dependent or -independent (in soft agar) growth. Other assays include selective inhibition of growth of H-*ras*-transformed cells as opposed to either their K-*ras*, *raf*- or *src*-transformed counterparts.

The most costly and time-consuming screening step involves the *in vivo* studies. The most commonly-used models are H-*ras*-transformed NIH3T3 or Rat1 cells grafted onto nude mice, followed by a variety of human tumour cell line xenografts and, less frequently to date, transgenic mice for H- or N-*ras*. Most treatment schedules have employed daily i.p. or subcutaneous (s.c.) treatments, with the exception of studies using Schering's SCH 59228 compound and its derivatives, which were given orally 4 times daily. Most treatments started on day one after implantation of the tumour, without leaving any period for the tumour to develop and establish a proper vasculature. Moreover, tumour weight is often substituted for tumour volume as an end point, with inhibition of tumour growth often only stated at one given time point. It is noticeable too that most compounds have been tested on a very limited number of tumour models. Therefore, these procedures depart from those most commonly used in evaluating the more classical anticancer agents, making it difficult to compare the real or relative efficacy of these original, novel compounds.

Sources of compounds as potential FPTase inhibitors

The chemical structures of most of the known FPTase inhibitors have been considered in detail in recent and comprehensive reviews [1–3]. This chapter, concentrating on the FPTase inhibitors which have demonstrated definite *in vivo* activity, therefore presents only a few of the essential structures (Fig. 3). In terms of sources of potential inhibitors of FPTase, various ones have been tapped by companies and academic groups, namely natural products, rationally-designed inhibitors including FPP, peptidic/peptidomimetic or bisubstrate derivatives, and finally non-peptidic derivatives mostly derived from high-throughput screening of large libraries of compounds.

Numerous natural-product inhibitors [33] have been discovered with reported IC_{50} values for FPTase varying from 0.05 to 40 μM. Where tested, good specificity against FPTase, as opposed to GGPTase I, has generally been shown. In the few cases where mechanistic studies have been carried out, FPP competition has been identified. However, in terms of biological activity, only the manumycin derivative UCF1 has shown some *in vivo* activity, inhibiting the growth of the K-*ras*-murine fibrosarcoma in BALB/c mice and of N-*ras* mutated HT1080 human fibrosarcoma and that of a human pancreatic cell line in nude mice [41]. Furthermore, UCF1 blocked Ras processing and MAPK activity, whilst it did not decrease Rap 1A geranylgeranylation in Ras-mutated Hep G2 human hepatoma [42], which is consistent with its being a specific inhibitor of FPTase.

FPP analogues have shown interesting activity against FPTase in enzyme assays, with IC_{50} values ranging from 10–1000 nM, and for certain compounds a good selectivity *versus* GGPTase I [1, 33] is shown. However, in terms of biological activity, they have proved disappointing, with some inhibition of Ras processing and of the growth of H-*ras*-transformed NIH3T3 being reported only *in vitro* (see review [33]).

Similarly, bisubstrate derivatives developed by Bristol Myers Squibb, notably BMS-186511, blocked Ras processing and cell growth in H-*ras*-transformed NIH3T3 cells, but were less effective in K-*ras*-transformed NIH3T3 cells, and ineffective against untransformed cells [43, 44].

Peptide derivatives (see reviews [1–3]) are currently receiving most interest, since peptide analogues or peptidomimetics designed on the basis of the last four amino acids of a Ras protein can inhibit the FPTase-catalysed farnesylation of Ras, at a concentration which does not interfere with GGPTase I. These products constitute the largest family of FPTase inhibitors studied to date, with several major pharmaceutical companies investing in this area of research. Since peptides and peptidomimetics contain a free carboxylic acid

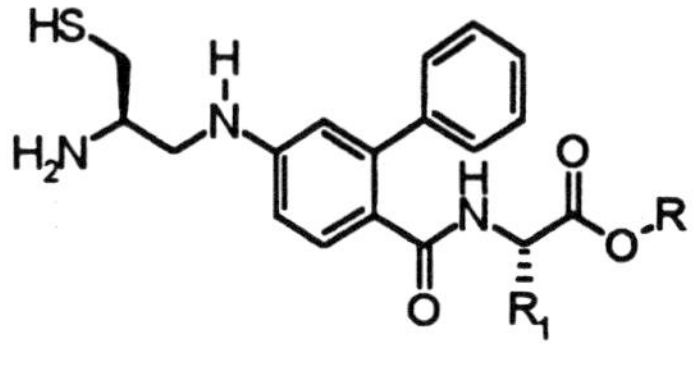

$R_1 = CH_2CH_2SMe$; R = H: FTI-276
$R_1 = CH_2CH_2SMe$; R = Me: FTI-277
$R_1 = CH_2CHMe_2$; R = Me: GGTI-286
$R_1 = CH_2CHMe_2$; R = H : GGTI-287

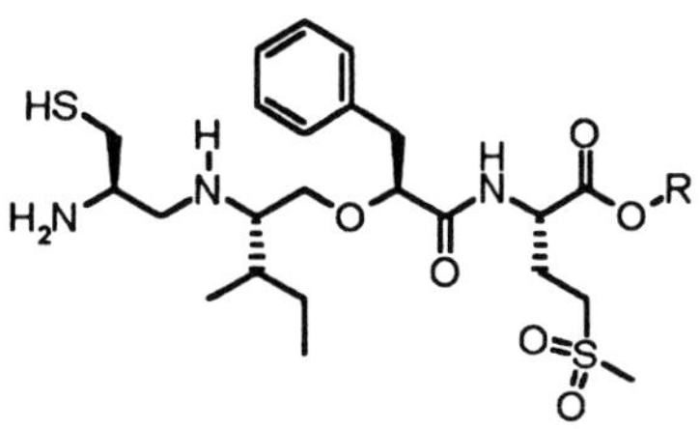

R = H : L-739,750
R = Me : L-739,749
R = iPr : L-744,832

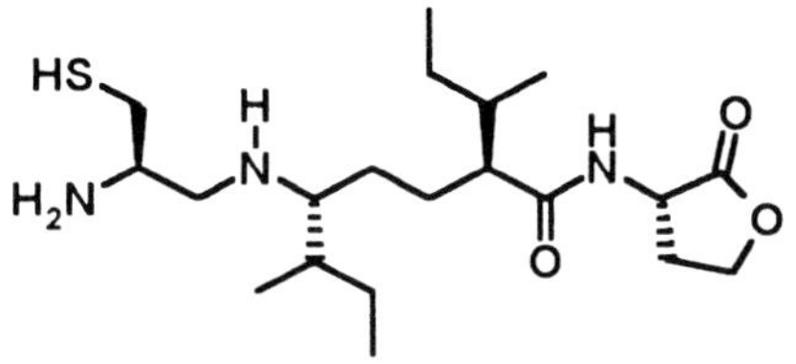

L-731,734

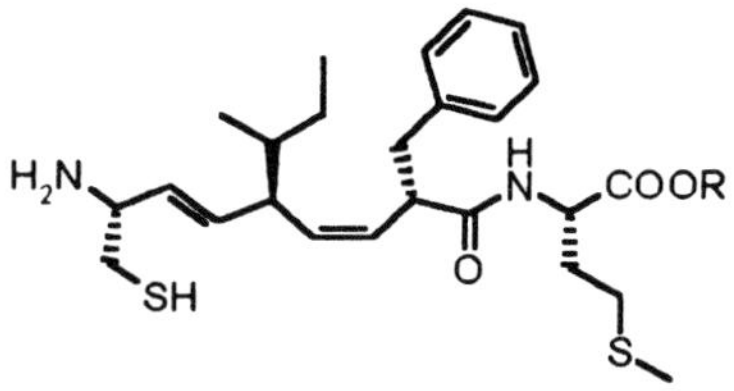

R = H : B-956 R = Me : B-1086

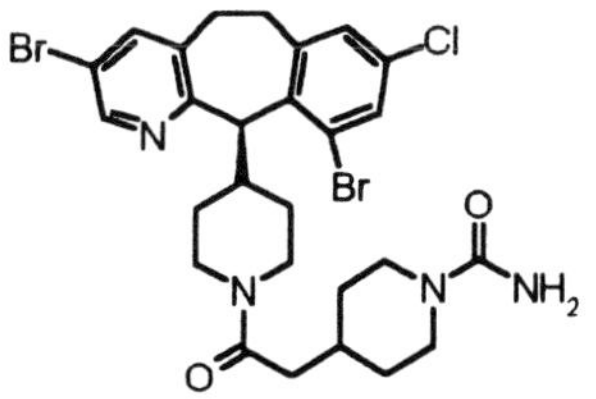

SCH 66336 R(+)

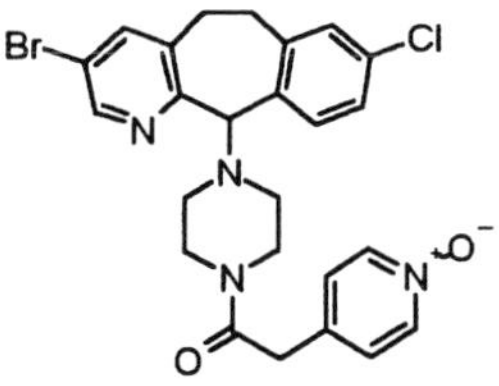

SCH 59228

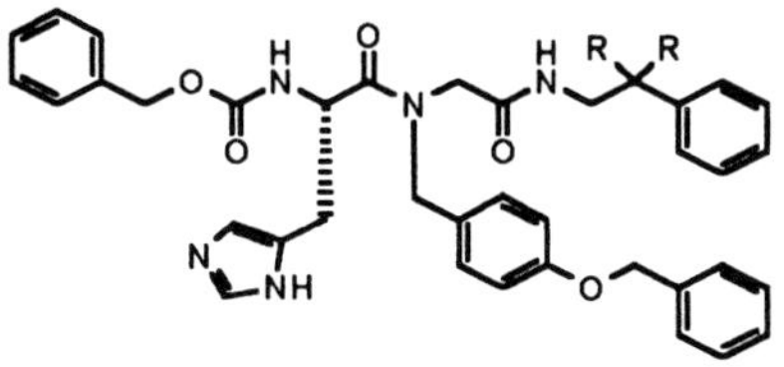

R = H : PD-161956
R = Me : PD-169451

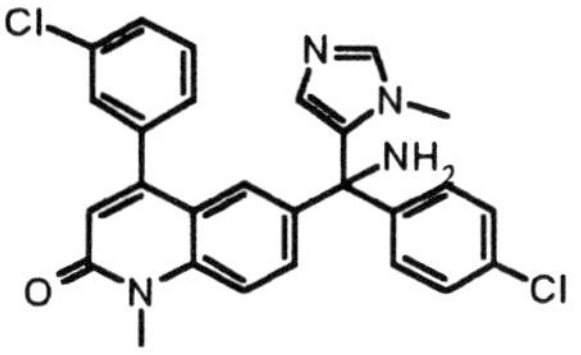

R115777 (+)

Figure 3. Structure of selected inhibitors of FPTase and GGPTase I.

group and permeate cells poorly, a widely-followed strategy has been to synthesise prodrugs of active peptides in the ester form (see structures displayed in Fig. 3).

In 1990 and 1991, the tetrapeptides CVIM [45] and CVFM [46] with IC_{50} values against FPTase of 150 and 60 nM, respectively, were described, which proved inactive against intact cells. In 1993, the first proof of principle at the cellular level of the relevance of targeting FPTase was provided, with the publication in the journal Science of two papers reporting benzodiazepine-based BZA-5B [47] and peptidomimetic L-731,734 [48] as inhibitors of both Ras processing and of *ras*-dependent cell growth in H-*ras*-transformed Rat1 cells, while being inactive against untransformed and *src*-transformed cells [47] or *raf*- and *mos*-transformed cells [48], respectively. In 1994, the methylester L-739,749 provided the first demonstration of *in vivo* activity of an inhibitor of Ras farnesylation [49]. *In vivo* activities are detailed in the next paragraph and they are critically analysed in the Perspectives section.

FPTase inhibitory compounds showing activity in vivo

Results obtained *in vivo* with FPTase inhibitors are summarised in Table 1. L-739,750 from Merck is a potent inhibitor of FPTase (IC_{50} value 2 nM) and appears selective relative to GGPTase (IC_{50} value >100 μM). Its methylester derivative L-739,749 demonstrated *in vivo* growth inhibition of *ras*-transfected Rat1 cells xenografted s.c. onto nude mice [49]. When injected i.p. at 20 mg/kg daily, from days 2 to 7 post-tumour implantation, L-739,749 induced a 66% decrease in tumour weight 5 days after the last treatment. Moreover, this compound did not affect the growth of either *raf*- or *mos*-transfected Rat1 cells. Furthermore, while the classical anticancer agent doxorubicin, used at its MTD of 2 mg/kg induced only a 33% reduction in tumour weight, with associated systemic toxicity, including weight loss, anorexia and inactivity, L-739,749 did not induce any visible side-effects, and microscopic *post mortem* examination of tissues like the retina, bone marrow and gastrointestinal tract failed to reveal any evidence of abnormalities.

Its isopropyl ester analogue L-744,832 [50], which has been reported to inhibit the processing of Ras proteins with an IC_{50} value between 0.1 and 1 μM, showed impressive activity against spontaneous mammary and salivary carcinoma in MMTV/H-*ras* transgenic mice. In this transgenic model, treatment was started later, when spontaneously occurring tumours had reached a volume of 50 to 350 mm^3. In 7 out of 7 mice treated daily with L-7344,832 at a dose of 40 mg/kg s.c., no detectable tumours could be found within 2 weeks of treatment. In this study, doxorubicin, used at its MTD (2 mg/kg) for 9 days, only slowed the tumour growth rate from 16.7 to 7.5 mm^3/day, inducing no regressions. Following cessation of treatment with L-744,832, tumour regrowth or growth at another site occurred in 5 out of 5 mice. However, retreatment at a dose of 40 mg/kg daily, starting when the tumour reached a volume of 1200 mm^3, induced further tumour regression in 2 out of 3 mice, within 2 to 6 weeks. Similarly, when tumours were first treated at a late stage (having reached a volume of 1200 mm^3), 3 out of 5 mice responded within 10 days with a 50% decrease in tumour volume. No visible toxicity was detected after 11 weeks of treatment with L-744,832 at the dose of 40 mg/kg/day.

Table 1. *In vivo* activities of selected inhibitors of FPTase

Compound	IC_{50} for FPTase	IC_{50} for GGPTase I	*In vivo* Activity: Model used	Daily administration schedule	Maximal effect	Refs
L-739,749	200 nM	>100 µM	H-*ras* Rat1 xeno	20 mg/kg i.p. days 2–7 post-implant	66% inhibition of tumour growth	49
L-734,832	nr	nr	MMTV-H-*ras* transgenics (mammary & salivary tumours)	40 mg/kg s.c. for 2 weeks	No tumour detected in 7/7 mice	50
			MMTV-N-*ras* transgenics (mammary & salivary tumours)	40 mg/kg s.c. for 38.5 days	Tumour regression, no lymphoma developed	51
			MMTV-TGFα -TGFα/*neu* transgenics (mammary tumours)	40 mg/kg s.c. for days	Tumour regression, increased apoptosis	54
B 1086	nr	nr	EJ-1 (H-*ras* mut) bladder tumour xeno	100 mg/kg for 18 days	Complete growth inhibition	55
			HT-1080 (N-*ras* mut) sarcoma xeno	100 mg/kg for 10 days	Growth inhibition of 50%	
			HCT-116 (K-*ras* mut) colon tumour xeno	100 mg/kg for 14 days	Growth inhibition of <15%	
SCH 59228	95 nM	>38 µM	H-*ras*-NIH3T3 xeno	50 mg/kg q.i.d. for 20–21 days	Growth inhibition of 95%	56
			K-*ras*-NIH3T3 xeno		Growth inhibition of 63%	
			DLD-1 (K-*ras* mut) colon tumour xeno		Growth inhibition of 43%	

(continued on the next page)

Table 1. (continued)

Compound	IC_{50} for FPTase	IC_{50} for GGPTase I	*In vivo* Activity: Model used	Daily administration schedule	Maximal effect	Refs
SCH 66336	1.9 nM	>50 μM	MIA PaCa-2 (K-*ras* mut) pancreatic tumour xeno	40 mg/kg q.i.d. for 20–21 days	Growth inhibition ranging from 70-100%.	57
			HCT-116 (K-*ras* mut) colon tumour xeno			58
			H-*ras*-NIH3T3 xeno			
PD 161956	4 nM	18 μM	H-*ras*-F xeno	150 mg/kg i.p. for 14 days	Growth inhibition of 88%	60
FTI-276	0.5 nM	50 nM	H-*ras*-NIH3T3 xeno	50 mg/kg i.p. for 10 days	Growth inhibition of 50%	62
			Calu-1 (K-*ras* mut) lung tumour xeno	50 mg/kg i.p. for 32 days	Complete growth inhibition	

xeno: xenografted s.c. onto nude mice
mut: mutated
nr: not reported

Moreover, histological examination of 28 tissues and organs did not reveal any clinical manifestations of toxicities. Systemic toxicity and some animal deaths were observed only at doses in excess of 100 mg/kg/day. However, no detailed toxicity data have been published.

Similar results were obtained using transgenic mice containing the MMTV/N-*ras* construct, which develop spontaneous lymphomas and mammary carcinomas with an average latency of 12 months. When detectable tumours reached a size of 200–700 mm^3, L-744,832 at 40 mg/kg s.c. was given daily for 38.5 days before sacrifice. Under these conditions, this compound induced tumour regression, with the tumour growth rate reaching a negative value (–0.7 mm^3/day *versus* 28.2 mm^3/day in the control group). Moreover, at the end of the treatment period, 0 of 6 mice in the treated group, *versus* 3 of 6 in the control group, developed lymphoma. Interestingly, in L-744,832-treated tumours, whilst H-Ras processing was completely inhibited, N-Ras processing was only partially affected; moreover, the apoptotic rate was higher in mammary tumours of treated mice than in the control group with figures of 2.9 *versus* 1.1% respectively [51].

The potential influence of the genetic background of v-H-*ras* transgenic mice and notably their p53 status, on tumour response to the FPTase inhibitor L-744,832 was studied. In $p53^{+/+}$ mice, upon treatment, tumour cell apoptosis rose from 0.3 to 4.5%, whilst the percentage of cells in the G_1 phase increased nonsignificantly from 93.8 to 95.4%. In $p53^{-/-}$mice, though, the tumour cell apoptotic rate rose from 0.3 to 2.5% and the percentage of cells in the G_1 phase increased from 84.8 to 94.5% upon treatment. Still, independently of p53 status, L-744,832 treatment induced dramatic tumour regression, with the tumour growth rate ranging from 12 to 33 mm^3/day in vehicle-treated *versus* a negative value of –10 to –13 mm^3/day in L-744,832-treated tumours [52]. Whilst L-744,832 has been reported to induce a block in G_1 via an increased expression of p21 in p53 wild-type cells, similar antiproliferative activity was still observed in p53-mutated cell lines, without any G_1 block [53], pointing out that the mechanism of action of FPTase inhibitors might not be determined only by the *ras*-status of the targeted tumour, but also by its genetic background and notably its p53 status. The overall conclusion of these studies is that FPTase inhibitors most probably work through/via an alteration of the fine balance between apoptosis and cell growth, with the molecular mechanisms differing according to the genetic background of the tumours studied.

MMTV-TGFα and MMTV-TGFα/*neu* transgenic mice, with wild-type *ras*, develop mammary tumours. Using this model system, L-744,832, administered daily at 40 mg/kg s.c., induced significant regression of tumours [54] with a tumour growth rate of –7 mm^3/day, against 19 mm^3/day in vehicle treated animals. Moreover, the apoptotic fraction increased from 0.2 to 2.6% in treated *versus* control tumours, combined with a G_1 phase arrest. As expected, MAPK activity was downregulated in tumour biopsies of treated *versus* control mice. Altogether, these results provide proof of principle that FPTase inhibitors could be useful against tumours such as breast carcinomas, which are characterised by a deregulation at the level of growth factor receptors, without any mutation of the *ras* gene.

The peptidomimetic B956 from Eisai is characterised by an IC_{50} value of 10 nM against FPTase. In an anchorage-independent *in vitro* cell growth assay, this compound preferentially inhibited the proliferation of cell lines with H-*ras* mutations (IC_{50} values of between 0.2 and 1 μM), rather than those harbouring either a N-*ras* mutation (IC_{50} values ranged between 2 and 10 μM), a K-*ras* mutation (IC_{50} values between 2 and 100 μM), or wild-type *ras* (IC_{50} values within the range of 10–100 μM). Similarly, *in vivo*, B1086, a methylester derivative of B956, induced complete inhibition of the growth of EJ-1 bladder tumours (mutated in H-*ras*) at a dose of 100 mg/kg daily i.p., with treatment starting on day 1 following s.c. tumour cell implantation. At this same dose, though, B1086 only slowed the growth of HT-1080 tumours (mutated in N-*ras*) and had merely a marginal effect on the growth of HCT-116 tumours (mutated in K-*ras*) [55]. These results suggest that the ranking in sensitivity of tumours to B1086 is H $\rightarrow$ N $\rightarrow$ K-*ras* mutations.

Non-peptidic derivatives identified via high-throughput screening at Schering-Plough and derived from the antidepressant methiotepine, surprisingly have been shown to inhibit FPTase, and are competitive with the Ras substrate. Recent derivatives have shown *in vivo* activity, employing a 4-times-daily dosage based on the relatively short half-life (1.4 h) of this compound in mouse plasma. SCH 59228 administered orally 4-times-daily for 20 to 21 days at the dose of 50 mg/kg, inhibited growth of H-*ras*-transformed NIH3T3 and BALB/3 T3 cells grafted onto nude mice by 95 and 91%, respectively, whilst inhibiting the growth of K-*ras*-transformed NIH3T3 and DLD-1 cells by 63 and 43%, respectively [56]. Therefore, against this limited panel of tumours, this SCH 59228 compound appeared more potent against cells presenting a constitutively activated H-*ras* as opposed to K-*ras*. Similarly, compound SCH 66336 inhibited in a dose-dependent fashion the growth of K-*ras*-mutated human pancreatic MIA PaCa-2 and human colon HCT-116 cells and proved most potent against mutated H-Ras-transformed NIH3T3 cells xenografted onto nude mice [57, 58]. Average tumour growth inhibition ranging from 70–100% at the dose of 40 mg/kg 4 times daily in a panel of tumours including lung, colon and pancreatic cell lines [57] was reported with SCH 66336. SCH 66336 entered phase-I trials during the autumn of 1997 [59] and publication of the results of these studies are keenly awaited.

PD 169451, from Parke-Davis, a selective inhibitor of FPTase over GGPTase (4500 fold) with IC_{50} values of 4 nM *versus* 18 μM, respectively, inhibited H-Ras processing in H-*ras*-F (murine fibroblast) cells in culture at 50 nM, and inhibited growth of this cell line *in vivo* by 88% on day 15 at the dose of 150 mg/kg i.p. administered daily for 14 consecutive days [60]. Interestingly though, a pentapeptide derivative, PD 169451, has been shown to be competitive vis-à-vis the FPP substrate of the FPTase enzyme [60].

One widely-published on product is FTI-276, the only compound issuing from academic research (namely from the laboratories of S. Sebti and

Hamilton, originally at the University of Pittsburgh [61]) which has shown some *in vivo* activity. FTI-276 is a potent (IC_{50} value 0.5 nM) FPTase inhibitor and relatively specific vis-à-vis GGPTase (IC_{50} value 50 nM). Its methyl ester derivative, FTI-277, has been shown to interfere potently with H-Ras processing in H-*ras*-transformed NIH3T3 cells with an IC_{50} value of 0.1 μM, while displaying no effect on the processing of the geranylgeranylated protein Rap 1A. Moreover, in the same model, FTI-277 started to block the constitutive activation of MAPK at a concentration between 0.3 and 1 μM, whilst concentrations 100 times higher were required to obtain the same degree of inhibition of MAPK in K-*ras4B*-transformed NIH3T3 cells [61].

When administered daily to mice at a dose of 50 mg/kg, FTI-276 selectively inhibited the growth of H-*ras*- NIH3T3 cells by 50% over a 10-day period, while showing no inhibitory effect on *raf*-transformed NIH3T3 cells. The effect of FTI-276 on the growth of two human lung carcinomas, one characterised by a K-*ras* mutation (Calu-1) and the other presenting no *ras* mutation (NCI-H810), xenografted onto nude mice, was studied [62]. Treatments involved a dose of 50 mg/kg daily i.p. and started 32 days after s.c. implantation. FTI-276, over a period of 14 days of treatment, had no effect on the growth rate of fast-growing NCI-H810 tumours, while the inhibitory effect on slowly growing Calu-1 tumours was spectacular, with growth inhibition being almost complete over the 36 days of treatment. In this study too, the authors stressed the absence of visible toxicity following 36 days of daily treatment with FTI-276 [62].

A divergent and somewhat interesting family of products are the GGPTase inhibitors synthesised by this same academic research group. As pointed out earlier, K-Ras is the most frequently mutated member of the Ras family encountered in clinical tumours [11]. It has been shown that K-Ras is an alternate substrate for GGPTase I [63]. Compounds designed selectively as inhibitors of the farnesylation of H-Ras, have proved generally less potent inhibitors, both *in vivo* and *in vitro,* of the growth of various tumour cell lines displaying K-*ras*- as compared to H-*ras*-mutations [55, 64, 65], even if this rule is not absolute. It was therefore interesting that a CAAX peptidomimetic inhibitor of GGPTase I, GGTI-287, was reported as displaying a limited selectivity for GGPTase I (IC_{50}: 7 nM) as opposed to FPTase (IC_{50}: 21 nM) *in vitro* [64]. Moreover, its methyl ester GGTI-286 proved a more potent inhibitor of the processing of Rap1A and K-Ras than of H-Ras in cells, and markedly inhibited the activation of MAPK by K-Ras, characteristics which were not shared by the potent and selective inhibitor of FPTase, FTI-277 [64]. Indeed, a subsequent derivative, namely GGTI-297 and its methyl ester GGTI-298, have been claimed as specific GGPTase I inhibitors with IC_{50} values of 56 and 203 nM against GGPTase and FPTase, respectively [66]. GGTI-298 has been reported to induce a G_0/G_1 block in A549 (human lung adenocarcinoma) cells, mutated in K-*ras*, whilst FTI-277 caused a G_2/M enrichment. Moreover, at a concentration of 30 μM, whilst GGTI-298 induced apoptosis, FTI-277 did not [67]. Unfortunately though, an appropriately wide range of FTI-277 concentrations were not studied, so any definitive conclusions vis-à-vis the apoptotic-inducing properties of those two types of compounds must remain open. It was further reported that a G_0/G_1 block was systematically observed following exposure to GGTI-298 using a panel of nine human tumour cell lines, which parallelled an induction of p21. More variable results, though, were obtained with FTI-277, since, depending on the cell line, there was either no effect on the cell cycle distribution, an arrest in G_0/G_1 or one in G_2/M [68]. These data could be taken to imply that the mechanism of action of FPTase inhibitors depends on the genetic background of the tumours, notably their p53 status, which in turn may well influence their effects on the cell cycle.

Furthermore, whilst FTI-277 has been reported as inhibiting Ras processing in H- and N-*ras*-transformed NIH3T3 cells, efficient inhibition of K-Ras processing in NIH3T3 cells was reported to require both FTI-277 and GGTI-298 [69]. In A549 and Calu-1 tumour cells in culture, processing of H- and N-Ras was efficiently inhibited by FTI-277, whilst processing of K-Ras was inhibited by a mixture of both FTI-277 and GGTI-298 in Calu-1 cells, although A549 cells were resistant to this combined treatment. However, against nude mice xenografts, FTI-277 and GGTI-298 alone (70 mg/kg daily i.p.), inhibited the growth of Calu-1 and A549 tumours by 60 and 80%, respectively, and a combination of both compounds (35 mg/kg each) did not show any superior antitumour activity [69]. Therefore, a confusing lack of correlation seems to exist between the processing data obtained *in vitro* and results of *in vivo* studies.

Perspectives/critical look at the field of FPTase inhibitors

Since the beginning of the 1990s considerable progress has been made in terms of the synthesis of potent, selective and stable inhibitors of FPTase. Cell membrane-permeable products, mostly esters of carboxylic acids, have been synthesised, and their cellular activities have been demonstrated. Evidence of antitumour efficacy *in vivo* has been obtained against Ras-dependent tumours xenografted onto nude mice and against stochastically-arising tumours in transgenic mice, as discussed above. To date (June 1999), three compounds (SCH 66636, R115777 and L-778,123) have been reported officially as being evaluated in clinical trials. However, there are still a number of unanswered questions and contentious issues.

The experimental systems studied

In cell culture, as well as *in vivo*, the favoured model has involved cells engineered to express activated Ras, namely H-*ras*-transformed fibroblastic cell lines, NIH3T3 and Rat1. The parameters followed *in vitro* have been mostly processing of H-Ras and Rap 1A and inhibition of cellular proliferation. A contentious issue appears to be the role of the enzyme GGPTase I, since K-*ras* and not H-*ras* is the most frequently mutated *ras* gene in human cancer. It appears that K-Ras farnesylation is not only difficult to inhibit, due to its high affinity for FPTase, but that K-Ras can also be geranylgeranylated instead of being farnesylated [70]. This is reflected in the fact that FPTase inhibitors generally are less effective against K- *versus* H-*ras* mutated cell lines [55, 64, 65]. Therefore, selecting for FPTase inhibitors devoid of GGPTase I activity might be considered disadvantageous in this respect. However, whilst inhibitors of GGPTase I, unlike FPTase inhibitors, have been reported [64, 68] to inhibit efficiently the prenylation of K-Ras *in vitro*, both types of compounds have been shown to be equally effective against K-*ras* mutated tumours *in vivo*. Unfortunately, the few GGPTase I inhibitory compounds presently available are relatively unselective and could even best be described as dual inhibitors of FPTase and GGPTase I. The issue as to whether GGPTase I might be a target of choice or an off-limit target is therefore still awaiting clarification and truly selective GGPTase inhibitors are called for to answer this question.

Therefore, in the future, model systems will probably have to include a panel of human tumour cell lines, or even human tumour biopsies, differing in their *ras* status. However, the *ras*-transformed murine fibroblast models have been invaluable in establishing the proof of principle that targeting FPTase was possible, so that potent inhibitors were synthesised, with *in vitro* as well as *in vivo* activities against these models. On the down side, however, it is likely that a bias was introduced by selecting for highly specific and highly potent inhibitors of H-*Ras* prenylation and function. Turning to human tumour xenografts data, though, the results appear to be highly variable from one tumour type to the other, failing to correlate well with *in vitro* results and mostly appearing independent of the *ras* status of the tumour studied.

In vivo results of xenografted tumours have been obtained generally by starting treatment on day 1 post-tumour implantation, without permitting the establishment of a proper tumour vasculature, and without waiting for tumours to reach an evaluatable size. These procedures depart from those used classically in antitumour drug screening *in vivo*. Similarly, instead of measuring tumour size as a function of time, several individual ways of monitoring effects on tumour growth have been employed, such as weighing tumours at a specific time without providing any rationale for choosing the single time point used. One might speculate that these steps had to be taken to show some efficacy of this new class of compounds. One often-heard argument to overcome this criticism is that one should not expect more from FPTase inhibitors, since they are cytostatic rather than cytotoxic agents, an issue discussed further below. The transgenic mouse expressing activated human H-*ras* or N-*ras*, is actually a very attractive *in vivo* model which is characterised by spontaneously arising, vascularised tumours in immunocompetent animals, and is probably closer to tumours encountered in patients. Truly, as detailed above, the most spectacular results in terms of tumour regression have been obtained with these transgenic mice models. However, results obtained with transgenic mice expressing K-*ras* mutations, which might be even closer to the human pathology, are eagerly awaited, although, these models are both costly and time-consuming. The 'unbeliever' might therefore ask the question if these models are not in fact a 'last chance' model to show some sort of antitumour activity with compounds which otherwise would prove, at best, minimally active against conventional nude mouse tumour xenografts.

Mode of action of FPTase inhibitors: inhibition of growth, induction of apoptosis?

It has been widely speculated that FPTase inhibitors would have cytostatic rather than cytotoxic effects. However, the complete regression of large tumours observed, notably in transgenic mice, seems to support some sort of cytotoxic action. The nature of the mechanism leading to tumour regression, be it apoptosis and/or necrosis, is not yet resolved, with conflicting results being obtained with different test compounds. In cultured A549 cells, FTI-277 at the highest tested concentration of 10 μM has been reported not to induce apoptosis, unlike the dual inhibitor GGTI-286 at the same concentration [67]. However, the question remains unanswered as to the effects of higher FTI-277 concentrations. Induction of apoptosis in cells in culture, with four different FPTase inhibitors, has been reported in the presence of a low serum concentration of 0.1%, whilst no effect was observed in the presence of the more conventionally used 10% serum [71]. Interpretation of these results is clouded, though, by their being obtained under non-physiological low concentrations of growth factors. Whilst clearly irrelevant to a clinical/physiological setting, they are, though, interesting mechanistically. On the other hand, compound L-744,832 clearly induced apoptosis in mammary tumours of H- and N-*ras* transgenic mice [52]. Moreover, several lines of evidence seem to link Ras to the regulation of apoptosis [16, 72, 73]. Globally, the sum of the evidence available seems to point to apoptosis as a mechanism of cell death resulting from exposure to FPTase inhibitors. Indeed, it might well be that FPTase inhibitors alter the delicate balance between the anti- and pro-apoptotic pathways downstream of Ras. Irrespective of the issue of the mode of action of FPTase inhibitors, the identity of the substrate protein(s) affected by inhibition of FPTase remains to be defined.

What are the targets of FPTase?

It is important to remember that H-Ras and Rap 1A, whose processing is followed to study the cellular effects of FPTase or GGPTase I inhibitors, respectively, are merely surrogate markers. Observing an upward band shift by Western blotting, i.e., a change in the processing of H-Ras, by no means provides firm evidence that H-Ras is the relevant target of the ***enzyme*** FPTase, which is actually inhibited. A puzzling – and rarely discussed – point too is the wide discrepancy *in vitro* between concentrations needed maximally to inhibit the processing of H-Ras and those required to achieve inhibition of proliferation. For instance, for a compound like FTI-277, we have shown that maximal inhibition of H-ras processing in DLD-1 colon adenocarcinoma cells occurs at concentrations as low as 0.1 μM, whilst a mere 50% inhibition of growth requires 20 μM of this product (unpublished data). These data do not fit well with the concept of Ras activity being central to cell proliferation in transformed cells. Therefore the question of the identity of the protein(s) whose processing, and therefore activity, is blocked by FPTase inhibitors is still open. Rho B is currently the most popular candidate. Rho B belongs to the Ras family of small G-proteins, and together with Rac controls the shape of the actin cytoskeleton [74] and therefore cell morphology, one of the hallmarks of transformation being a change in morphology. Moreover, Rho [75, 76] and Rac [77] activity has been shown to be necessary for Ras transformation and, whilst Rho proteins are geranylgeranylated [78], Rho B is both geranylgeranylated and farnesylated [79]. The FPTase inhibitor L731,734 has been shown preferentially to inhibit the processing of Rho B over Ras [80]. Moreover, it has been reported that the effects of L-731,734 on cell morphology preceded those on cell growth [80], which is compatible with a preferential inhibition of Rho B over Ras. L-739,749 inhibited RhoB farnesylation, whilst in parallel the amount of geranylgeranylated Rho B increased in cells [81]. In line with these results, the growth stimulatory function of constitutively activated Rho B^{V14} in Rat1 cells was abolished by L-739,749. Altogether these results suggest an anti-transforming gain of function for Rho B upon switching from the farnesylated to the geranylgeranylated form. However, here too, the relationship observed is purely correlative, not causative, and other prenylated proteins could well be involved. Moreover, the spectrum of small G-proteins evaluated in processing studies is very limited and limiting: most widely studied being H-Ras, Rap 1A, those rarely studied being Rho B, N-Ras and K-Ras, and even less often, Rho A. This fact can be explained, at least in part, by the limited availability of high-quality commercial antibodies.

Use of the FPTase inhibitors in the clinic

One of the first questions to be considered is that of the potential ***side-effects*** associated with this new class of compounds, even though an absence of toxicity at therapeutic doses has been emphasised in all the experimental *in vivo* studies reported to date. Ras proteins are part of normal signalling pathways in all cells, and yet all the reports dealing with FPTase inhibitors *in vitro* have shown a clear selectivity for inhibiting Ras processing and cellular growth in *ras*-transformed over *raf*- or *mos*-transformed and untransformed cells. A dominant negative effect exerted by cytosolic oncogenic Ras accumulated upon treatment with FPTase inhibitors, over oncogenic Ras still associated with the plasma membrane, has therefore been postulated to occur in transformed, but not in untransformed cells [82]. However, some important substrates of FPTase, such as rhodopsin kinase and transducin, are involved in the visual signalling pathway and the lamins A and B are involved in the control of nuclear membrane assembly. Even so, no major side-effects or gross toxicities in mice treated with any of these potential inhibitors have been reported, even when periods of daily treatment lasted for up to 11 weeks [50]. This observation might be explained by a higher affinity of these protein substrates for the FPTase enzyme, as compared to H-ras for instance, making them more resistant to inhibitors of farnesylation. However, the inhibitory effects on tumour growth of all FPTase inhibitors has beeen shown clearly to be dependent on ***continuous*** treatment. Therefore, success in the clinic might well depend on long-term therapy, raising the question of potentially adverse effects of such chronic administration.

Preliminary reports of the phase-I clinical trial of R115777 at 125 mg twice daily orally for 21 of 28 days indicated grade-2 adverse effects such as nausea, vomiting, headache and fatigue [59]. Since then, doses have been escalated up to 420 mg and the toxicities encountered were mostly haematologic in nature. Two patients with stable disease (one with parotid carcinoma, one with a prostate tumour) have been treated for over 6 months [83]. SCH 66336 entered phase-I clinical trials in late 1997, and the toxicities encountered were mostly gastrointestinal [84]. The MTD was determined as 200 mg twice daily on continuous treatment and stable disease was noted in 2 patients (one with a thyroid tumour and one with pseudomyxoma peritonei). Similar toxicities and MTD values were determined using a 2-week on, 2-week off schedule of administration. Stable disease was observed in patients with colon cancer [85]. Farnesylation of prelamin A in patients' buccal mucosa cells was assessed as a surrogate marker. Amongst peptidomimetics, preliminary results of a phase-I clinical trial of Merck's L-778,123, a compound of yet undisclosed structure, have been presented orally [86]. Severe grade-4 thrombocytopenia and grade-3 neurological toxicities, including confusion and electroencephalogram abnormalities, were noted at the highest administered dose of 1120 mg/m^2. In parallel, inhibition of farnesylation of the protein hDJ2 was followed in peripheral blood monocytes as a surrogate marker. With limonene at the MTD of 8 g/m^2 a day, given orally in 21-day cycles, the limiting toxicities were nausea, vomiting and diarrhoea [34], whilst for its derivative perillyl alcohol, nausea and vomiting were described [35]. No data are yet available as to whether such toxicities are cumulative, a critical factor if continuous administration is to be considered. Indeed, as seen in the case of anti-HIV treatments, even relatively mild side-effects might hinder compliance if the drug is to be given on a daily basis.

Another important question to be answered is the one of ***which clinical endpoint*** (survival, disease stabilisation, tumour regression) should be used to assess the efficacy of this new family of anticancer agents? A second is to know whether these products are expected to show efficacy as single agents or will they need to be used in ***combinations***? Such combinations might involve

Table 2. A selection of combined evaluations of conventional cytotoxic agents and FPTase inhibitors

FPTase inhibitor	Conventional cytotoxic agents tested	Model used	Effects identified	Refs
FTI-277 L-739,749	X-irradiation	*In vitro.* Panel of tumour cell lines	Decrease in survival fraction	84
L-744,832	Taxol, desoxyepothilone Doxorubicin, cisplatin, 5FU, vinblastine	*In vitro.* HTP-177 lung carcinoma cell line	Synergy Additivity	85
SCH 66336	Cyclophosphamide, 5FU, vincristine	*In vivo.* MDA-MB -468 breast tumour xeno	Increased tumour growth inhibition	58
	Paclitaxel	*In vivo.* NCI-H460 lung tumour xeno	Increased tumour growth inhibition	98
	Paclitaxel	*In vitro.* Panel of tumour cell lines	Synergy or additivity	98

xeno: xenografted s.c. onto nude mice

inhibitors of both FPTase and GGPTase I, as earlier considered by Sebti [69], even though preliminary data were not overencouraging, or alternatively combinations with classical antitumour agents (Tab. 2). Combinations with radiotherapy have been described *in vitro*. Using clonogenic assays, FPTase inhibitors FTI-277 and L-744,832 markedly decreased the surviving fraction following irradiation in H-*ras*-mutated cell lines, whilst a combination of FTI-277 and of GGTI-298 was necessary in K-*ras*-mutated cell lines [87]. *In vitro*, synergy between the FTI inhibitor L-744,832 and the microtubule-stabilising agents, taxol and desoxyepothilone A, have also been reported, whilst mere additivity has been observed with either doxorubicin, cisplatin, 5-fluorouracil (5-FU) or vinblastine [88]. Similarly, considering compound SCH 66336 at 40 mg/kg combined *in vivo* with cyclophosphamide, 5-FU or vincristine, efficacy in terms of tumour growth inhibition increased from 60% to 80% in the HTB-177 human lung carcinoma model [58]. Comparable results have been obtained with recently disclosed compounds: FTI-2148 from the group of Sebti in combination with cisplatin or gemcitabine against A549 (lung) xenografts [89], Abbott's A-197574 with cyclophosphamide against PANC-01 (pancreas) xenografts [90], Janssen's R115777 in combination with taxol or cisplatin against LoVo xenografts [91] and Rhône-Poulenc Rorer's RPR130401 with irinotecan against HCT-116 (colon) xenografts [92]. The bottom line of these experiments appears to be that *in vivo*, no antagonism has been identified and some potential additivity between several classical clinically-useful anticancer agents and inhibitors of FPTase has been noted.

Considering taxol and microtubule-interacting agents in general, a potential for synergy exists, since a cluster of evidence seems to link the microtubule-based cytoskeleton and Ras proteins, notably K-Ras. K-Ras has been shown in an elegant biochemical study to be associated with microtubules [93]. The

FPTase inhibitors SCH 56582 and B1088 have been shown to revert morphological changes induced by the expression of v-K-Ras in the NRK (normal rat kidney) cell line [94]. Moreover, these two compounds synergised strongly with vincristine to induce a block at the G_2/M phases of the cycle. Therefore, if the paradigm of the relative lack of toxicity of FPTase inhibitors holds in the clinic, combinations with classical drugs, notably with taxol, will probably be considered quite early in development programmes. Meanwhile, a phase-I clinical trial of a combination of R115777 with 5-FU has been reported as ongoing [95] and the main toxicity encountered was myelosuppression. Interestingly, this approach of studying combinations in early phase-III clinical trials is also being considered now for other 'unconventional' antitumour agents, for example, with flavopiridol, an inhibitor of cyclin-dependent kinase 2 [96] and certain angiogenic inhibitors [97].

FPTase inhibitors: tomorrow's blockbuster drugs or merely exquisite laboratory tools?

In conclusion, progress in the search for potent inhibitors of FPTase which might eventually be used in the clinic has been impressive in less than one decade: beginning in the 1990s with the description of the first tetrapeptide inhibitors, in 1993 there were the first reports of cellular activity, and in 1994 the first *in vivo* results in experimental animals, followed in 1997 by the initiation of phase-I clinical trials. Interestingly, this research has been led by the pharmaceutical industry, with the notable exception of the groups of Sebti and Hamilton at the University of Pittsburgh. Ironically, the compounds which were actually first being tested in the clinic were not rationally-designed FPTase inhibitors, but rather those compounds either not directly inhibiting the FPTase enzyme, like limonene, lovastatin and phenyl acetate, or compounds issuing from high-throughput screening like, for example, Janssen's compound R115777 and Schering-Plough's compound SCH 66336. It is hoped that publication in peer-reviewed journals of the results of these phase-I trials in the not so distant future will help clarify the issue of toxicity of this novel family of compounds.

If successful in the clinic, this new class of drugs, which will likely have to be taken daily over a long period of time to maintain remissions or prevent recurrences, is potentially a blockbuster for the pharmaceutical industry. If, however, the clinical potential of this class of compounds is not realised, they will merely have proved to be an exquisite research tool for dissecting cell functions involving prenylated proteins, but surely at a prohibitive cost.

Acknowledgments
The assistance of Stephane Gras and Marie Lamothe in producing the figures is gratefully acknowledged.

References

1 Halazy S, Gotteland JP, Lamothe M, Perrin D, Hill BT (1997) Rationally designed FPTase inhibitors. *Drugs Future* 22: 1133–1146

2 Williams TM (1998) Inhibitors of protein farnesylation. *Expert Opin Ther Patents* 8: 553–569

3 Leonard DM (1997) Ras farnesyltransferase: a new therapeutic target. *J Med Chem* 40: 2971–2990

4 Hunter T (1997) Oncoprotein networks. *Cell* 88: 333–346

5 Campbell SL, Khosravi-Far R, Rossman KL, Clark GJ, Der C (1998) Increasing complexity of Ras signaling. *Oncogene* 17: 1395–1413

6 Eyster KM (1998) Introduction to signal transduction. *Biochem Pharmacol* 55: 1927–1938

7 Koera K, Nakamura K, Nakao K, Miyoshi J, Toyoshima K, Hatta T, Otani H, Aiba A, Katsuki M (1997) K-Ras is essential for the development of the mouse embryo. *Oncogene* 15: 1151–1159

8 Johnson L, Greenbaum D, Cichowski K, Mercer K, Murphy E, Schmitt E, Bronson RT, Umanoff H, Edelmann W, Kucherlapati R et al (1997) K-ras is an essential gene in the mouse with partial functional overlap with N-ras. *Gene Develop* 11: 2468–2481

9 Yan J, Roy S, Apolloni A, Lane A, Hancock JF (1998) Ras isoforms vary in their ability to activate Raf-1 and phosphoinositide 3-kinase. *J Biol Chem* 273: 24054–24056

10 Barbacid M (1987) Ras genes. *Annu Rev Biochem* 56: 779–827

11 Bos JL (1989) *ras* oncogenes in human cancer: a review. *Cancer Res* 49: 4682–4689

12 Shirazawa S, Furuse M, Yokoyama N, Sasazuki T (1993) Altered growth of human colon cancer cell lines disrupted at activated Ki-*ras*. *Science* 26: 85–88

13 Plattner R, Anderson MJ, Sato KY, Fasching CL, Der CJ, Stanbridge EJ (1996) Loss of oncogenic ras expression does not correlate with loss of tumorigenicity in human cells. *Proc Natl Acad Sci USA* 93: 6665–6670

14 Watanabe M, Nobuta A, Tanaka J, Asaka M (1996) An effect of K-*ras* gene mutation on epidermal growth factor receptor signal transduction in Panc-1 pancreatic carcinoma cells. *Int J Cancer* 67: 264–268

15 Rak J, Mitsuhashi Y, Bayko L, Filmus J, Shirazawa S, Sasasuki T, Kerbel RS (1995) Mutant ras oncogenes upregulate VEGF/VPF expression: implications for induction and inhibition of tumor angiogenesis. *Cancer Res* 55: 4575–4580

16 Rak J, Mitsuhashi Y, Erdos V, Huang S, Filmus J, Kerbel RS (1995) Massive programmed cell death induced by three-dimensional growth conditions: suppression by mutant c-*H-ras* oncogene expression. *J Cell Biol* 131: 1587–1598

17 Lowe SW, Ruley HE, Jacks T, Housman DE (1993) p53-dependent apoptosis modulates the cytotoxicity of anticancer agents. *Cell* 74: 957–967

18 Zachos G, Spandidos DA (1997) Expression of ras proto-oncogenes: regulation and implications in the development of human tumors. *Crit Rev Oncol Hematol* 26: 65–75

19 Cox AD, Der CJ (1992) The ras/cholesterol connection: implications for ras oncogenicity. *Crit Rev Oncogen* 3: 365–400

20 Gutierrez L, Magee AI, Marshall CJ, Hancock JF (1989) Post-translational processing of p21ras is two step and involves carboxyl-methylation and carboxy-terminal proteolysis. *EMBO J* 8: 1093–1098

21 Kashani-Sabet M, Funato T, Florenes VA, Fodstad O, Scanlon KJ (1994) Suppression of the neoplastic phenotype *in vivo* by an anti-*ras* ribozyme. *Cancer Res* 54: 900–902

22 Li M, Lonial H, Citarella R, Lindh D, Colina L, Kramer R (1996) Tumor inhibitory activity of anti-Ras ribozymes delivered by retroviral gene transfer. *Cancer Gene Therapy* 3: 221–229

23 Cowsert LM (1997) *In vitro* and *in vivo* activity of antisense inhibitors of ras: potential for clinical development. *Anti-Cancer Drug Des* 12: 359–371

24 Shayesteh L, Lu Y, Kuo W-L, Baldocchi R, Godfrey T, Collins C, Pinkel D, Powell B, Mills GB, Gray JW (1999) PIK3CA is implicated as an oncogene in ovarian cancer. *Nat Genet* 21: 99–101

25 Abrams SI, Horan Hand P, Tsang K, Schlom J (1996) Mutant *ras* epitopes as targets for cancer vaccines. *Semin Oncol* 23: 118–134

26 Gjersten MK, Bakka A, Breivik J, Saeterdal I, Gedde-Dahl IIIT, Stokke KT, Solheim BG, Egge TS, Soreide O, Thorsby E et al (1996) *Ex vivo* peptide vaccination in patients with advanced pancreatic cancer: results of a phase-I/II study. *Int J Cancer* 65: 450–453

27 Thibault A, Samid D, Tompkins AC, Figg WD, Cooper MR, Hohl RJ, Trepel J, Liang B, Patronas N, Venzon DJ et al (1996) Phase I study of lovastatin, an inhibitor of the mevalonate pathway, in

patients with cancer. *Clin Cancer Res* 75: 2932–2938
28 Larner J, Jane J, Laws E, Packer R, Myers C, Shaffrey M (1998) A phase-I-II trial of lovastatin for anaplastic astrocytoma and glioblastoma multiforme. *Amer J Clin Oncol* 21: 579–583
29 Liu L, Shack S, Stetler-Stevenson WG, Hudgins WR, Samid D (1994) Differenciation of cultured human melanoma cells induced by the aromatic fatty acids phenylacetate and phenylbutyrate. *J Invest Dermatol* 103: 335–340
30 Samid D, Ram Z, Hudgins WR, Shack S, Liu L, Walbridge S, Oldfield EH, Myers CE (1994) Selective activity of phenylacetate against malignant gliomas: resemblance to fetal brain damage in phenylketonuria. *Cancer Res* 54: 891–895
31 Shack S, Chen L, Miller AC, Danesi R, Samid D (1995) Increased susceptibility of *ras*-transformed cells to phenylacetate is associated with inhibition of $p21_{ras}$ isoprenylation and phenotypic reversion. *Int J Cancer* 63: 124–129
32 Thibault A, Samid D, Cooper MR, Figg WD, Tompkins AC, Patronas N, Headlee DJ, Kohler DR, Venzon DJ, Myers CE (1995) Phase I study of phenylacetate administered twice daily to patients with cancer. *Cancer* 75: 2932–2938
33 Perrin D, Halazy S, Hill BT (1996) Inhibitors of the Ras signal transduction pathway as potential antitumour agents. *J Enzyme Inhib* 11: 77–95
34 Vigushin DM, Poon GK, Boddy A, English J, Halbert GW, Pagonis G, Jarman M, Coombes RC (1998) Phase I clinical trial of perillyl alcohol administered daily. *Cancer Chemother Pharmacol* 42: 111–117
35 Ripple GH, Gould MN, Stewart JA, Tutsch KD, Arzoomanian RZ, Alberti D, Feierabend C, Pomplun M, Wildind G, Bailey HH (1998) Phase I clinical trial of perillyl alcohol administered daily. *Clin Cancer Res* 4: 1159–1164
36 Nagase T, Kawata S, Nakajima H, Tamura S, Yamasaki E, Fukui K, Yamamoto K, Miyagawa JI, Matsumura I, Matsuda Y et al (1999) Effect of farnesyltransferase overexpression on cell growth and transformation. *Int J Cancer* 80: 126–133
37 Tanimoto H, Mehta KD, Parmley TH, Shigemasa K, Parham GP, Clarke J, OBrien TJ (1997) Expression of the farnesyltransferase beta-subunit gene in human ovarian carcinoma: correlation to K-ras mutation. *Gynecol Oncol* 66: 308–312
38 Laezza C, Di Marzo V, Bifulco M (1998) v-K-ras leads to preferential farnesylation of p21(ras) in FRTL-5 cells: multiple interference with the isoprenoid pathway. *Proc Natl Acad Sci USA* 95: 13 646–13 651
39 Moomaw JF, Casey PJ (1992) Mammalian protein geranylgeranyltransferase. *J Biol Chem* 267: 17 438–17 443
40 Brown MS, Goldstein JL (1993) Mad bet for Rab. *Nature* 366: 14–15
41 Ito T, Kawata S, Tamura S, Igura T, Nagase T, Miyagawa JI, Yamazaki E, Ishiguro H, Matsuzawa Y (1996) Suppression of human pancreatic pancreatic cancer growth in Balb/c nude mice by manumycin, a farnesyl-protein transferase inhibitor. *Jpn J Cancer Res* 87: 113–116
42 Nagase T, Kawata S, Tamura S, Matsuda Y, Inui Y, Yamasaki E, Ishiguro H, Ito T, Matsuzawa Y (1996) Inhibition of cell growth by a farnesyl protein transferase inhibitor: a preferential suppression of Ras farnesylation. *Int J Cancer* 65: 620–626
43 Patel DV, Gordon EM, Schmidt RJ, Weller HN, Young MG, Zahler R, Barbacid M, Carboni JM, Gullo-Brown JL, Hunihan L et al (1995) Phosphonyl acid-based bisubstrate analog inhibitors of Ras farnesyl protein transferase. *J Med Chem* 38: 435–442
44 Manne V, Yan N, Carboni JM, Tuomari AV, Ricca CS, Gullo-Brown J, Andahazy ML, Schmidt RJ, Patel D, Zahler R et al (1995) Bisubstrate inhibitors of farnesyltransferase: a novel class of specific inhibitors of ras transformed cells. *Oncogene* 10: 1763–1779
45 Reiss Y, Goldstein JL, Seabra MC, Casey PJ, Brown MS (1990) Inhibition of purified p21ras farnesyl protein transferase by Cys-AAX peptides. *Cell* 62: 81–88
46 Goldstein JL, Brown MS, Stradley SJ, Reiss Y, Gierasch LM (1991) Nonfarnesylated tetrapeptide inhibitors of protein farnesyl transferase. *J Biol Chem* 266: 15 575–15 578
47 James GL, Goldstein JL, Brown MS, Rawson TE, Somers TC, McDowell RS, Crowley CW, Lucas BK, Levinson AD, Marsters JC (1993) Benzodiazepine peptidomimetics: potent inhibitors of Ras farnesylation in animal cells. *Science* 260: 1937–1942
48 Kohl NE, Mosser SD, Jane deSolms S, Giuliani EA, Pompliano DL, Graham SL, Smith RL, Scolnick EM, Oliff A, Gibbs JB (1993) Selective inhibition of *ras*-dependent transformation by a farnesyltransferase inhibitor. *Science* 260: 1934–1937
49 Kohl NE, Wilson FR, Mosser SD, Giuliani E, Jane deSolms S, Conner MW, Anthony NJ, Holtz

WJ, Gomez RP, Lee T et al (1994) Protein farnesyltransferase inhibitors block the growth of *ras*-dependent tumors in nude mice. *Proc Natl Acad Sci USA* 91: 9141–9145
50 Kohl NE, Omer CA, Conner MW, Anthony NJ, Davide JS, Jane deSolms S, Giuliani E, Gomez R, Graham SL, Hamilton K et al (1995) Inhibition of farnesyltransferase induces regression of mammary and salivary carcinomas in *ras* transgenic mice. *Nat Med* 1: 792–796
51 Mangues R, Corral T, Kohl NE, Symmans WF, Lu S, Malumbres M, Gibbs JB, Oliff A, Pellicer A (1998) Antitumor effect of a farnesyl protein transferase inhibitor in mammary and lymphoid tumors overexpressing N-ras in transgenic mice. *Cancer Res* 58: 1253–1259
52 Barrington RE, Subler MA, Rands E, Omer CA, Miller PJ, Hundley JE, Koester SK, Troyer DA, Bearss DJ, Conner MW et al (1998) A farnesyltransferase inhibitor induces tumor regression in transgenic mice harbouring multiple oncogenic mutations in both cell cycle control and apoptosis. *Mol Cell Biol* 18: 85–92
53 Sepp-Lorenzino L, Rosen N (1998) A farnesyl-protein transferase inhibitor induces p21 expression and G1 block in p53 wild type tumor cells. *J Biol Chem* 273: 20 243–20 251
54 Norgaard P, Law B, Joseph H, Page DL, Shyr Y, Mays D, Pietenpol JA, Kohl NE, Oliff A, Coffey RJ et al (1999) Treatment with farnesyl-protein transferase inhibitor induces regression of mammary tumors in transforming growth factor (TGF) α and TGFα/*neu* transgenic mice by inhibition of mitogenic activity and induction of apoptosis. *Clin Cancer Res* 5: 35–42
55 Nagasu T, Yoshimatsu K, Rowell C, Lewis MD, Garcia AM (1995) Inhibition of human tumor xenograft growth by treatment with the farnesyltransferase inhibitor B956. *Cancer Res* 55: 5310–5314
56 Liu M, Bryant MS, Chen J, Lee S, Yaremko B, Li Z, Dell J, Lipari P, Malkowski M, Prioli N et al (1999) Effects of SCH 59228, an orally bioavailable farnesyl protein transferase inhibitor, on the growth of oncogene-transformed fibroblasts and a human colon carcinoma xenograft in nude mice. *Cancer Chemother Pharmacol* 43: 50–58
57 Njoroge FG, Vibulbhan B, Pinto P, Bishop R, Bryant MS, Nomeir AA, Lin CC, Liu M, Doll RJ, GirijavallabhanV et al (1998) Potent, selective, and orally bioavailable tricyclic pyridiyl acetamide n-oxide inhibitors of farnesyl protein transferase with enhanced *in vivo* antitumor activity. *J Med Chem* 41: 1561–1567
58 Liu M, Bryant MS, Chen J, Lee S, Yaremko B, Lipari P, Malkowski M, Ferrari E, Nielsen L, Prioli N et al (1998) Antitumor activity of SCH 66336, an orally bioavailable tricyclic inhibitor of farnesyl protein transferase, in human tumor xenograft models and wap-ras transgenic mice. *Cancer Res* 58: 4947–4956
59 Lavelle F (1998) American Association for *Cancer Res* 1998: promises and prospects for the next century. *Expert Opin Invest Drugs* 7: 1015–1021
60 McNamara DJ, Dobrusin EM, Leonard DM, Shuler KR, Kaltenbronn JS, Quin JI, Bur S, Thomas CE, Doherty AM, Scholten JD et al (1997) C-terminal modifications of histidyl-N-benzylglycinamides to give improved inhibition of ras farnesyltransferase, cellular activity, and anticancer activity in mice. *J Med Chem* 40: 3319–3322
61 Lerner EC, Qian Y, Blaskovich MA, Fossum RD, Vogt A, Sun J, Cox AD, Der CJ, Hamilton AD, Sebti SM (1995) Ras CAAX peptidomimetic FTI-277 selectively blocks Ras signaling by inducing cytoplasmic accumulation of inactive Ras-Raf complexes. *J Biol Chem* 270: 26 802–26 806
62 Sun J, Qian Y, Hamilton AD, Sebti SM (1995) Ras CAAX peptidomimetic FTI-276 selectively blocks tumor growth in nude mice of a human lung carcinoma with K-ras mutation and p53 deletion. *Cancer Res* 55: 4243–4247
63 James GL, Goldstein JL, Brown MS (1995) Polylysine and CVIM sequences of K-RasB dictate specificity of prenylation and confer resistance to benzodiazepine peptidomimetic *in vitro*. *J Biol Chem* 270: 6221–6226
64 Lerner EC, Qian Y, Hamilton AD, Sebti SM (1995) Disruption of oncogenic K-Ras4B processing and signaling by a potent geranylgeranyltransferase I inhibitor. *J Biol Chem* 270: 26 770–26 773
65 Sepp-Lorenzino L, Ma Z, Rands E, Kohl NE, Gibbs JB, Oliff A, Rosen N (1995) A peptidomimetic inhibitor of farnesyl:protein transferase blocks the anchorage-dependent and-independent growth of human tumor cell lines. *Cancer Res* 55: 5302–5309
66 Qian Y, Vogt A, Vasudevan A, Sebti SM, Hamilton AD (1998) Selective inhibition of type-I geranylgeranyltransferase *in vitro* and in whole cells by CAAL peptidomimetics. *Bioorgan Med Chem* 6: 293–299
67 Miquel K, Pradines A, Sun J, Qian Y, Hamilton AD, Sebti SM, Favre G (1997) GGTI-298 induces G_0-G_1 block and apoptosis whereas FTI-277 causes G_2-M enrichment in A549 cells. *Cancer Res*

57: 1846–1850
68 Vogt A, Sun J, Qian Y, Hamilton AD, Sebti SM (1997) The geranylgeranyltransferase-I inhibitor GGTI-298 arrests human tumor cells in G0/G1 and induces $p21_{(WAF1/CIP1/SDI1)}$ in a p53-independent manner. *J Biol Chem* 272: 27 224–27 229
69 Sun J, Qian Y, Hamilton AD, Sebti SM (1998) Both farnesyltransferase and geranyltransferase inhibitors are required for inhibition of oncogenic K-ras prenylation but each alone is sufficient to suppress human tumor growth in nude mouse xenografts. *Oncogene* 16: 1467–1473
70 Zhang FL, Kirschmeier P, Carr D, James L, Bond RW, Wang L, Patton R, Windsor WT, Syto R, Zhang R et al (1997) Characterization of Ha-Ras, N-Ras, Ki-Ras4A, and Ki-Ras4B as *in vitro* substrates for farnesyl protein transferase and geranylgeranyl protein transferase type I. *J Biol Chem* 272: 10 232–10 239
71 Suzuki N, Urano J, Tamanoi F (1998) Farnesyltransferase inhibitors induce cytochrome c release and caspase 3 activation preferentially in transformed cells. *Proc Natl Acad Sci USA* 95: 15 356–15 361
72 Lebowitz PF, Sakamuro D, Prendergast GC (1997) Farnesyl transferase inhibitors induce apoptosis of Ras-transformed cells denied substratum attachment. *Cancer Res* 57: 708–713
73 Kauffmann-Zeh A, Rodriguez-Viciana P, Ulrich E, Gilbert C, Coffer P, Downward J, Evan G (1997) Suppression of c-myc-induced apoptosis by Ras signalling through PI(3)K and PKB. *Nature* 385: 544–548
74 Hall A (1998) Rho GTPases and the actin cytoskeleton. *Science* 279: 509–513
75 Prendergast GC, Khosravi-Far R, Solski PA, Kurzawa H, Lebowitz PF, Der CJ (1995) Critical role of Rho in cell transformation by oncogenic Ras. *Oncogene* 10: 2289–2296
76 Qiu R-G, Chen J, McCormick F, Symons M (1995) A role for Rho in Ras transformation. *Proc Natl Acad Sci USA* 92: 11 781–11 785
77 Qiu R-G, Chen J, Kirn D, McCormick F, Symons M (1995) An essential role for Rac in Ras transformation. *Nature* 374: 457–459
78 Lebowitz PF, Davide JP, Prendergast GC (1995) Evidence that farnesyltransferase inhibitors suppress Ras transformation by interfering with Rho activity. *Mol Cell Biol* 15: 6613–6622
79 Armstrong SA, Hannah VC, Goldstein JL, Brown MS (1995) CAAX geranylgeranyl transferase transfers farnesyl as efficiently as geranylgeranyl to RhoB. *J Biol Chem* 270: 7864–7868
80 Prendergast GC, Davide JP, Jane deSolms S, Giuliani EA, Graham SL, Gibbs JB, Oliff A, Kohl NE (1994) Farnesyltransferase inhibition causes morphological reversion of *ras*-transformed cells by a complex mechanism that involves regulation of the actin cytoskeleton. *Mol Cell Biol* 14: 4193–4202
81 Lebowitz PF, Casey PJ, Prendergast GC, Thissen JA (1997) Farnesyltransferase inhibitors alter the prenylation and growth-stimulating function of RhoB. *J Biol Chem* 272: 15 591–15 594
82 Gibbs JB (1991) Ras C-terminal processing enzymes-new drug targets? *Cell* 65: 1–4
83 Hudes GR, Schol J, Baab J, Rogatko A, Bol C, Horak C, Langer C, Goldstein LJ, Szarka C, Meropol NJ et al (1999) Phase I Clinical and pharmacological trial of the farnesyltransferase inhibitor R115777 on a 21-day dosing schedule. *Proc Am Soc Clin Oncol* 18: 156a
84 Eskens F, Awada A, Verweij J, Cutler DL, Hanauske A, Piccart M (1999) Phase I and pharmacologic study of continuous oral SCH 66336, a novel farnesyl transferase inhibitor, in patients with solid tumors. *Proc Am Soc Clin Oncol* 18: 156a
85 Hurwitz HI, Colvin OM, Petros WP, Williams R, Conway D, Adams DJ, Casey PJ, Calzetta A, Mastorides P, Statkevitch P et al (1999) Phase I and pharmacokinetic study of SCH 66336, a novel FPTI, using a 2-week on, 2-week off schedule. *Proc Am Soc Clin Oncol* 18: 156a
86 Soignet S, Yao S-L, Britten C, Spriggs D, Pezzuli D, McCreery H, Mazina K, Deutsch P, Lee Y, Lobell R et al (1999) Pharmacokinetics and pharmacodynamics of the farnesyl protein transferase inhibitor (L-778,123) in solid tumors. *Proc Am Ass Cancer Res* 40: 517
87 Bernhard EJ, McKenna WG, Hamilton AD, Sebti SM, Qian Y, Wu JM, Muschel RJ (1998) Inhibiting Ras prenylation increases the radiosensitivity of human tumor cell lines with activating mutations of *ras* oncogenes. *Cancer Res* 58: 1754–1761
88 Moasser MM, Sepp-Lorenzino L, Kohl NE, Oliff A, Balog A, Su DS, Danishefsky SJ, Rosen N (1998) Farnesyltransferase inhibitors cause enhanced mitotic sensitivity to taxol and epothilones. *Proc Natl Acad Sci USA* 95: 1369–1374
89 Sun J, Marfurt J, Blaskovich MA, Bailey RD, Qian Y, Hamilton AD, Sebti SM (1999) Effective combination therapy with the non-thiol farnesyltransferase inhibitor FTI-276 and taxol, gemcitabine or cisplatin for human tumor xenografts in nude mice. *Proc Amer Assoc Cancer Res* 40:

522
90 Rosenberg SH, Alder J, Bauch J, Cherian SP, Cohen J, Ewing P, Hamilton AD, Henry KJ, Marsch K, Mitten M et al (1999) *In vitro* and *in vivo* activity of the farnesyltransferase inhibitor A-197574. *Proc Amer Assoc Cancer Res* 40: 522
91 Skrzat SG, Bowden CR, End DW (1999) Interaction of the farnesyl protein transferase inhibitor (FTI) R115777 with cytotoxic chemotherapeutics *in vitro* and *in vivo*. *Proc Amer Assoc Cancer Res* 40: 523
92 Vrignaud P, Bissery M-C, Mailliet P, Lavelle F (1999) *In vivo* combination of RPR 130401, a non-peptidomimetic farnesyltransferase inhibitor, with chemotherapy. *Proc Amer Assoc Cancer Res* 40: 523
93 Thissen J, Gross JM, Subramanian K, Meyer T, Casey PJ (1997) Prenylation-dependent association of Ki-Ras with microtubules. Evidence for a role in subcellular trafficking. *J Biol Chem* 272: 30 362–30 370
94 Suzuki N, Del Villar K, Tamanoi F (1998) Farnesyltransferase inhibitors induce dramatic morphological changes of KNRK cells that are blocked by microtubule interfering agents. *Proc Natl Acad Sci USA* 95: 10 499–10 504
95 Peeters M, Van Cutsem E, Marse H, Plamer P, Walraven V, Willems L (1999) Phase-I combination trial of the farnesyltransferase inhibitor (FTI) R115777 with a 5FU/LV regimen in advanced colorectal (CRC) or pancreatic (PC) cancer. *Proc Am Soc Clin Oncol* 18: 223a
96 Bible KC, Kaufmann SH (1997) Cytotoxic synergy between flavopiridol (NSC 649890, L86-8275) and various antineoplastic agents: the importance of sequence of administration. *Cancer Res* 57: 3375–3380
97 Kakeji Y, Teicher BA (1997) Preclinical studies of the combination of angiogenic inhibitors with cytotoxic agents. *Invest New Drugs* 15: 39–48
98 Shi B, Gurnani M, Yaremko B, Lee S, Chen J, Lipari P, Ferrari E, Malkowski M, Liu M, Gerald Hajian G et al (1999) Enhanced efficacy of the farnesyl protein transferase inhibitor SCH 66336 in combination with paclitaxel. *Proc Amer Assoc Cancer Res* 40: 524

New Approaches to Drug Development
ed. by P. Jollès

The future of obesity treatment

John Wilding

University Hospital Aintree, Liverpool L9 7AL, UK

Summary. Obesity is rapidly becoming a worldwide epidemic, with significant consequences in terms of clinical burden and economic costs in treating its complications, so effective new approaches are urgently needed. Development of new drugs in this therapeutic area requires a detailed understanding of the physiology underlying body weight regulation. Recently several significant advances have been made in this area, including the identification of the appetite regulating hormone, leptin, and a detailed understanding of its targets in the central nervous system (CNS), such as neuropeptide Y (NPY) and the melanocortin-4 receptor. The observation that some humans with severe childhood-onset obesity have defects in these regulatory systems has confirmed their relevance in humans as well as in animal models, and the search is now on to produce drugs which act on these and other CNS targets such as glucagon-like peptide 1 and the orexins to help reduce body weight. Other recently identified targets outside the central nervous system include agents acting to inhibit digestive enzymes, specifically pancreatic lipase in the form of orlistat (which has recently been licensed for obesity treatment), and looking to the future, the possibility of altering energy expenditure by modulating the newly identified uncoupling proteins is being considered. It should be remembered however, that pharmacotherapy for obesity is unlikely to provide a 'magic bullet', and that diet and lifestyle changes are likely to remain the cornerstone of treatment for the foreseeable future

Obesity: the scale of the problem

On a worldwide scale, an epidemic of obesity is rapidly developing; for example, in the United States, over 50% of adults are overweight and nearly a quarter are clinically obese [1]. Europe is rapidly catching up; for example, in the UK, 17% of men and 20% of women are obese, as defined by having a body mass index greater than 30 kg/m^2 [2]. The reasons for this are multifactorial, and include a genetic predisposition and lifestyle changes, notably a reduction in physical activity, combined with the ready availability of palatable food with a high fat content. The health consequences of this are significant, and include a global epidemic of diabetes, which is likely to affect 220 million people worldwide by the year 2010 [3], an increased risk of cardiovascular and respiratory disease and a host of other problems, including osteoarthritis, gout, malignancy and psychiatric diseases such as depression [4]. Overall, the burden of these obesity-related diseases is estimated to be between 2 and 7% of health care costs in the developed world [1].

Unfortunately, the range of available therapy for this problem remains limited. Diet and behavioural therapy are effective in the short term for many but only succeed in the long term for a minority of patients; current drug therapy produces an average weight loss of 5–10%, but only 30% of patients achieve this long term. Surgery is effective, but carries significant morbidity, and may only be appropriate for the most severely overweight subjects [5], [6]. Hence there is a real need for safe, effective drugs to combat this growing clinical problem.

Physiology of weight control

The number of potential therapeutic targets for agents to control body weight has increased dramatically in recent years, as our understanding of the mechanisms controlling energy balance has improved. Most of the targets are within the central nervous system (CNS), as this is where the complex systems regulating appetite reside; however, some peripheral targets and systems linking the two are also of interest. Significant advances in this area include the discovery of the fat-derived hormone leptin and its receptor [7,8] and the understanding of the melanocortin-4 (MC-4) receptor and Agouti-Related Protein (AGRP) as causes of obesity in the yellow obese mouse [9, [10]. The importance of these observations has been underlined by the identification of several genetic mutations in the same genes in children with early-onset obesity [11-13], and successful treatment of the first patient with leptin deficiency with recombinant human leptin. The identification of several other neurotransmitters such as glucagon-like peptide-1 (GLP-1), cocaine- and amphetamine-related transcript (CART) and orexins as potential physiological regulators of appetite, and of two new uncoupling proteins, is also significant [14-18].

Controlling the energy balance equation

Regulation of energy balance in an organism is at face value a straightforward task which should match intake to expenditure, whilst taking into account requirements for growth in an immature organism. A complex system has evolved to regulate this process in mammals which involves tight regulation of intake and expenditure. This system can be considered to have four principal components:

1. A series of short-term feedback signals indicating the end of a meal (satiety)
2. A system indicating the status of long-term energy stores
3. A central integrating system that detects these signals and regulates both energy intake and energy expenditure (usually in a reciprocal fashion).
4. A peripheral component that may modulate energy expenditure.

These regulatory processes are of course subject to modulation by a number of factors, including psychological and social background, the taste, smell and palatability of foods and intercurrent illness. The 'set point' for the system may also be varied by altering factors such as physical fitness, and also per-

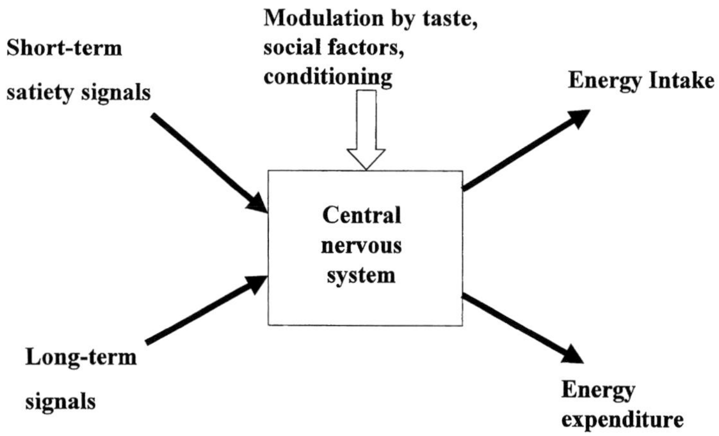

Figure 1. Components of the systems that regulate body weight. Most of the integration takes place in the central nervous system, which responds to external metabolic and hormonal signals, and internal cues to regulate energy intake and expenditure

haps with increasing age. Each of these components will now be discussed in more detail.

Short-term feedback

Examples of short-term feedback signals include neuronal satiety signals from the stomach indicating that it is full, mediated by stretch receptors in the stomach wall, and transmitted to the CNS via the vagus nerve, and hormonal signals from the gut, such as cholecystokinin (CCK) and bombesin, which may act either locally by influencing the vagus nerve, or directly on the CNS via the circulation [19-22]. Pancreatic hormones, such as insulin and glucagon, that are released after a meal have also been found to have receptors within the CNS in areas that are accessible to the circulation and may also play a role in inhibiting food intake after a meal [23,24].

Long-term energy stores

The principal peripherally generated signal of long-term energy stores identified to date is leptin. This fat-derived hormone is the product of the ob gene, and was originally identified as mutations caused a syndrome of severe obesity in mice [7]. It is generally produced in proportion to the amount of body fat. Leptin is thought to cross the blood-brain barrier and enter the cerebrospinal fluid (CSF) via a number of routes, utilising a truncated form of its receptor

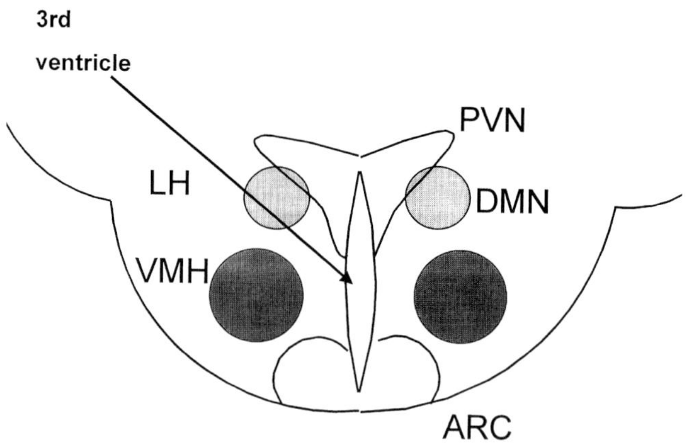

Figure 2. Appetite regulating areas of the hypothalamus. Fibres originating in the arcuate nucleus (ARC) project to the paraventricular (PVN) and dorsomedial (DMN) nuclei. The ventromedial hypothalamus (VMH) and lateral hypothalamic areas were traditionally regarded as satiety and feeding centres respectively.

(ObRs) as a transport protein; the leptin receptor (ObRb) is localised to the hypothalamic arcuate nucleus (see below). Animals and humans which fail to produce leptin (such as the ob/ob mouse) or have a defect in the leptin receptor (such as the db/db mouse or Zucker rat) are characterised by severe obesity, insulin resistance, dyslipidaemia, and in some cases, diabetes [25](Fig. 2).

In normal animals, circulating leptin levels fall during starvation, and this activates a number of systems within the hypothalamus (including increased synthesis and release of the orexigenic peptide neuropeptide Y (NPY) to attempt to restore energy balance by increasing the drive to eat and by reducing energy expenditure. However, if normal animals are fed highly palatable diets, leptin levels rise in parallel with their increasing obesity, suggesting that the system can easily be overridden in some situations; interestingly, central sensitivity to leptin is maintained, suggesting that the defect may lie with the leptin transport system [26].

Other proposed regulators of long-term energy balance include insulin, corticosteroids (both of which also regulate leptin sensitivity and expression) and circulating nutrients such as certain amino-acids and glucose.

The Central Integrating System

The critical importance of the hypothalamus in the regulation of energy balance was originally noted in the 1950s, when experimental lesions of the ventromedial nucleus were found to produce uncontrollable hyperphagia and weight gain, whereas lesions of the lateral hypothalamic area produced weight loss and cachexia. These regions remain important in appetite control, but

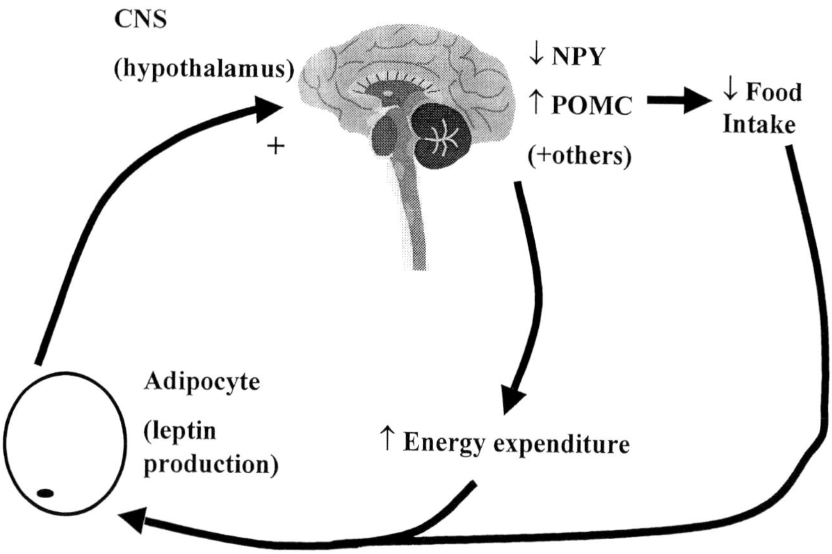

Figure 3. Leptin signalling pathways as targets for obesity treatment.

many other areas, most notably the arcuate nucleus and the paraventricular and dorsomedial nuclei, have recently been found to be important (Fig. 3).

A wide range of neurotransmitters are present in the hypothalamus and other regions of the brain concerned with appetite control, such as the brainstem and amygdala. A few, such as NPY, catecholamines, orexins, growth hormone releasing hormone (GHRH) and galanin act to increase appetite, but most inhibit food intake. These neurotransmitter systems are highly complex. They are influenced by a wide range of peripheral metabolic and neuronal signals, such as the short-term and long-term signals described above, and probably also by higher centres. In general, factors which stimulate energy intake tend to reduce peripheral energy expenditure principally by reducing sympathetic nervous system (SNS) activity to brown adipose tissue (BAT), and also altering the release of other hormones such as glucocorticoids, thyroid hormone, growth hormone and pancreatic hormones concerned with metabolic regulation. Some aspects of this system have been well characterised, and include the peptide neurotransmitter NPY – when given into the CNS, this is a potent stimulus to food intake and inhibits energy expenditure. Given repeatedly to normal animals it results in obesity, with many associated features such as insulin resistance, hypertension and hyperlipidaemia [27]. Hypothalamic NPY is synthesised in the arcuate nucleus of the hypothalamus, and neurons from this nucleus project to the paraventricular and dorsomedial nuclei. The physiological signals regulating arcuate hypothalamic NPY synthesis include insulin and leptin, which inhibit NPY production, and glucocorticoids, that stimulate it. The main physiological role of hypothalamic NPY appears to be as part of the adaptive response to starvation: when rodents are deprived of food, circulating leptin and insulin levels fall and this acts a a potent stimulus to NPY synthesis, which promotes food intake and conserves energy by reducing sympathetic outflow to BAT [28]. Another important role of this system is to switch off luteinising hormone (LH) production (and thus reproductive capacity) which could be considered an important response to starvation. The role of this system in response to energy excess is less clear, but there is some evidence that the NPY system is underactive under such conditions in models such as dietary obesity, which might suggest that further inhibition would have little effect on energy intake and energy expenditure [29,30]. This question will only be answered, however, when specific antagonists of the NPY appetite system are tested in such models, and eventually in humans.

An example of an inhibitory system is GLP-1. This peptide is synthesised in the brainstem, and appears to act principally in the hypothalamic paraventricular nucleus, but also in other brain regions, such as the amygdala. Recent evidence suggests that it is a mediator of post-prandial satiety [14]. The reasoning behind this statement is based on experimental work that shows that although administration of GLP-1 inhibits energy intake under most circumstances when food intake is stimulated (including after NPY administration), it does not inhibit food intake in deprived rats, indicating that it is only effective when the 'satiety' system has been activated. A GLP-1 agonist might be of potential therapeutic use in obese patients with impaired glucose tolerance or early diabetes, as this peptide is also a powerful stimulus to pancreatic insulin secretion and delays gastric emptying in addition to its satiating effect [31].

The number of potential targets for new drugs in this area is huge (Tab. 1), but there are also potential problems; first, because of the built-in redundancy in the system, which could limit the effectiveness of a single agent, but also

Table 1. Potential targets for obesity treatment. Note that some factors will have more than one potential site of action

Enhance satiety	Inhibit nutrient absorption	Reduce appetite	Increase thermogenesis
CCK receptor agonists	Orlistat	Leptin	β3 adrenoceptor agonists
GLP-1 receptor agonists		Leptin receptor agonists	Activators of uncoupling proteins
Bombesin receptor agonists		NPY Y5 receptor antagonists	Protein Kinase A activators
Serotonin/nor-adrenaline reuptake inhbitors		Melanocortin-4 receptor agonists	Protein tyrosine Phosphatase B1 inhibitors
		Orexin receptor antagonists	Serotonin/nor-adrenaline reuptake inhibitors
		Galanin receptor antagonists	Leptin

because of the possibility of significant interaction with other neuroendocrine regulatory processes, such as the effect of NPY on reproductive hormones.

Peripheral metabolic regulation

A fourth area which may provide many new targets for drug discovery is in the field of regulation of peripheral metabolism. This can occur at several different levels.

1. Reduction of substrate intake – the most studied of these is inhibition of fat absorption with the drug orlistat (inhibition of glucose absorption does not seem to influence weight, presumably because the unabsorbed products are absorbed further along the gut). This agent is an inhibitor of pancreatic and intestinal lipases, and at therapeutic doses, inhibits absorption of approximately 30% of dietary fat. In clinical trials, weight loss of just under 10% is achieved over a six-month period, and largely maintained for up to two years [32].
2. Modulation of energy expenditure. Some (but not all) obese subjects have reduced energy expenditure once their increased lean body mass has been taken into consideration. This has been attributed in part to a polymorphism in the β3-adrenoceptor, but a number of other possibilities including dysregulation of uncoupling proteins (UCP) 1–3 are being investigated. These proteins are another group of potential targets for treatment.
3. Peripheral metabolism. Some recently identified molecular targets include the protein tyrosine phosphatase-1B (PTB-1B) gene and the enzyme protein kinase A. PTB-1B is thought to be an important component of the insulin signalling pathway, possibly by dephosphorylating the activated insulin receptor. Rats in which the PTB-1B gene is disrupted are insulin-sensitive and appear resistant to the development of dietary-induced obesity. Inhibitors of this enzyme might therefore be useful in the treatment of obesity and the associated insulin resistance [33]. The enzyme protein kinase A is also important in several intracellular signalling pathways, and disruption of this enzyme also renders mice resistant to dietary obesity [34]. Another group of molecular targets include the peroxisome proliferator-activated receptors (PPARs); drugs which act on these systems are already in clinical use, in the form of fibrate drugs, which activate PPARα, and the insulin-sensitising drugs, the thiazolidenediones, which act via PPARγ. These nuclear receptors are important in regulating adipocyte differentiation and proliferation. It is of interest that the PPARγ agonists currently in use produce some weight gain in animal models, but in humans result in a redistribution of body fat, with increased subcutaneous adipose tissue, and decreases in abdominal fat [35]. This is of potential importance as it is abdominal fat that is thought to pose the greatest threat to health in terms of diabetes and cardiovascular risk. It is therefore possible that in the future, drugs acting on

such systems may be used to produce a more favorable fat distribution in terms of metabolic risk factors.

Known genetic defects that lead to obesity

Amongst this morass of information, the identification of specific genetic defects that are associated with the development of obesity in both animals and man can provide important clues as to which systems are most physiologically relevant; the use of targeted gene knockout can also be invaluable in this respect (Tab. 2).

Table 2. Some examples of naturally occurring and experimental genetic defects associated with obesity or altered energy balance in animal models and humans

Gene	Species	Phenotype
Leptin	Mouse (ob/ob mouse), man	Severe obesity, hyperphagia, (reduced thermogenesis in mice, but not humans), insulin resistance, infertility
Leptin receptor	Mouse (db/db), rat (Zucker, ZDF), man	Severe obesity, hyperphagia, infertility, insulin resistance, diabetes
Agouti	Mouse	Obesity, insulin resistance, yellow coat colour
Carboxypeptidase E mutation	Mouse (fat gene)	Obesity, POMC processing defect
PC-1 gene mutation	Man	Obesity, insulin processing defect, POMC processing defect
MC-4 receptor knockout	Mouse	Obesity, insulin resistance
UCP-1 knockout	Mouse	Obesity; reduced thermogenesis, hyperphagia
Protein tyrosine phosphatase 1B gene knockout	Mouse	Resistance to dietary obesity; insulin sensitive
Protein kinase A knockout	Mouse	Resistance to dietary obesity
NPY knockout	Mouse	Phenotypically normal, but reduced leptin responsiveness
NPY Y1 receptor knockout	Mouse	Moderate Obesity

It can be seen from the table that whilst some defects, such as those in leptin or its receptor or in the MC-4 receptor are clearly associated with significant obesity, others, perhaps surprisingly, such as the NPY knockout, are relatively normal. This may reflect developmental differences, but also again illustrates the degree of 'redundancy' built into the system, with alternative pathways and systems taking over if one component of the system fails.

Past and current approaches

The earliest attempts to use drugs to treat obesity used the centrally acting sympathomimetic drugs available at the time, and included diethylpropion, phentermine, fenfluramine and dexfenfluramine. These agents have largely fallen out of use, or been withdrawn. This was either because of unacceptable CNS side-effects, addictive potential or, in the case of fenfluramine and dexfenfluramine, an association with the development of primary pulmonary hypertension and valvular heart disease similar to that in patients with carcinoid tumours [36].

Two newer agents have recently come into clinical use. The first is sibutramine, a centrally-acting inhibitor of nor-adrenaline and serotonin reuptake, that has appetite-suppressant, satiating and thermogenic properties in animal studies and in humans [37]. The second is orlistat, an intestinal and pancreatic lipase inhibitor, which inhibits the absorption of about 30% of dietary fat [32]. Both have been found to produce weight loss of an average of 8–10% when combined with dietary restriction, and to improve several obesity-related co-morbidities and risk factors.

New molecular targets

From the physiology outlined above, it can be seen that there are many potential new agents or molecular targets for antiobesity treatments, but also that there is a high degree of redundancy in the system; current agents produce weight loss in the region of 10% with optimal dietary intervention; new agents seem unlikely to exceed this on the basis of current evidence, but it may be that combination treatment, using different agents or different components of the system, will be more effective. The effects may be additive, as is often seen in combinations of different agents to other conditions, such as diabetes or hypertension. It must however always be remembered that any pharmacological intervention to treat obesity is unlikely to be effective in the absence of changes to diet and lifestyle, and such drugs should complement, rather than be substitutes for these important therapeutic modalities.

References

1 WHO (1998) *Obesity – Preventing and managing the global epidemic.* WHO, Geneva
2 Anon Health Survey for England 1996 (1998) HMSO, London
3 WHOStudy Group (1997) *Prevention of diabetes mellitus.* 844. WHO, Geneva,
4 Jung RT (1997) Obesity as a disease. *Brit Med Bull* 53: 307–321
5 Sjostrom L, Larsson B, Backman L, Bengtsson C, Bouchard C, Dahlgren S (1992) Swedish obese subjects (SOS). Recruitment for an intervention study and a selected description of the obese state. *Int J Obes Relat Metab Disord* 16: 465–479
6 Pories WJ (1995) Surgery for obesity – an update of a randomized trial – invited commentary. *Obes Surg* 5: 152–152

7 Zhang YY, Proenca R, Maffei M, Barone M, Leopold L, Friedman JM (1994) Positional cloning of the mouse obese gene and its human homolog. *Nature* 372: 425–432
8 Tartaglia LA, Dembski M, Weng X, Deng NH, Culpepper J, Devos R, Richards GJ, Campfield LA, Clark FT, Deeds J et al (1995) Identification and expression cloning of a leptin receptor, ob-r. *Cell* 83: 1263–1271
9 Lu D, Willard D, Patel IR, Kadwell S, Overton L, Kost T, Luther M, Chen W, Woychik RP, Wilkison WO et al (1994) Agouti protein is an antagonist of the melanocyte-stimulating hormone receptor. *Nature* 371: 799–802
10 Rossi M, Kim MS, Morgan DG, Small CJ, Edwards CM, Sunter D, Abusnana S, Goldstone AP, Russell SH, Stanley SA et al (1998) A C-terminal fragment of Agouti-related protein increases feeding and antagonizes the effect of alpha-melanocyte stimulating hormone *in vivo*. *Endocrinology* 139: 4428–4431
11 Jackson RS, Creemers JWM, Ohagi S, RaffinSanson ML, Sanders L, Montague CT, Hutton JC, Orahilly S (1997) Obesity and impaired prohormone processing associated with mutations in the human prohormone convertase 1 gene. *Nat Genet* 16: 303–306
12 Montague CT, Farooqi IS, Whitehead JP, Soos MA, Rau H, Wareham NJ, Sewter CP, Digby JE, Mohammed SN, Hurst JA et al (1997) Congenital leptin deficiency is associated with severe early-onset obesity in humans. *Nature* 387: 903–908
13 Rau H, Reaves BJ, O'Rahilly S, Whitehead JP (1999) Truncated human leptin (delta133) associated with extreme obesity undergoes proteasomal degradation after defective intracellular transport. *Endocrinology* 140: 1718–1723
14 Turton MD, O'Shea D, Gunn I, Beak SA, Edwards CMB, Meeran K, Choi SJ, Taylor GM, Heath MM, Lambert PD et al (1996) A role for glucagon-like peptide-1 in the central control of feeding. *Nature* 379: 69–72
15 Thim L, Kristensen P, Larsen PJ, Wulff BS (1998) CART, a new anorectic peptide. [Review] [10 refs]. *Int J Biochem Cell Biol* 30: 1281–1284
16 Sakurai T, Arnemiya T, Ihsii M, Matsukai I, Chemelli RM, Tanaka H, Williams SC, Richardson JA, Kozlowski GP, Wilson S et al (1998) Orexins and orexin receptors: a family of hypothalamic neuropeptides and G protein coupled receptors that regulate feeding behaviour. *Cell* 92: 573–585
17 Boss O, Samec S, PaoloniGiacobino A, Rossier C, Dulloo A, Seydoux J, Muzzin P, Giacobino JP (1997) Uncoupling protein-3: a new member of the mitochondrial carrier family with tissue-specific expression. *FEBS Lett* 408: 39–42
18 Fleury C, Neverova M, Collins S, Raimbault S, Champigny O, LeviMeyrueis C, Bouillaud F, Seldin MF, Surwit RS, Ricquier D et al (1997) Uncoupling protein-2: a novel gene linked to obesity and hyperinsulinemia. *Nat Genet* 15: 269–272
19 Barrachina MD, Martinez V, Wei JY, Tache Y (1997) Potentiating interaction between leptin and cck to reduce food intake in lean mice. *Gastroenterology* 112: A1131–A1131
20 Dourish CT, Rycroft W, Iversen SD (1989) Postponement of satiety by blockade of cholecystokinin (CCK-B) receptors. *Science* 245: 1509–1511
21 Lieverse RJ, Jansen JBMJ, Masclee AAM, Lamers CBHW (1994) Satiety effects of cholecystokinin in humans. *Gastroenterology* 106: 1451–1454
22 Stuckey JA, Gibbs J, Smith GP (1985) Neural disconnection of gut from brain blocks bombesin induced satiety. *Peptides* 6: 1249–1252
23 Schwartz MW, Figlewicz DP, Baskin DG, Woods SC, Porte D (1992) Insulin in the bra*In*: A hormonal regulator of energy balance. *Endocrin Rev* 13: 387–413
24 Smith GP, Gibbs J (1987) The effect of gut peptides on hunger, satiety, and food intake in humans. *Ann N Y Acad Sci* 499: 132–136
25 Pelleymounter MA (1997) Leptin and the physiology of obesity. *Curr Pharmaceut Design* 3: 85–97
26 VanHeek M, Compton DS, France CF, Tedesco RP, Fawzi AB, Graziano MP, Sybertz EJ, Strader CD, Davis HR (1997) Diet-induced obese mice develop peripheral, but not central, resistance to leptin. *J Clin Invest* 99: 385–390
27 Zarjevski N, Cusin I, Rohner-Jeanrenaud F, Jeanrenaud B (1993) Chronic intracerebroventricular neuropeptide Y administration to normal rats mimics hormonal and metabolic changes of obesity. *Endocrinology* 133: 1753–1758
28 Wang Q, Bing C, Al-Barazanji K, Mossakowaska DE, Wang X, McBay D, Neville WA, Taddayon M, Pickavance L, Dryden S et al (1997) Interactions between Leptin and Hypothalamic Neuropeptide Y Neurones in th Control of Food Intake and Energy Homeostasis in the Rat.

Diabetes; *in press*
29 Wilding JPH, Gilbey SG, Mannan M, Aslam N, Ghatei MA, Bloom SR (1992) Increased neuropeptide Y content in individual hypothalamic nuclei, but not neuropeptide Y mRNA, in diet induced obesity in rats. *J Endocrinol* 132: 299–304
30 Widdowson PS, Upton R, Henderson L, Buckingham R, Wilson S, Williams G (1997) Reciprocal regional changes in brain NPY receptor density during dietary restriction and dietary-induced obesity in the rat. *Brain Res* 774: 1–10
31 Todd JF, Edwards CM, Ghatei MA, Mather HM, Bloom SR (1998) Subcutaneous glucagon-like peptide-1 improves postprandial glycaemic control over a 3-week period in patients with early type 2 diabetes. *Clin Sci* 95: 325–329
32 Sjostrom L, Rissanen A, Andersen T, Boldrin M, Golay A, Koppeschaar HPF, Krempf M (1998) Randomised placebo-controlled trial of orlistat for weight loss and prevention of weight regain in obese patients. *Lancet* 352: 167–172
33 Elchebly M, Payette P, Michaliszyn E, Cromlish W, Collins S, Loy AL, Normandin D, Cheng A, Himms-Hagan J, Chan CC et al (1999) Increased insulin sensitivity and obesity resistance in mice lacking the protein tyrosine phosphatase-1B gene. *Science* 283: 1544–1548
34 Cummings DE, Brandon EP, Planas JV, Motamed K, Idzerda RL, McKnight GS (1996) Genetically lean mice result from targeted disruption of the RII beta subunit of protein kinase A [see comments]. *Nature* 382: 622–626
35 Adams M, Montague CT, Prins JB, Holder JC, Smith SA, Sanders L, Digby JE, Sewter CP, Lazar MA et al (1997) Activators of peroxisome proliferator-activated receptor gamma have depot-specific effects on human preadipocyte differentiation. *J Clin Invest* 100: 3149–3153
36 Finer N (1997) Present and future pharmacological approaches. *Br Med Bull* 53: 409–432
37 Bray GA, Ryan DH, Gordon D, Heidingsfelder S, Cerise F, Wilson K (1996) Double-blind randomized placebo-controlled trial of sibutramine. *Obes Res* 4: 263–270

New Approaches to Drug Development
ed. by P. Jollès

Antithrombotic therapy in acute coronary syndromes

M. Buerke and H.J. Rupprecht

II. Department of Medicine, Johannes Gutenberg-University, D-55101 Mainz, Germany

Summary. Current therapy of acute coronary syndromes (i.e., unstable angina and non-Q-wave myocardial infarction, Q-wave myocardial infarction) consists of thrombolytic, anti-platelet, and anticoagulant therapy.

In most cases of acute coronary syndromes, the pathogenesis is a mural thrombus formation on a ruptured or eroded atherosclerotic plaque. Both platelets and thrombin play an essential role in the pathophysiology of acute coronary syndromes.

Aspirin and heparin are essential treatments for patients with acute coronary syndromes. Novel thrombin and platelet inhibitors have been developed and demonstrated useful effects for improving both acute and long-term clinical outcomes in acute coronary syndromes.

Tissue plasminogen activator is the compound for effective thrombolytic therapy. New developments like reteplase and TNK-tPA can be administered as bolus injection and result in rapid reperfusion. Combination of thrombolytic therapy and glycoprotein IIb/IIIa inhibitors seem to accelerate and improve reperfusion. Clopidogrel as anti-aggregatory compound demonstrated profound effects following stent implantation as well as in patients with aspirin intolerance. Administration of glycoprotein IIb/IIIa inhibitors like abciximab, eptifibatide, tirofiban results in reduction of cardiovascular events in patients with unstable angina and following coronary intervention. Low-molecular-weight heparins like enoxiparin and dalteparin seem to be more effective than heparin, and their use is evolving in patients with unstable angina. Anti-thrombin therapy with hirudin results in slight reduction of cardiovascular events in combination with tolerable safety profile.

It has yet to be determined which combination of agents and procedural strategies most significantly reduces mortality and serious events in patients with acute coronary syndromes.

Pathophysiology

The term acute coronary syndromes (ACS) summarizes a spectrum of clinical syndromes ranging from unstable angina to non-Q-wave and Q-wave myocardial infarction (MI). The diagnosis is based on a history of chest pain at rest or accelerating symptoms of previous stable angina to myocardial infarction with typical signs of electrocardiography and elevated cardiac enzymes.

Unstable angina is a complex condition. The most important pathophysiological mechanism of ischaemia in acute coronary syndromes is a primary reduction of myocardial oxygen supply due to plaque disruption with associated thrombosis and vasoconstriction [1, 2]. Plaque rupture results in platelet deposition, activation of blood coagulation and thrombus formation. However, there is less thrombus formation in unstable angina than in myocardial infarction, which is associated with an occlusive thrombus. Thrombus formation associated with unstable angina consists mostly of platelets, whereas the thrombus formation associated with acute myocardial infarction consists to a large extent of red blood cells.

The process of atherogenesis, with lipid accumulation, cell proliferation, or extracellular matrix synthesis takes years. Pathological analyses have demonstrated that vulnerable plaques are composed of a lipid-rich core separated from the vessel lumen by a fibrous cap [1, 2]. Plaques that are likely to disrupt have a high concentration of cholesteryl esters. Infiltrating foam cells are able to weaken the fibrous cap and physical forces will cause plaque disruption. Wall stress, cap condition, location, size and consistency of the atheromatous core are responsible for plaque rupture near the plaque shoulder. Following plaque rupture, subendothelial adhesive proteins, extracellular matrix, α-granule adhesive proteins and von Willebrand factor are exposed. Platelets adhere via glycoproteins Ib and $\alpha_2\beta_1$, become activated and release proaggregatory mediators like serotonin, thromboxane A_2 and adenosine diphosphate (ADP) as well as proadhesive proteins including fibrinogen, von Willebrand factor and thrombospondin [3]. The extrinsic coagulation cascade is activated to generate thrombin, known to stimulate further platelet aggregation and endothelial activation. Disruption of an unstable plaque changes the geometry of the plaque and can result in acute occlusion or subocclusion. Further, activated platelets release mediators like serotonin or thromboxane A_2, which cause vasoconstriction. This is responsible for the clinical manifestations of unstable angina or other acute coronary syndromes. In unstable angina, a relatively small erosion or fissuring of an atherosclerotic plaque may lead to an acute change in plaque structure with reduction in coronary blood flow and exacerbation of angina. Episodes of thrombotic vessel occlusion at the site of plaque injury may occur. Since this thrombus is usually labile, temporary vascular occlusion will last only 10 to 20 min. In non-Q-wave MI, severe plaque damage will result in persistent thrombotic occlusion for up to 1 h. However, spontaneous thrombolysis, loss of vasoconstriction, or presence of collateral circulation may prevent the formation of transmural necrosis by limiting the duration of myocardial ischaemia. In Q-wave MI, larger plaque fissures may result in the formation of a persistent thrombus, leading to myocardial ischaemia for more than 1 h with subsequent transmural necrosis. The coronary lesion responsible for the infarction and the other acute coronary syndromes is often only a moderate stenosis (Fig. 1).

Therapy

The treatment of acute coronary syndromes should result in pain relief with morphine and antianginal therapy. Further, stabilisation of the thrombotic process with antithrombotic therapy is important to prevent myocardial infarction and death. If myocardial infarction develops, a standard procedure such as thrombolytic therapy or primary angioplasty should be used. Longer-term therapy should include treatment of cardiac risk factors such as hypertension, dyslipidemia and smoking.

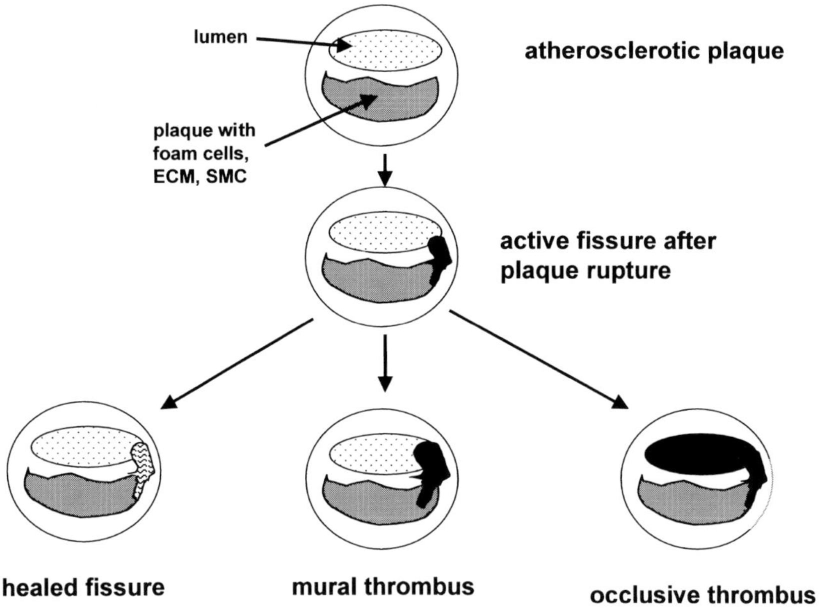

Figure 1. Mechanisms of acute coronary syndromes.

Reperfusion therapy

Thrombolytic therapy

Thrombolytic therapy has clearly been shown to improve survival and left ventricular function in patients with acute myocardial infarction. All patients with ST segment elevation indicating an acute myocardial infarction (MI) who present within 12 h from the onset of symptoms should be considered for myocardial reperfusion therapy [4]. Definite contraindications for thrombolytic therapy are active bleeding and recent stroke, trauma, or major surgery. For patients with relative contraindications to thrombolytic therapy, primary angioplasty can be considered.

The benefit of thrombolytic therapy or primary angioplasty is independent of age, gender and most of the baseline characteristics. However, the biggest benefits occur in patients who are treated early, and those with an anterior myocardial infarction. Patients treated with different thrombolytic agents (streptokinase (SK), tissue plasminogen activator (t-PA), urokinase or anistreplase) demonstrated a profound reduction of myocardial necrosis and mortality [5–11]. Benefit was confirmed across subgroups including patients with ST elevation or bundle branch block, diabetics, patients with prior myocardial infarction and patients receiving therapy within 12 h from onset of symptoms. However, patients treated within the first hour have the highest benefit [12].

In the GISSI-2/International trial [13] patients were randomly assigned to t-PA or SK. Mortality at 30 days was 8.9% and 8.5%, respectively. In the ISIS-3 trial [14] 41 299 patients were treated with three different thrombolytic agents (duteplase SK and anistreplase). However, there were no differences in mortality (10.3%, 10.5%, and 10.6%). Subsequently, the Global Use of Streptokinase and t-PA for occluded coronary arteries [15] evaluated 41 021 patients randomly assigned to accelerated alteplase t-PA over 90 min with intravenous heparin, streptokinase with intravenous or with subcutaneous heparin, or the combination of t-PA and SK with intravenous heparin. t-PA treatment with heparin resulted in a significant relative risk reduction of 30-day mortality by 15% compared to SK treatment. The benefit of accelerated t-PA in the GUSTO-I trial was related to improvement in rapid and complete restoration of coronary blood flow [16] (Fig. 2).

Common Thrombolytic Drugs

	Streptokinase	Urokinase	Alteplase	Reteplase
Fibrin specifity	minimal	moderate	moderate	moderate
Plasminogen binding	indirect	direct	direct	direct
Allergic reaction	yes	no	no	no
Administration (min)	60	60	90	2 boli
Dose	1.5 Mio Units	2 Mio Units	100mg	20mg
Plasma clearance (min)	15-25	15-20	4-8	

Figure 2.

Many new thrombolytic agents are being developed [17]. The mutants of native t-PA are reteplase (r-PA), lanoteplase (n-PA) and TNK (a triple mutant). Other agents that are being developed include recombinant staphylokinase and vampire bat plasminogen activator.

Many of these agents have a prolonged half-life and can be adminstered as a bolus injection. TNK, staphylokinase, and vampire bat plasminogen activator have remarkable fibrin specificity.

Reteplase (r-PA)

Reteplase represents the first of the third generation thrombolytics to become commercially available. Reteplase is a deletion mutant of t-PA and is given as two 10 MU boluses since it has a longer half life. Intravenous heparin is used in conjunction with r-PA. Two angiographic trials [18, 19] indicate the potential of faster coronary thrombolysis for r-PA compared with conventional or accelerated t-PA. The INJECT trial (International Joint Evaluation of Coronary Thrombolysis) compared r-PA with streptokinase in over 6000 patients and demonstrated a small, but not statistically meaningful benefit of r-PA [20]. More recently, the GUSTO-3 trial compared r-PA and accelerated t-PA in over 15 072 patients. No benefit for r-PA over t-PA was demonstrated [21].

Lanoteplase (n-PA)

Lanoteplase was tested in the InTIME trial (Intravenous n-PA for Treating Infarcting Myocardium Early) and suggested a higher potential for n-PA at higher doses [22]. This molecule is a deletion mutant but one of the kringle domains is preserved.

TNK

TNK a triple mutant, has been assessed angiographically at escalating doses in the TIMI-10 trial [23], and has achieved at least as good a potency rate as accelerated t-PA. The safety of several doses of TNK tissue-plasminogen activator (t-PA), given as a single bolus to patients with acute myocardial infarction was tested in the ASSENT-1 trial [24]. The overall safety profile of a single bolus of 30 to 50 mg TNK-tPA is comparable to that of accelerated r-tPA observed in other large trials. Currently a large phase-III mortality trial comparing weight-adjusted single-bolus TNK-tPA with accelerated recombinant tissue plasminogen activator (rt-PA) is ongoing (ASSENT-2).

Recombinant staphylokinase

Recombinant staphylokinase is resistant to plasminogen activator inhibitor (PAI-1) and has almost complete fibrin-specificity. In small clinical trials recombinant staphylokinase demonstrated higher potency compared to accelerated t-PA [25].

Saruplase

Saruplase (recombinant single-chain urokinase-type plasminogen activator) is a naturally occurring glycoprotein which is converted into urokinase by plasmin. In a comparative study with streptokinase, the use of saruplase was associated with higher coronary artery patency rates, earlier reperfusion, less fibrinogen consumption and fewer bleeding complications [26].

Recombinant vampire bat plasminogen activator

Recombinant vampire bat plasminogen activator also has promise because of its plasminogen activator inhibitor (PAI-1) resistance and fibrin-specificity, but it has not yet been assessed in large angiographic or clinical trials.

Antithrombotic therapy

Heparin

A number of trials have evaluated the use of standard heparin in unstable angina [9, 27–30]. Heparin has been shown to be more effective than aspirin for relieving angina [27, 31]. In the RISC study (Research Group on Instability in Coronary Artery Disease) 796 men with unstable angina or non-Q-wave MI were randomized to receive placebo, aspirin (75 mg per day), heparin (5000 units, every 6 h i.v.), placebo, or aspirin plus heparin [9]. Aspirin and heparin had the lowest event rate (risk of death or myocardial infarction) during the initial 5 days. However, heparin is known to induce thrombocytopenia due to inhibition of adenylate cyclase (heparin-induced thrombocytopenia type I, HIT I) or heparin-induced anti-platelet antibodies (HIT II).

Low-molecular-weight heparin

Subcutaneous injection and consistent anticoagulation are some of the advantages of low-molecular-weight heparin.

In the FRISC study 1506 patients with unstable angina or non-Q-wave MI were randomized to receive either placebo or low-molecular-weight heparin (dalteparin) [32]. Death and myocardial infarction were reduced from 4.8% to 1.8% in the first 6 days of treatment with low-molecular-weight heparin. In the FRIC trial (Low Molecular Weight Heparin (Fragmin) in the Treatment of Unstable Coronary Artery Disease) the efficacy of subcutaneous fragmin (120 units per kg, 12-hourly over at least 48 h) was compared with heparin [33]. Neither in the first phase (until day 6) nor in the second phase (days 6 to 45) was a significant difference noted in the incidence of death or nonfatal myocardial infarction between fragmin-treated patients and heparin-treated patients.

In the ESSENCE study (Enoxaparin in Unstable Angina and Non-Q-Wave MI), 3171 patients with unstable angina or non-Q wave MI were randomized to receive intravenous heparin (dose-adjusted) plus aspirin, or subcutaneous enoxaparin (1 mg per kg, every 12 h) in combination with aspirin. After 30 days there was a 15% relative risk reduction for the composite endpoint recurrent angina, death and nonfatal myocardial infarction with enoxaparin therapy. The incidence of major haemorrhage was similar in the two treatment groups.

Several trials have demonstrated that low-molecular-weight heparins in unstable angina are at least as effective as standard heparin. However, they are easier to administer and do not require aPTT measurements. Low-molecular-weight heparin therapy in unstable angina is associated with similar bleeding complications compared to the standard heparin therapy.

Hirudin

Heparin has a number of limitations, including a limited effect on fibrin or platelet-bound thrombin. Further heparin requires antithrombin III, can be inactivated by platelet factor 4 and thrombospondin, requires monitoring and is associated with thrombocytopenia.

Hirudin is a 6S-amino acid peptide isolated from leech saliva. The direct antithrombins do not require antithrombin III as a cofactor and are little affected by plasma proteins or platelet factor 4. These agents also have a more consistent effect on the aPTT than heparin and are able to inactivate clot-bound thrombin.

Topol et al. randomized 166 patients with unstable angina to receive two doses of heparin or escalating doses of hirudin [34]. All patients had angiographic thrombus and stenoses of at least 60%. A repeat angiogram after 72 to 120 h showed a reduction in stenosis severity in the hirudin-treated patients. However there was no statistically significant difference.

The GUSTO IIa trial, investigation of hirudin compared to heparin with thrombolysis in acute myocardial infarction, was interrupted prematurely due to increased incidence of cerebral vascular bleeding complications. Later the GUSTO IIb trial was performed. In the GUSTO-IIb trial, patients with symptoms of acute myocardial infarction and electrocardiographic ST segment elevation were treated with thrombolytic therapy (t-PA or streptokinase) and randomly assigned to receive hirudin or heparin [35]. Among SK-treated patients, death or reinfarction at 30 days occurred more often in those treated additionally with heparin (14.4%) rather than hirudin (8.6%, $p = 0.004$). Among t-PA-treated patients, the rates were 10.9% with heparin and 10.3% with hirudin, suggesting that the beneficial treatment effect of hirudin was limited to the SK-treated patients.

The randomized, double-blind phase-III r-Hirudin for Improvement of Thrombolysis study (HIT III) compared a recombinant hirudin with heparin [36]. Seven thousand patients with acute myocardial infarction and a duration of symptoms of less than 6 h were randomized to receive intravenous heparin or hirudin infused over 48 to 72 h. In a pilot phase, patients received front-loaded alteplase for thrombolysis. After enrollment of 302 patients, the trial was stopped after an increased rate of intracranial bleeding was observed in the hirudin group (3.4%) compared with the heparin group (0%). The overall stroke rate was 3.4% in the hirudin group and 1.3% in the heparin group.

The TIMI 9 trial evaluated whether the direct antithrombin hirudin is more effective than an indirect-acting antithrombin, heparin, as supplementary therapy for thrombolysis in myocardial infarction [37]. Heparin and hirudin as supplementary therapy to t-PA and streptokinase had an equal effect in preventing unsatisfactory outcomes in patients with acute myocardial infarction. Similar rates of major bleeding were observed for patients in the heparin and hirudin groups. Therefore, the therapeutic range of hirudin as a supplement to thrombolysis may be smaller than previously thought.

The Organization to Assess Strategies for Ischemic Syndromes (OASIS) study-2 randomized 10 141 patients with unstable angina or suspected acute myocardial infarction without ST elevation, who were randomly assigned heparin or hirudin for 72 h in a double-blind trial [38]. At 7 days the primary endpoint (cardiovascular death or new myocardial infarction) was not statistically different in the heparin group (4.2%) compared to the hirudin group (3.6%). The secondary endpoints (number of patients with cardiovascular death, new myocardial infarction, or refractory angina) at 7 days were significantly different in the heparin treated group (6.7%) compared to the hirudin group (5.6%). These differences were primarily observed during the 72 h treatment period. There was no excess of life-threatening episodes or strokes. However, with hirudin there was an excess of major bleeding requiring transfusion (1.2% vs 0.7% with heparin; $p = 0.01$).

Hirudin seems to be a potent direct antithrombin with some beneficial effects in acute myocardial infarction and unstable angina. However, the therapeutic benefit seems to be limited to the period of infusion and hirudin treatment is restricted by its narrow therapeutic range. On the other hand, hirudin is the only effective drug which can be administered following heparin-induced thrombocytopenia type II in patients with unstable angina.

Hirulog

Hirulog is a synthetic 20-amino-acid peptide. It has a shorter half-life than hirudin. The TIMI-7 trial randomized patients with unstable angina to receive four different dosages of hirulog infused for 72 h [39]. Infusion of hirulog resulted in a dose-dependent reduction incidence of death or myocardial infarction.

In the Hirulog Angioplasty Study, 4098 patients undergoing angioplasty received either heparin or hirulog. Hirulog reduced the primary end point death, myocardial infarction, reocclusion, bypass surgery, intra-aortic balloon pumping, or repeat angioplasty from 14.2% to 9.1% ($p < 0.05$). Hirulog treatment was associated with lower bleeding rates.

In the HELVETICA trial, 1141 patients with unstable angina who were scheduled for angioplasty received one of three treatments: (1) a bolus dose of 10 000 IU of heparin followed by an intravenous infusion of heparin for 24 h and subcutaneous placebo twice daily for three days, (2) a bolus dose of 40 mg of hirudin followed by an intravenous infusion of hirudin for 24 h and subcutaneous placebo twice daily for three days, or (3) the same hirudin regimen except that 40 mg of hirudin was given subcutaneously instead of placebo twice daily for three days [40]. Although significantly fewer early cardiac events occurred with hirudin than with heparin, hirudin had no apparent benefit with longer-term follow-up.

Antiplatelet therapy

Aspirin

Acetylsalicylic acid potently inhibits platelet aggregation by inhibiting the cyclooxygenase irreversibly, resulting in reduced synthesis of thromboxane A_2 [41] (Fig. 3). Aspirin has been shown to prevent the development of myocardial infarction and death in patients with unstable angina [42, 43] and to reduce mortality and reinfarction in patients with acute infarction treated with or without thrombolytic therapy [44]. Aspirin also reduces vascular events for at least 2 years in patients with previous infarction, unstable angina, or stable angina [45].

Four studies have shown that aspirin reduces mortality [9, 27, 42, 43]. The dosage used in these studies has varied from 75 mg per day to 325 mg four

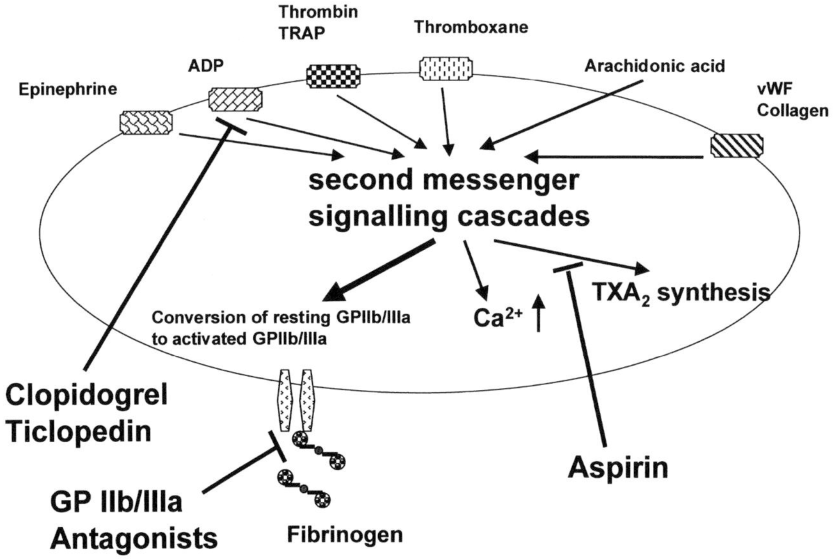

Figure 3. Platelet activation and inhibition of aggregation.

times a day. The effect of aspirin is rapid. In normal volunteers, 300 mg of aspirin loading dose and 100 mg as continuing dose inhibited *ex vivo* platelet aggregation with collagen and thromboxane generation [46]. All patients with a diagnosis of unstable angina should receive aspirin as soon as possible unless there is active bleeding or documented hypersensitivity. The initial dose of aspirin should be 150 to 325 mg due to the possibility of decreased intestinal blood flow during ischaemia and to ensure complete inhibition of thromboxane A_2 production.

Ticlopidine and clopidogrel

Ticlopidine and clopidogrel are thienopyridine derivatives, and both are prodrugs. ticlopidine and clopidogrel are selective antagonists of ADP-induced platelet aggregation (Fig. 3). Ticlopidine has a delayed onset of action (2 to 3 days for maximal antiplatelet effect), Therefore, ticlopidine is not recommended as initial therapy for patients with unstable angina. However, it could be used in patients with contraindications against aspirin. Patients must be closely monitored with blood tests because of possible neutropenia. It has been demonstrated that administration of ticlopidine (250 mg twice daily) in patients with unstable angina or non-Q-wave MI resulted in a reduction of cardiovascular mortality and nonfatal myocardial infarction (13.6% to 7.3%) after 6 month. Since clopidogrel and ticlopidine seem to have the same active drug and the risk of neutropenia is much smaller for clopidogrel, it is the drug of choice.

In the CAPRIE trial (Clopidogrel Versus Aspirin in Patients at Risk of Ischemic Events) clopidogrel (75 mg daily) was compared with aspirin (325 mg per day) in 19 000 patients with symptomatic peripheral vascular disease, recent MI, or a recent ischemic stroke. There was a modest reduction of vascular death, myocardial infarction, or ischemic stroke with clopidogrel as compared with aspirin (5.32% *versus* 5.83%, $p = 0.04$). In the subgroup of patients with recent MI a little benefit (5.03% with clopidogrel *versus* 4.84% with aspirin) was observed. Clopidogrel had a similar safety profile compared to aspirin [47].

Platelet glycoprotein IIb/IIIa antagonists

Binding of fibrinogen to the glycoprotein IIb/IIIa receptor is the final pathway for platelet aggregation. There are 50 000 to 80 000 glycoprotein IIb/IIIa receptors on the surface of each platelet. Fibrinogen cross-linking and formation of a platelet thrombus can lead to vascular occlusion (Fig. 3). The effects of thrombin, thromboxane A_2, collagen, ADP and catecholamine-induced platelet aggregation can be prevented by specific inhibitors of the glycoprotein IIb/IIIa receptors (Fig. 3). However, these inhibitors do not prevent shape change and granule release following platelet activation. Since these inhibitors are highly effective and have steep dose-response curves, close monitoring is necessary to prevent major bleeding complications.

Abciximab

The antibody (Abciximab, c7E3) binds with high affinity for a long time to platelets. The Evaluation of 7E3 for the Prevention of Ischemic Complication (EPIC) trial randomized 2099 patients with high-risk angioplasty (unstable angina, recent or evolving infarction, or high-risk angiographic lesions (grade B or C)) to receive a placebo, a bolus of abciximab, or a bolus plus a 12-hour infusion of abciximab [48]. Bolus plus infusion of abciximab resulted in a 35% reduction compared with a single bolus in the combined end point (i.e., death, nonfatal infarction, emergency revascularization, or stent placement). Further, after 6 months there was a lower incidence of major ischemic events or the need for revascularization of the angioplasty-treated artery and there was a reduction in the need for repeat angioplasty or bypass surgery. However, the benefit was accompanied by increased major bleeding for the bolus plus infusion of abciximab (15.4% *versus* 7.0% with heparin).

The EPILOG trial (The Evaluation in PTCA to Improve Long-Term Outcome with Abciximab Glycoprotein IIb/IIIa Blockade) was conducted with a lower dose of weight-adjusted heparin and was stopped prematurely after an interim analysis showed that death and myocardial infarction were reduced from 8.1% in heparin-treated patients to 2.6% in abciximab-treated patients [49]. Interestingly, there was no increase in bleeding with abciximab.

In the CAPTURE trial (Chimeric 7E3 Antiplatelet Therapy in Unstable Refractory Angina), 1050 patients received either placebo or abciximab prior to angioplasty. After thirty days the combined end point (death, myocardial infarction, or the need for intervention) was significantly reduced (10.8% *versus* 16.4% in placebo-treated patients) [50] (Fig. 4).

In the EPISTENT study, 2399 patients with ischemic heart disease and suitable coronary-artery lesions were randomly assigned to stenting plus placebo, to stenting plus abciximab, or balloon angioplasty plus abciximab. The main outcomes that occurred less with abciximab were death and large myocardial infarction: 7.8% in the placebo group, 3.0% for stent plus abciximab ($p < 0.001$), and 4.7% for balloon angioplasty plus abciximab ($p = 0.01$) [51]. Abciximab (c7E3) binds also to the vitronectin receptor, which might be one of the reasons for superior benefit when compared with other GPIIb/IIIa inhibitors.

Tirofiban

Tirofiban is a small nonpeptide antagonist of the glycoprotein IIb/IIIa receptor that mimics the tripeptide arginine-glycine-alanine (RGD) sequence in fibrinogen. It is nonimmunogenic and has a high degree of selectivity for the platelet fibrinogen receptor, producing an acute effect within 5 min of administration. The effects are reversible in 4 to 6 h.

In the RESTORE trial (Randomized Efficiency Study of Tirofiban for Outcome and Restenosis) 2600 patients with unstable angina or postmyocar-

GP IIb/IIIa Antagonists in Unstable Angina and non Q-wave MI trials

Trial	Agent	Patients	Death or non fatal MI Placebo	Death or non fatal MI GP IIb/IIIa-Antagonist
PRISM	Tirofiban	3231	7.0	5.7
PRISM Plus	Tirofiban	1570	11.9	8.7
PARAGON	Lamifiban	2282	11.7	11.3
PURSUIT	Eptifibatide	10948	15.7	14.2

GP IIb/IIIa Antagonists in PTCA trials

Trial	Agent	Patients	Death or non fatal MI Placebo	Death or non fatal MI GP IIb/IIIa-Antagonist
EPIC	Abciximab	2099	10.1	7.0
EPILOG	Abciximab	2792	9.1	4.0
CAPTURE	Abciximab	1265	9.0	4.8
EPISTENT	Abciximab	2399	10.2	5.2
IMPACT-II	Eptifibatide	4010	8.4	7.1
RESTORE	Tirofiban	2139	6.3	5.1

Figure 4.

dial infarction undergoing angioplasty received either tirofiban or placebo. After 30 days there was no significant difference for the composite primary end point (death, myocardial infarction, bypass surgery, repeat angioplasty, or bailout stenting) [52]. However, there were significant reductions at 48 h (8.7% *versus* 5.4%) and at 7 days (10.1% *versus* 7.4%). In the PRISM trial, 3231 patients with unstable angina or non-Q-wave MI were treated with either tirofiban or heparin [53]. Treatment with tirofiban, resulted in a significant reduction of all components of the composite end point after 48 h (refractory ischaemia (5.3% to 3.6%), myocardial infarction (1.3% to 0.9%), and death (0.6% to 0.4%). Interestingly, the PRISM trial demonstrated that aspirin plus tirofiban significantly reduced mortality compared with aspirin plus heparin [53].

The PRISM-Plus trial was conducted in high-risk patients with unstable angina [55]. There were three treatment arms comparing tirofiban (0.6 mg per kg per minute for 30 min followed by 0.15 mg per kg per minute), heparin, and tirofiban plus heparin. The tirofiban arm was prematurely stopped because of a higher death rate after 7 days, However, increased mortality was not observed after 48 h, 30 days, or 6 months (Fig. 4).

Integrilin

Integrilin is a cyclic peptide inhibitor of the glycoprotein IIb/IIIa receptor that exerts rapid platelet inhibition. It has a short half-life, with platelet aggregation returning to baseline within 2 to 4 h.

In the IMPACT-II trial (Integrilin to Minimize Platelet Aggregation and Prevent Coronary Thrombosis-II), 4010 patients with unstable angina undergoing angioplasty were treated with either Integrilin or placebo (IMPACT-II, 1997). Integrilin treatment (135 mg per kg bolus followed by 0.75 mg per kg per hour) resulted in a significant reduction in ischemic events or emergency revascularization after 24 h. Among patients who underwent emergency coronary stenting, clinical events were significantly less in those treated with integrilin. In the PURSUIT trial, patients with unstable angina receiving integrilin had a significant reduction in the combined endpoint death and myocardial infarctions at 30 days [56] (Fig. 4).

Lamifiban

Lamifiban is a synthetic, nonpeptide, highly selective glycoprotein IIb/IIIa antagonist with a pharmacodynamic half-life of approximately 4 h.

In the Canadian Lamifiban Study, patients with unstable angina were randomized to receive double-blind placebo or four different dosages (1, 2, 4, or 5 mg per minute) of lamifiban infused for 72 to 120 h (Theroux et al., 1996). Lamifiban reduced the risk of death, nonfatal myocardial infarction, or the need for urgent revascularization during the period of infusion from 8.1% to 3.3% (for all dosages).

Lamifiban at a dosage of 1 or 5 mg per kg per minute has been tested in the PARAGON trial (Platelet IIb/IIIa Antagonists for the Reduction of Acute Coronary Syndrome Events in a Global Organization Network) in combination with or without heparin [57]. The 30-day death and reinfarction rates were not significantly different within the groups. Six-month follow-up data showed that death and nonfatal myocardial infarction were reduced by approximately 25% by the low-dosage lamifiban regimen compared with heparin treatment (Fig. 4).

Several oral glycoprotein IIb/IIIa inhibitors like Xemilofiban or lefradafiban have a longer half-life; however, preliminary results of clinical trials with patients with acute coronary syndromes were disappointing with regard to efficacy and safety (OPUS TIMI-16 study, FROST-study).

Therefore, aspirin and inhibitors of the glycoprotein IIb/IIIa receptors are effective means to achieve short and long-term beneficial effects in patients with ACS.

Agents in development

Argatroban

Argatroban is a synthetic, competitive thrombin inhibitor that also inhibits factor-Xa-induced platelet aggregation. It has a half-life of 35 to 40 min and has been tested in phase-II trials of patients with unstable angina.

Tissue factor inhibitors

Tissue factor inhibitors might be also an useful strategy, since tissue factor is an integral protein of the vascular endothelium. It is an important cofactor for proteolytic activity of factor VII.

Hirudisins

Hirudisins are drugs that are combinations of hirudin and IIb/IIIa receptor antagonists. Residues 32 to 35 of hirudin have been replaced by integrilin, and hirudin has been combined with antibodies against the IIb/IIIa receptor.

Pentasaccharide

The antithrombotic activity is due to the binding of the pentasaccharide to antithrombin III (ATIII) and the increase in the factor Xa inhibitory activity of the complex.

Local antithrombotic therapy

In experimental studies the efficacy of gene therapy is tested in vein grafts to prevent acute occlusion. Vein grafts are particularly well suited for gene transfer in the clinic because direct access to vein is available during surgical preparation for grafting. Gene therapy with tissue plasminogen activator, thrombomodulin, nitric oxide synthase or prostacyclin synthase is intended to reduce early thrombosis as well as to reduce neointima formation,

Further local-delivery catheter systems are developed to administer the different antithrombotic agents during cardiovascular interventions to avoid systemic anticoagulation with severe bleeding complications.

Thrombus formation after intracoronary stent implantation is also a new target for local antithrombotic drug release. Stents coated with slow-release polymers containing argatroban or heparin are tested for antithrombotic potency following coronary interventions.

References

1 Fuster V, Badimon L, Badimon JJ et al (1992) The pathogenesis of coronary artery disease and the acute coronary syndromes 1. *N Engl J Med* 326: 242–250
2 Fuster V, Badimon L, Badimon JJ et al (1992) The pathogenesis of coronary artery disease and the acute coronary syndromes 2. *N Engl J Med* 326: 310–318
3 Libby P (1995) Molecular bases of the acute coronary syndromes. *Circulation* 91: 2844–2850
4 Fibrinolytic therapy trialists' (FTT) collaborative group (1994) Indications for fibrinolytic thera-

py in suspected acute myocardial infarction: Collaborative overview of early mortality and major morbidity results from all randomized trials of more than 1000 patients. *Lancet* 343: 311–322
5 Granger CB, Califf RM, Topol EJ (1992) Thrombolytic therapy for acute myocardial infarction, a review. *Drugs* 44: 293–325
6 AIMSTrial Study Group (1988) Effect of intravenous APSAC on mortality after acute myocardial infarction: Preliminary report of a placebo-controlled clinical trial. *Lancet* 1: 545–549
7 Wilcox RG, von der Lippe G, Olsson CG et al (1988) Trial of tissue plasminogen activator for mortality reduction in acute myocardial infarction (ASSET). *Lancet* 2: 525–530
8 The ISAMStudy Group (1986) A prospective trial of intravenous streptokinase in acute myocardial infarction (ISAM). Mortality, morbidity, and infarct size at 21 days. *N Engl J Med* 314: 1465–1471
9 RISCGroup (1990) Risk of myocardial infarction and death during treatment with low dose aspirin and intravenous heparin in men with unstable coronary artery disease. *Lancet* 336: 827–830
10 ISIS (International Studies of Infarct Survival) Pilot Study Investigators (1987) Randomized factorial trial of high-dose intravenous streptokinase, of oral aspirin and of intravenous heparin in acute myocardial infarction. *Eur Heart J* 8: 634–642
11 Gruppo Italiano per lo Studio della Streptochinasi nell'Infarto Miocardico (GISSI) (1986) Effectiveness of intravenous thrombolytic treatment in acute myocardial infarction. *Lancet* 1: 397–401
12 Boersma E, Maas AC, Deckers JW, Simoons ML (1996) Early thrombolytic treatment in acute myocardial infarction: Reappraisal of the golden hour. *Lancet* 348: 771–775
13 Gruppo Italiano Per Lo Studio Della Sopravvivenza Nell'Infarto Miocardico (1990) GISSI-2: A factorial randomized trial of alteplase *versus* streptokinase and heparin *versus* no heparin among 12 490 patients with acute myocardial infarction. *Lancet* 336: 65–71
14 ISIS 3 (Third International Study of Infarct Survival) Collaborative Group (1992) *Isis* 3: A randomized comparison of streptokinase vs tissue plasminogen activator vs anistreplase and of aspirin plus heparin vs aspirin among 41 299 cases of suspected acute myocardial infarction. *Lancet* 339: 753–770
15 The GUSTO Investigators (1993) An international randomized trial comparing four thrombolytic strategies for acute myocardial infarction. *N Engl J Med* 329: 1615–1622
16 The GUSTO Angiographic Investigators (1993) The effects of tissue plasminogen activator, streptokinase, or both on coronary-artery patency, ventricular function, and survival after acute myocardial infarction. *N Engl J Med* 329: 1615–1622
17 Verstraete M, Lijnen Hr Collen D (1995) Thrombolytic agents in development. *Drugs* 50: 29–42
18 Smalling RW, Bode C, Kalbfleisch J et al (1995) More rapid, complete, and stable coronary thrombolysis with bolus administration of reteplase compared with alteplase infusion in acute myocardial infarction. *Circulation* 91: 2725–2732
19 Bode C, Smalling RW, Berg G et al (1996) Randomized comparison of coronary thrombolysis achieved with double-bolus Reteplase (Recombinant Plasminogen Activator) and front-loaded, accelerated Alteplase (Recombinant Tissue Plasminogen Activator) in patients with acute myocardial infarction. *Circulation* 94: 891–898
20 International Joint Efficacy Comparison of Thrombolytics (1995) Randomised, double-blind comparison of reteplase double-bolus administration with streptokinase in acute myocardial infarction (INJECT): Trial to investigate equivalence. *Lancet* 346: 329–336
21 Topol EJ, for the GUSTO-3 Investigators (1997) An international, multicenter, randomized comparison of reteplase and tissue plasminogen activator for acute myocardial infarction. *N Engl J Med*; *in press*
22 den Heijer P, Vermeer F, Ambrosioni E, Sadowski Z, Lopez-Sendon JL, von Essen R, Beaufils P, Thadani U, Adgey J, Pierard L et al (1998) Evaluation of a weight-adjusted single-bolus plasminogen activator in patients with myocardial infarction: a double-blind, randomized angiographic trial of lanoteplase *versus* alteplase. *Circulation* 98(20): 2117–25
23 Cannon CP, McCabe CH, Gibson CM et al (1997) TNK-tissue plasminogen activator in acute myocardial infarction. *Circulation* 95: 351–356
24 Van de Werf F, Cannon CP, Luyten A, Houbracken K, McCabe CH, Berioli S, Bluhmki E, Sarelin H, Wang-Clow F, Fox NL et al (1999) Safety assessment of single-bolus administration of TNK tissue-plasminogen activator in acute myocardial infarction: the ASSENT-1 trial. The ASSENT-1 Investigators. *Amer Heart J* 137(5): 786–91

25 Vanderschueren S, Barrios L, Kerdsinchai P et al (1995) A randomized trial of recombinant staphylokinase *versus* alteplase for coronary artery potency in acute myocardial infarction. *Circulation* 92: 2044–2049
26 The PRIMI study group (1989) Randomized double-blind trial of recombinant prourokinase against streptokinase in acute myocardial infarction. *Lancet* 1: 863–869
27 Théroux P, Ouimet H, McCans J et al (1988) Aspirin, heparin, or both to treat acute unstable angina. *N Engl J Med* 319: 1105–1111
28 Williams DO, Kirby MG, McPherson K et al (1986) Anticoagulant treatment of unstable angina. *Brit J Clin Pract* 40: 114–116
29 Theroux P, Waters D, Qiu S et al (1993) Aspirin *versus* heparin to prevent myocardial infarction during the acute phase of unstable angina. *Circulation* 88: 2045–2048
30 Gurfinkel EP, Manos EJ, Mejail RI et al (1995) Low molecular weight heparin *versus* regular heparin or aspirin in the treatment of unstable angina and silent ischemia. *J Amer Coll Cardiol* 26: 313–318
31 Neri Serneri GG, Modesti PA, Gensini GF et al (1995) Randomised comparison of subcutaneous heparin, intravenous heparin, and aspirin in unstable angina. *Lancet* 345: 1201–1204
32 Fragmin During Instability in Coronary Artery Disease (FRISC) Study Group (1996) Low-molecular-weight heparin during instability in coronary artery disease. *Lancet* 347: 561–568
33 Klein W, Buchwald A, Hillis WS, Monrad S, Sanz G, Turpie AG, van der Meer J, Olaisson E, Undeland S, Ludwig K (1997) Fragmin in unstable angina pectoris or in non-Q-wave acute myocardial infarction (the FRIC study). Fragmin in Unstable Coronary Artery Disease. *Amer J Cardiol* 80(5A): 30E–34E
34 Topol EJ, Fuster V, Harrington RA et al (1994) Recombinant hirudin for unstable angina pectoris: A multicenter, randomized angiographic trial. *Circulation* 89: 1557–1566
35 Metz BK, White HD, Granger CB, Simes RJ, Armstrong PW, Hirsh J, Fuster V, MacAulay CM, Califf RM, Topol EJ (1998) Randomized comparison of direct thrombin inhibition *versus* heparin in conjunction with fibrinolytic therapy for acute myocardial infarction: results from the GUSTO-IIb Trial. Global Use of Strategies to Open Occluded Coronary Arteries in Acute Coronary Syndromes (GUSTO-IIb) Investigators. *J Amer Coll Cardiol* 31(7): 1493–8
36 Neuhaus KL, von Essen R, Tebbe U, Jessel A, Heinrichs H, Maurer W, Doring W, Harmjanz D, Kotter V, Kalhammer E et al (1994) Safety observations from the pilot phase of the randomized r-Hirudin for Improvement of Thrombolysis (HIT-III) study. *Circulation* 90(4): 1638–42
37 Antman EM, Hirudin in acute myocardial infarction (1996) Thrombolysis and Thrombin Inhibition in Myocardial Infarction (TIMI) 9B trial, *Circulation* 94: 911–921
38 OASIS-2 Investigators (1999) Effects of recombinant hirudin (lepirudin) compared with heparin on death, myocardial infarction, refractory angina, and revascularisation procedures in patients with acute myocardial ischaemia without ST elevation: a randomised trial. Organisation to Assess Strategies for Ischemic Syndromes. *Lancet* 353: 429–38
39 Fuchs J, Cannon CP, the TIMI 7 Investigators (1995) Hirulog in the treatment of unstable angina: Results of the Thrombin Inhibition in Myocardial Ischemia (TIMI) 7 Trial. *Circulation* 92: 727–733
40 Serruys PW, Herrman JP, Simon R, Rutsch W, Bode C, Laarman GJ, van Dijk R, van den Bos AA, Umans VA, Fox KA et al (1995) A comparison of hirudin with heparin in the prevention of restenosis after coronary angioplasty Helvetica Investigators. *N Engl J Med* 333(12): 757–63
41 Roth GJ, Majerus PW (1975) The mechanism of the effect of aspirin on human platelets: I: Acetylation of a particular fraction protein. *J Clin Invest* 56: 624–632
42 Lewis HD, Davis JW, Archibald DG et al (1983) Protective effects of aspirin against acute myocardial infarction and death in men with unstable angina: Results of a Veterans Administration cooperative study. *N Engl J Med* 309: 396–403
43 Cairns JA, Gent M, Singer J et al (1985) Aspirin, sulfinpyrazone, or both in unstable angina: Results of a Canadian multicenter trial. *N Engl J Med* 313: 1369–1375
44 ISIS 2 (Second international study of infarct survival) Collaborative Group (1988) Randomized trial of intravenous streptokinase, oral aspirin, both, or neither among 17 187 cases of suspected acute myocardial infarction. *Lancet* 11: 349–360
45 Antiplatelet Trialists' Collaboration (1994) Collaborative overview of randomised trials of antiplatelet therapy – I: Prevention of death, myocardial infarction, and stroke by prolonged antiplatelet therapy in various categories of patients. *Brit Med J* 308: 81–106
46 Buerke M, Pittroff Walter Meyer J, Darius H (1995) Aspirin therapy: Optimized platelet inhibition

with different loading and maintenance doses. *Am Heart J* 130: 465–472
47 CAPRIE Steering Committee (1996) A randomised, blinded, trial of clopidogrel *versus* aspirin in patients at risk of ischaemic events (CAPRIE). *Lancet* 348: 1329–1339
48 EPIC Investigators (1994) Use of a monoclonal antibody directed against the platelet glycoprotein IIb/IIIa receptor in high-risk coronary angioplasty. *N Engl J Med* 330: 956–961
49 The EPILOG Investigators (1997) Platelet glycoprotein IIb/IIIa receptor blockade and low-dose heparin during percutaneous coronary revascularization. *N Engl J Med* 336: 1689–1696
50 The CAPTURE Investigators (1997) Platelet glycoprotein IIb/IIIa receptor blockade and low-dose heparin during percutaneous coronary revascularization. *N Engl J Med* 336: 1689–1696
51 The EPISTENT Investigators (1998) Randomized placebo-controlled and balloon-angioplasty-controlled trial to assess safety of coronary stenting with use of platelet glycoprotein-llb/llla Blockade. *Lancet* 352: 87–92
52 The RESTORE Investigators (1997) Effects of platelet glycoprotein IIb/IIIa blockade with tirofiban on adverse cardiac events in patients with unstable angina or acute myocardial infarction undergoing coronary angioplasty. The RESTORE Investigators. Randomized Efficacy Study of Tirofiban for Outcomes and REstenosis. *Circulation* 96(5): 1445–53
53 White HD, on behalf of the Platelet Receptor Inhibition in Ischemic Syndrome Management (PRISM) Study Investigators (1997) The Platelet Receptor Inhibition in Ischemic Syndrome Management (PRISM) Study. Presented at the American College of Cardiology 46th Annual Scientific Sessions, Anaheim, CA, March 1997
54 PRISM-PLUS Study Investigators (1998) Inhibition of the platelet glycoprotein IIb/IIIa receptor with tirofiban in unstable angina and non-Q-wave myocardial infarction. Platelet Receptor Inhibition in Ischemic Syndrome Management in Patients Limited by Unstable Signs and Symptoms (PRISM-PLUS) Study Investigators. *N Engl J Med* 338(21): 1488–1497
55 IMPACT-II Study Group (1997) Randomised placebo-controlled trial of effect of eptifibatide on complications of percutaneous coronary intervention: IMPACT-II. Integrilin to Minimise Platelet Aggregation and Coronary Thrombosis-II. *Lancet* 349: 1422–1428
56 PURSUIT Trial Investigators (1998) Inhibition of platelet glycoprotein IIb/IIIa with eptifibatide in patients with acute coronary syndromes. The PURSUIT Trial Investigators. Platelet Glycoprotein IIb/IIIa in Unstable Angina: Receptor Suppression Using Integrilin Therapy. *N Engl J Med* 339(7): 436–43
57 PARAGON Study Group (1998) International, randomized, controlled trial of lamifiban (a platelet glycoprotein IIb/IIIa inhibitor), heparin, or both in unstable angina. The PARAGON Investigators. Platelet IIb/IIIa Antagonism for the Reduction of Acute coronary syndrome events in a Global Organization Network. *Circulation* 97(24): 2386–95

New Approaches to Drug Development
ed. by P. Jollès

Strategies to delay the onset of Alzheimer's disease

Michael H. Hüll, Bernd L. Fiebich, Klaus Lieb

Dept. of Psychiatry, University of Freiburg, D-79104 Freiburg, Germany

Summary. Several processes are implicated in the neuropathology of Alzheimer's disease (AD), such as the deposition of amyloid, the formation of paired helical filaments and the proinflammatory activation of microglial and astroglial cells. Proinflammatory activation of glial cells has been a focus of research for a mere ten years now. However, the availability of and broad experience with anti-inflammatory drugs has led to several ongoing clinical trials to verify the capacity of anti-inflammatory drugs to ameliorate the deterioration in AD. The enzymatic cleavage of the amyloid-precursor-protein or the hyperphosphorylation of tau as well as the subsequent aggregation of the resulting products are further targets for drugs intended to delay the neuropathological destruction observed in AD.

Pathogenetic processes in Alzheimer's disease (AD)

Genetic factors

AD is a common disease in elderly patients. With increasing age the probability rises to find several members of one family affected by AD. Autosomal dominant gene mutations as causes of AD are extremely rare. Up to now, only about 200 families have been described worldwide, in which an autosomal dominant form of AD was clearly linked to different mutations in three different genes [1]. These genes were identified as coding for the presenilin-1 protein on chromosome-14, for the presenilin-2 protein on chromosome-1 and for the amyloid precursor protein (APP) on chromosome-21. All three genes seem to interfere with the processing of APP. Except for a few patients with mutations in the presenilin-2 gene, all individuals carrying pathogenic mutations develop AD before the age of 65 [2].

Mutations in the amyloid-precursor-protein (APP) gene

Mutations in the APP-gene on chromosome-21 were the first mutations described in families with an autosomal dominant transmission of AD. APP harbors a 40–43 amino acid sequence, the so-called βA4 peptide, which is consistenly found in plaques in the brains of AD patients. Once the βA4 peptide is produced by a yet unknown process, it has a strong tendency to form protein aggregates and amyloid deposits. One of these APP mutations, an exchange of two amino acids near the N-terminus of βA4-(codon 670 and 671 of APP, "swedish double mutation"), has been shown to favor the generation of amyloidogenic peptides from APP. This observation underlines the hypothesis that a metabolic defect in the processing of APP with a subsequent overproduction of amyloidogenic βA4 is a major pathogenetic pathway in AD. Another pathogenic mutation in the APP gene at codon 717 does not lead to a quantitative overproduction of βA4. However, an altered proteolytic process has been shown with a shift to slightly longer βA4 fragments (42/43 amino acids) which accelerate amyloid formation in comparison to shorter fragments of 40 amino acids [2, 3]. APP mutations as causes of AD have been reported in less than 20 families affected wordwide. Several other mutations in the APP-gene cause a familiar form of cerebral haemorrhages.

Mutations in the Presenilin gene

Mutations in the gene coding for the presenilin-1 protein on chromosome-14 and for the presenilin-2 protein on chromosome-1 have been found as autosomal dominant causes of AD. More than 90% of all genetically determined cases of AD are caused by the mutation in the presenilin-1 protein. Up to now the presence of presenilin mutations have been reported in less than 180 families worldwide. The presenilin proteins interfere with the NOTCH signalling pathway, a signal transduction pathway involved in organogenesis. Interestingly, the NOTCH protein is proteolytically cleaved near its transmembrane domain in its action of signalling. Presenilins are presumably either directly involved in this cleavage or connected to the intracellular trafficking bringing the NOTCH-protein in contact with the unknown protease. The formation of βA4 depends on a proteolytic processing of APP similar to the proteolytic processing of NOTCH. Therefore, the role of presenilins in AD may be connected to the altered proteolytic processing of APP [4,5]. Cells transfected with the mRNA for presenilins show an enhanced production of amyloidogenic βA4 [6].

Apolipoprotein E4 allels

The apolipoprotein-E4 (apoE4) allele is one of three different allelic forms of the apolipoprotein-E. ApoE4 is associated with an increased risk of developing AD. The risk of carriers homozygous for the apoE4 allele is increased more than threefold compared to individuals without an apoE4 allele. However, the apoE4 allele is neither a necessary nor a determining factor for the development of AD. Genetic analysis of the apoE polymorphism is not useful as a diagnostic marker in AD [7, 8]. The pathogenetic pathway influenced by apoE4 is not known so far. Several hypotheses link the apoE polymorphism with either amyloid metabolism or tau pathology [9, 10]. Interestingly, the synthesis of both proteins, apoE4 and APP, is elevated under inflammatory conditions. The presence of apoE in amyloid plaques has been correlated to the activation of microglial cells [11, 12]

Head trauma

In epidemiological studies severe head trauma has been shown to be a risk factor for the development of AD. A *post mortem* study of victims of traffic accidents, who survived the initial few weeks after trauma, showed a high incidence of amyloid plaques, which were presumably formed within the few weeks between trauma and death [13]. Besides the deposition of amyloid, microglial cells are activated by severe head trauma.

Age as a risk factor

Age itself is an essential "risk factor" of AD. The risk of developing AD increases with age. Dementia affects 5–10% of the global population older than 65 years and AD accounts for more than 60% of all cases in which dementia is diagnosed. In Germany, 10% of the population is older than 65 years. From a total population of 80 million people, approximately 0.6 million suffer from AD [14].

Neuropathology of AD

Major neuropathological hallmarks of AD are amyloid plaques, neurofibrillary degenerations and loss of synapses. The correlation between these different types of lesions and the cognitive decline is complex. Earlier studies showed a correlation between the number of amyloid plaques and the loss of cognitive functions. However, more detailed studies attributed this finding to a "floor" and "ceiling" effect. In the absence of any amyloid plaque, there are usually no cognitive deficits. On the other hand, in the presence of extreme high numbers of amyloid plaques, cognition is always severely impaired. However, in between these two extreme situations, a correlation between cognition and plaque number cannot be found [15]. The formation of neurofibrillary degenerations in the soma of neurons, so-called tangles, follows a predictable course. The formation of tangles starts in the transentorhinal cortex, then spreads to the hippocampus and out into the isocortex. These stages of tangle formation show some correlation with the cognitive decline [16]. The loss of synapses as determined by electron microscopy is most closely correlated with the cognitive decline [17]. However, it is unclear so far how the loss of synapses, amyloid deposition and neurofibrillary degeneration are related to each other.

During the last couple of years, an inflammatory activation of microglial and astroglial cells combined with the synthesis of cytokines, prostaglandins and acute-phase-proteins has been identified as another pathogenetic process in AD.

Deposition of amyloid

The major component of amyloid plaques is the βA4 protein, a fragment derived from the APP. *In vitro* studies of the neurotoxicity of βA4 show inconsistent effects depending on the cell culture system [18]. High-density cultures of primary neurons are usually quite resistant to βA4 toxicity. In many neuronal cell culture systems, a delayed cell death is measurable after 48 h of incubation with 1–100 μm βA4 [19]. The degree of toxicity of amyloidogenic peptides derived from the βA4 sequence seems to depend on the aggregation of these peptides into fibrils of 7–9 nm. The dynamics of aggregation and the degree of toxicity is also extremely heterogenous between different peptides derived from the full 1–42 amino acid sequence of βA4 [18]. Several lines of evidence support the hypothesis that βA4 in cell culture, like other amyloidogenic peptides, induces a necrotic cell death by a common pathway involving the generation of free radicals [20, 21]. *In vivo* injection of βA4 is not associated with major neurodegeneration [22]. Although the loss of synapses is not markedly accentuated within amyloid deposits, pathological processes initiated in amyloid deposits may alter distant synaptic cell contacts [23, 24].

Neurofibrillary pathology

Another neuropathological hallmark of AD is the formation of tangles in the neuronal cell soma and neuritic degeneration of neuronal processes. Both intracellular lesions consist of paired helical filaments (PHF) which disrupt the function of the neuronal cytoskeleton. PHF in the neuronal processes and PHF of perinuclear tangles both consist of hyperphosphorylated tau proteins. The microtubuli of the cytoskeleton are stabilized by tau which belongs to the so-called microtubule-associated proteins [25, 26]. The binding of tau to microtubuli is diminished by phosphorylation and restored by dephosphorylation [27]. The formation of tangles starts in the neuronal somata in the transentorhinal cortex and expands into the hippocampus and further into the neocortex [28].

Neuritic degenerations of neuronal cell processes are found throughout the cortex in AD. Neuritic degenerations are not confined to amyloid plaques and are also found in areas without amyloid deposition. Diffuse amyloid plaques usually contain only few degenerative neurites [29]. In contrast, a corona of degenerative neurites is always present in primitive and classic amyloid plaques [30]. In hippocampal cell cultures, βA4 can induce hyperphosphorylation of tau [31]. The intracellular accumulation of PHF and the associated disruption of the normal cytoskeleton is accompanied by a shrinkage of the dendritic tree. Therefore, neurofibrillary degeneration may facilitate the loss of synapses.

Inflammatory processes

Activated microglial cells, interleukin-1α (IL-1α) and interleukin-6 (IL-6) can be found already in diffuse amyloid plaques, before neurofibrillary changes are detectable [32, 33]. Diffuse amyloid plaques which are sporadically present in non-demented elderly persons are not associated with an activation of microglial cells. However, activated microglial cells are frequently found in diffuse plaques in demented persons [34, 35]. Therefore, the activation of microglial cells constitutes an important element in the progression of pathology in AD. The extracellular stimuli which activate microglial cells in AD have not been characterized completely. However, most *in vitro* studies point to an involvement of βA4 and its fragments, either acting by itself or in combination with cytokines [36–39]. The predominant role of fragments of APP in the activation of microglial cells is further underlined by experiments with transgenic animals. The overexpression of a known pathogenic mutant of APP (the swedish mutation) in mice elicits microglial activation in amyloid deposits [40].

In 1992, Araujo and Cotman showed that βA4 induces IL-1 in rat microglial and astroglial cells [41]. Induction of IL-1 by βA4 could also be shown in human microglial cells [42]. In astrocytes, βA4 itself seems to be only a weak stimulator of IL-6 synthesis [43]. Thus, to explain the synthesis of IL-6 in AD, one must postulate additional signal transduction pathways activated by other stimuli and acting synergistically with the pathways activated by βA4. In human astrocytoma cells (U373MG), research showed that fibrillar (aged) βA4 markedly enhanced the IL-1-induced IL-6 synthesis, whereas the removal of extracellular calcium abolished this synergistic effect [44]. Therefore, a βA4-induced influx of calcium has been proposed as the mechanism of augmentation of IL-1-induced IL-6 synthesis [44]. We have thoroughly characterized the different intracellular signal transduction pathways leading to IL-1-induced transcription of IL-6 in the human astrocytoma cell line U373MG. We especially demonstrated the role of the transcription factor NF-κB, which may be a suitable target to disrupt the proinflammatory activation of glial cells [45, 46]. Furthermore, we identified several other potential stimulators of IL-6 synthesis, such as prostaglandin-E2 or ceramides, which may also act as costimulators together with βA4 [45, 47]

In correspondence with the known function of IL-6, acute-phase proteins, such as α_1-antichymotrypsin and c-reactive protein, are found in amyloid plaques and elevated concentrations of these acute-phase proteins have been measured in brain extracts of AD patients [48-51]. The acute-phase protein α_1-antichymotrypsin has been implicated in the formation of amyloid fibrils [52]. Furthermore, several acute-phase proteins are potent protease inhibitors. Therefore acute-phase proteins may interfere with extracellular degradation of amyloid and growth of amyloid plaques.

Microglial cells activated by βA4 produce neurotoxic substances [37, 53]. In neuronal cell cultures, IL-1 has been shown to enhance the neuronal toxicity of βA4 [54]. IL-6 has diverse effects on neuronal cells and leads to cellular deterioration after long-term exposure [55, 56]. Transgenic animals over-expressing IL-6 show a marked alteration of dendritic morphology, cerebral gliosis, a loss of cholinergic innervation of the hippocampus, a reduction of long-term potentiation and a progressive decline in learning and memory capacities [56-59]. Therefore, IL-6 synthesis may contribute to the pathology in AD.

Activated microglial cells and interleukin-1-stimulated astroglial cells are capable of synthesizing cyclooxygenase-2 (COX-2). Some *post mortem* studies indicate an increased expression of COX-2 in the brains of AD patients, but findings in general are contradictory [60-62]. COX-2 is the key enzyme in the synthesis of prostaglandin E2 and other prostanoids. Interestingly, we showed that prostaglandin E2 is able to induce IL-6 in astrocytic cells [47].

Most interestingly, recent investigations show a widespread activation of microglia outside amyloid plaques in AD. High numbers of activated microglial cells are not only found in amyloid plaques but also in hippocampal areas without amyloid deposition [63]. This activation of microglial in the hippocampus shows a much better correlation to the extent of tangle formation along the entorhinal-hippocampal pathway than to amyloid deposition [64]. Therefore, microglial activation does not only follow amyloid deposition in AD. In conclusion, the activation of microglia is a third element, in addition to amyloid plaques and neurofibrillary tangles, in the progression of AD. Microglial activation may be partly dependent on the other two types of lesions, but may be in part attributed to unknown processes.

Loss of cortical cholinergic inervation

A loss of cholinergic fibers in the cortex and a reduced number of cholinergic neurons in the Nucleus basalis Meynert are common findings in AD. Acetylcholine is an important neurotransmitter involved in vigilance, concentration and attention processes. The impact of acetylcholine on pure memory function is not well documented in animal models [65, 66]. However, the impairment of attention or other cognitive processes may add to the psychopathology in AD. The loss of cholinergic fibers is accentuated in areas functionally impaired in AD (temporal cortex, hippocampus). In primary sensory or motor cortex areas, the function and distribution of cholinergic fibers are spared [67].

Approaches altering the progression of AD

APP metabolism and aggregation

In finding drugs that reduce the progression of AD, one of the major research strategies has been the modulation of APP-processing pharmacologically. APP may be processed by so-called α-secretases to yield soluble forms of APP (sAPP). These sAPP are not harmful and may even have protective effects in the brain [68]. The enzymatic process leading to βA4 has not been completely elucidated. Intracellularly, βA4 fragments of APP are presumably generated in the endoplasmic reticulum or trans-Golgi network in neurons [69]. Despite enormous efforts, the enzymes responsible for the generation of βA4 have not been characterized at the protein level. Especially the enzyme responsible for the cleavage of APP in the transmembrane region which results in production of βA4, the so-called γ-secretase, remains an enigma. If specific inhibitors of this enzyme could be generated, they would be of utmost interest for the therapy of AD. Interestingly, as mentioned above, the known pathogenic mutations of presenilins seem to influence and enhance the activity of the γ-secretase. While the development of efficient γ-secretase inhibitors is still to come, the stimulation of the formation of sAPP has been shown to decrease production of βA4. Stimulation of sAPP production has been shown for m_1-muscarinergic agonists, mGlu$R_{1\alpha}$-glutamatergic agonists, $5HT_{2a}$ and $5HT_{2c}$ serotonin receptor agonists, vasopressin and bradykinin [70-73]. This stimulation of sAPP processing involves a stimulation of the phospho-inositol pathway and/or protein-kinase C (PKC). Further research in this field may allow development of other substances which may stimulate the formation of sAPP by acting directly on the intracellular processes involved.

Cholinesterase inhibitors

The development of cholinesterase inhibitors has been an important step in the treatment of AD symptoms, although the improvement in cognitive functioning has been shown to be only moderate. However, the rise in cortical acetylcholine levels brought about by inhibition of cholinesterase may also lead to the production of sAPP via stimulation of m_1-receptors. Moreover, some but not all cholinesterase inhibitors have been shown to stimulate sAPP production independent of activation of acetylcholine receptors, arguing for a second pathway affected by these substances [74]. Nevertheless, up to now none of the widely clinically used acetylcholine esterase inhibitors have been shown to alter the progression of the course of the disease. Cholinesterase inhibitors have several drawbacks, such as their indirect stimulation of m_1-receptors. First of all, a substantial rise of acetylcholine may not be elicited in the areas of amyloid deposition because of a loss of cholinergic terminals. Second, the higher availability of acetylcholine will not only lead to stimulation of m_1-receptors but also to stimulation of several other acetylcholine receptors with divergent results. The development of direct m_1-receptor agonists may overcome some of these obstacles. One newly developed m_1-receptor agonist, xanomeline, has been shown to stimulate the production of sAPP [75, 76].

However, the observation that stimulation of the production of sAPP leads to a reduction of the production of βA4 has to be tested for each substance and species-specific cell type. Activation of PKC with subsequent production of sAPP is not associated with a decrease of βA4 production in human neurons [77].

Apart from altering the protein processing of APP, several approaches have been suggested in order to reduce the extracellular aggregation of βA4. Sulfonated dyes, such as congo red, and other classes of chemical substances have been shown to interfere with amyloid aggregation [78]. However, long-term toxicity is a major concern in most of these substances. Another approach to influencing βA4 aggregation is to identify aggregation-promoting factors. One of these factors is the acute-phase reactant α_1-antichymotrypsin [79, 80]. We have shown that IL-1 is able to induce α_1-antichymotrypsin in human astroglioma cells [81]. This result has been replicated in primary human astrocytes [82]. Therefore, the amyloidogenic and the inflammatory pathways of further tissue destruction may interact with each other. We have shown that the novel non-steroidal anti-inflammatory drug (NSAID) tepoxalin inhibits IL-1-induced α_1-antichymotrypsin synthesis in human astroglioma cells [83].

Drug targets in PHF pathology

The hyperphosphorylation of tau-protein prevents the interaction of tau with microtubuli and precedes the self-aggregation of tau in PHF. Cell culture studies identified several kinases able to phosphorylize tau, most notably glycogen-synthase-kinase 3β [84]. An inhibition of this kinase by lithium reduces tau phosphorylation in cultured rat hippocampal neurons and rat embryos. However, the concentration necessary to inhibit glycogen-synthase-kinase 3β with lithium is greater than 5 mM and therefore way above the maximal non-toxic concentration in humans (appr. 1,2 mM) [85, 86]. Another way of influencing hyperphosphorylation of tau would be to activate endogenous phosphatases, which usually revert tau-phosphorylation and restore the capacity of micotubuli binding [27]. Glutamatergic stimulation has been shown to activate phosphatase-1 and phosphatase-2a and thus induces dephosphorylation of tau in hippocampal neurons [87]. However, glutamatergic overstimulation may also induce cell death. In addition, βA4 has been shown to potentiate glutamate-induced cell death [88]. Further research should reveal alternative pathways able to modulate intracellular tau phosphorylation.

Another process which may be inhibited pharmacologically is tau aggregation. *In vitro* studies have shown the potential of phenothiazines to inhibit self-aggregation of tau without interfering with the binding of tau to microtubuli [89]. Considerations for the development of drugs blocking either tau or βA4 aggregation are similar in principle. One should note, though, that in contrast to extracellular βA4 deposition, tau aggregation takes place within the cell. Therefore, drugs modulating tau aggregation have to be able to enter neurons on the one side without being toxic in their effective concentrations. Several other factors such as glycation or interaction with heparin have been suggested to play an important role in the dynamics of tau aggregation [84]. The elucidation of these factors may reveal further targets in tau pathology.

Inhibition of microglial activation

Several elements of signal transduction pathways are activated by βA4 in microglial cells including mitogen-activated kinase p38 (p38MAPK), extracellular signal regulated kinase (ERK1 and ERK2), c-jun-N-terminal-kinase (JNK), PKC and protein-tyrosin-kinases (PTKs) [90–92]. The inhibition of p38MAPK or ERK reduced the induction of tumour necrosis factor α by βA4 in microglial cells [93]. The inhibition of PKC or PTK abolished chemotaxis in microglial cells [94]. Different intracellular pathways may therefore be responsible for the complex action of βA4 on microglial cells. Inhibition of these pathways may be targets for future drug developments. In a recent animal study it was shown that indomethacin is capable of attenuating the activation of microglia induced by intraventricular infusion of βA4 [95].

Propentofyllin is an inhibitor of adenosine transporter and phosphodiesterase but has also major inhibitory effects on the activation of microglial cells [96]. In cultured microglial cells propentofyllin reduces the proliferation and oxygen radical production [97]. The effect of propentofyllin on the synthesis of cytokines in brain cells is not well known and may be independent of adenosine receptor stimulation. Stimulation of the adenosine A_{2b}-receptor in human astrocytoma cells leads to enhanced production of IL-6 [98] and stimulation of the adenosine A_{2a}-receptor in rat microglia induces COX-2 [99]. Propentofyllin has been tested in clinical trials for the treatment of vascular dementia and AD [100].

Non-steroidal anti-inflammatory drugs (NSAIDs) in AD

More than fourteen epidemiological studies have been conducted to identify risk factors or protective factors in the development of AD. One major finding was the reduced risk of developing AD in subjects treated with anti-inflammatory drugs [101]. In a neuropathological study, the activation of microglial cells in the brains of long-term users of NSAIDs was markedly reduced as compared to age-matched controls [102]. A pilot study with indomethacin given over a period of 6 months showed promising results, most notably a reduced progression of AD [103]. However, the statistical power of this pilot study with 28 participants and a duration of 6 months of active treatment was too weak to draw final conclusions. There was a high rate of drop-outs due to gastro-intestinal disturbances, a well-known side-effect of indomethacin. A novel NSAID, celecoxib [104], is currently tested in an ongoing trial for AD. We have shown that PGE_2 induces IL-6 synthesis in human astroglioma [47]. Therefore, inhibition of COX-2 with subsequent reduction of IL-6 synthesis may be one of the beneficial actions of NSAIDs in AD. We further demonstrated that the newly developed NSAIDs tenidap and tepoxalin do not only inhibit the enzymatic function of COX-2 but also the IL-1-induced synthesis of IL-6 [83, 105]. Novel NSAIDs which do not only inhibit COX-2 but also reduce cytokine synthesis directly may be promising in the treatment of AD.

Estrogen and estrogen derivates as protective agents in AD

Several observations suggest a protective effect of estradiol on the development of AD [106]. Women show a higher risk of developing AD after menopause and estradiol substitution reduces this risk [107]. Estradiol has been shown to suppress IL-6 synthesis which activates osteoclasts in osteoporosis [108]. These *in vitro* findings may explain the protective influence of estradiol on osteoporosis. The characterization of the interaction between estrogen and cytokines may help to understand the beneficial effects of estrogen in AD. There are currently several trials started or planned with estradiol or its derivate in AD [109].

Prednisolon

By the middle of 1999, data from a large trial conducted by the American Alzheimer's Disease Collaborative Study Unit on the effect of prednison in AD will be available [110]. Up to now, no serious side effects have been reported [111]. Several animal studies suggested that long-term elevated levels of steroids may impair memory function [112]. In healthy old persons, steroids reduce hippocampal glucose utilization and memory function [113, 114]. In contrast to estrogen, corticosterone potentiates the toxic properties of oxygen radicals, glutamate and βA4 in hippocampal cell cultures [115]. The effects of prednison on these processes take away from the beneficial effect of reducing inflammatory activation in AD.

References

1 The Ronald, Nancy Reagan Research Institute of the Alzheimer's Association, the National Institute on Aging Working Group (1998) Consensus report of the Working Group on: "Molecular and Biochemical Markers of Alzheimer's Disease". *Neurobiol Aging* 19: 109–116

2 Hardy J (1997) Amyloid, the presenilins and Alzheimer's disease. *Trends Neurosci* 20: 154–159

3 Maruyama K, Tomita T, Shinozaki K, Kume H, Asada H, Saido TC, Ishiura S, Iwatsubo T, Obata K (1996) Familial Alzheimer's disease-linked mutations at Val_{717} of amyloid precursor protein are specific for the increased secretion of Aβ42(43). *Biochem Biophys Res Commun* 227: 730–735

4 Zhou JH, Liyanage U, Medina M, Ho C, Simmons AD, Lovett M, Kosik KS (1997) Presenilin 1 interaction in the brain with a novel member of the Armadillo family. *Neuroreport* 8: 2085–2090

5 Hutton M, Hardy J (1997) The presenilins and Alzheimer's disease. *Hum Mol Genet* 6: 1639–1646

6 Borchelt DR, Thinakaran G, Eckman CB, Lee MK, Davenport F, Ratovitsky T, Prada CM, Kim G, Seekins S, Yager D et al (1996) Familial Alzheimer's disease-linked presenilin 1 variants elevate Aβ1-42/1-40 ratio *in vitro* and *in vivo*. *Neuron* 17: 1005–1013

7 Saunders AM, Hulette C, Welsh-Bohmer KA, Schmechel DE, Crain B, Burke JR, Alberts MJ, Strittmatter WJ, Breitner JCS, Rosenberg C et al (1996) Specificity, sensitivity, and predictive value of apolipoprotein-E genotyping for sporadic Alzheimer's disease. *Lancet* 348: 90–93

8 Breitner JCS (1996) APOE genotyping and Alzheimer's disease. *Lancet* 347: 1184–1185

9 Roses AD (1996) Apolipoprotein E alleles as risk factors in Alzheimer's disease. *Annu Rev Med* 47: 387–400

10 Mahley RW, Nathan BP, Pitas R (1996) Apolipoprotein E – Structure, function, and possible roles in Alzheimer's disease. *Ann.NY Acad.Sci.* 777: 139–145

11 Saitoh T, Kang D, Mallory M, DeTeresa R, Masliah E (1997) Glial cells in Alzheimer's disease: Preferential effect of APOE risk on scattered microglia. *Gerontology* 43: 109–118

12 Uchihara T, Duyckaerts C, He Y, Kobayashi K, Seilhean D, Amouyel P, Hauw J-J (1995) ApoE immunoreactivity and microglial cells in Alzheimer's disease brain. *Neurosci Lett* 195: 5–8

13 Roberts GW, Gentleman SM, Lynch A, Murray L, Landon M, Graham D (1994) βAmyloid protein deposition in the brain after severe head injury: implications for the pathogenesis of Alzheimer's disease. *J Neurol Neurosurg Psychiat* 57: 419–425

14. Förstl H (1998) Alzheimer's disease: the size of the problem, clinical manifestation and heterogeneity. In Alzheimer's Disease – From Basic Research to Clinical Applications. H.J. Gertz and T. Arendt, editors. Springer, Wien, New York. 1–8

15 Jellinger K, Bancher C, Fischer P, Lassmann H (1992) Quantitative histopathologic validation of senile dementia of the Alzheimer type. *Eur J Gerontol* 3: 146–156

16 Braak H, Braak E (1995) Staging of Alzheimer's disease-related neurofibrillary changes. Response to commentaries. *Neurobiol Aging* 16: 284–284

17 Terry RD, Masliah E, Salmon DP, Butters N, DeTeresa R, Hill R, Hansen LA, Katzman R (1991) Physical basis of cognitive alterations in Alzheimer's disease: synapse loss is the major correlate of cognitive impairment. *Ann Neurol* 30: 572–580

18 Iversen LL, Mortishire RJ, Pollack SJ, Shearman MS (1995) The toxicity *in vitro* of β-amyloid protein. *Biochem J* 311: 1–16

19 Hensley K, Carney JM, Mattson MP, Aksenova M, Harris M, Wu JF, Floyd RA, Butterfield DA (1994) A model for beta-amyloid aggregation and neurotoxicity based on free radical generation by the peptide: relevance to Alzheimer's disease. *Proc Natl Acad Sci USA* 91: 3270–3274

20 Behl C (1997) Amyloid beta-protein toxicity and oxidative stress in Alzheimer's disease. [Review] [153 refs]. *Cell Tiss Res* 290: 471–480

21 Mattson MP, Rydel RE (1996) Alzheimer's disease – Amyloid ox-tox transducers. *Nature* 382: 674–675

22 Winkler J, Connor DJ, Frautschy SA, Behl C, Waite J, Cole GM, Thal LJ (1994) Lack of long-term effects after β-amyloid protein injections in rat brain. *Neurobiol Aging* 15: 601–607

23 Masliah E, Terry RD, Mallory M, Alford M, Hansen LA (1990) Diffuse plaques do not accentuate synapse loss in Alzheimer's disease. *Amer J Pathol* 137: 1293–1297

24 Masliah E (1995) Mechanisms of synaptic dysfunction in Alzheimer's disease. *Histol Histopathol* 10: 509–519

25 Goedert M, Spillantini MG, Cairns NJ, Crowther RA (1992) Tau proteins of Alzheimer paired helical filaments: abnormal phosphorylation of all six brain isoforms. *Neuron* 8: 159–168

26 Goedert M, Wischik CM, Crowther RA, Walker JE, Klug A (1988) Cloning and sequencing of the cDNA encoding a core protein of the paired helical filament of Alzheimer's disease: identification as the microtubule-associated protein tau. *Proc Natl Acad Sci USA* 85: 4051–4055

27 Wang JZ, Grundke-Iqbal I, Iqbal K (1996) Restoration of biological activity of Alzheimer abnormally phosphorylated tau by dephosphorylation with protein phosphatase-2A, -2B and -1. *Brain Res.* Molecular Brain Research. 38: 200–208

28 Braak H, Braak E (1997) Frequency of stages of Alzheimer-related lesions in different age categories. *Neurobiol Aging* 18: 351–357

29 Yamazaki T, Yamaguchi H, Okamoto K, Hirai S (1991) Ultrastructural localization of argyrophilic substances in diffuse plaques of Alzheimer-type dementia demonstrated by methenamine silver staining. *Acta Neuropathol Berl.* 81: 540–545

30 Wisniewski HM, Terry RD (1973) Reexamination of the pathogenesis of the senile plaque. *Prog Neuropathol* 2: 1–26

31 Busciglio J, Lorenzo A, Yeh J, Yankner BA (1995) b-amyloid fibrils induce tau phosphorylation and loss of microtuble binding. *Neuron* 14: 879–888
32 Hüll M, Strauss S, Volk B, Berger M, Bauer J (1995) Interleukin-6 is present in early stages of plaque formation and is restricted to the brains of Alzheimer's disease patients. *Acta Neuropathol* 89: 544–551
33 Griffin WST, Sheng JG, Roberts GW, Mrak RE (1995) Interleukin-1 expression in different plaque types in Alzheimer's disease: Significance in plaque evolution. *J Neuropathol Exp Neurol* 54: 276–281
34 Mackenzie IRA, Hao CH, Munoz DG (1995) Role of microglia in senile plaque formation. *Neurobiol Aging* 16: 797–804
35 Mackenzie IRA (1994) Senile plaques do not progressively accumulate with normal aging. *Acta Neuropathol Berl.* 87: 520–525
36 Giulian D, Haverkamp LJ, Yu JH, Karshin W, Tom D, Li J, Kirkpatrick J, Kuo YM, Roher AE (1996) Specific domains of β-amyloid from Alzheimer plaque elicit neuron killing in human microglia. *J Neurosci* 16: 6021–6037
37 Meda L, Bernasconi S, Bonaiuto C, Sozzani S, Zhou D, Otvos L Jr, Mantovani A, Rossi F, Cassatella MA (1996) *Beta*-amyloid (25-35) peptide and IFN-gamma synergistically induce the production of the chemotactic cytokine MCP-1/JE in monocytes and microglial cells. *J Immunol* 157: 1213–1218
38 Araujo DM, Cotman CW (1992) β-Amyloid stimulates glial cells *in vitro* to produce growth factors that accumulate in senile plaques in Alzheimer's disease. *Brain Res* 569: 141–145
39 Murphy GMJ, Yang L, Cordell B (1998) Macrophage colony-stimulating factor augments beta-amyloid-induced interleukin-1, interleukin-6, and nitric oxide production by microglial cells. *J Biol Chem* 273: 20 967–20 971
40 Frautschy SA, Yang F, Irrizarry M, Hyman B, Saido TC, Hsiao K, Cole GM (1998) Microglial response to amyloid plaques in APPsw transgenic mice. *Amer J Pathol* 152: 307–317
41 Araujo DM, Cotman CW (1992) *Beta*-amyloid stimulates glial cells *in vitro* to produce growth factors that accumulate in senile plaques in Alzheimer's disease. *Brain Res* 569: 141–145
42 Walker DG, Kim SU, McGeer PL (1995) Complement and cytokine gene expression in cultured microglial derived cells from postmortem human brains. *J Neurosci* 40: 478–493
43 Del Bo R, Angeretti N, Lucca E, Grazia M, Simoni De, Forloni G (1995) Reciprocal control of inflammatory cytokines, IL-1 and IL-6, and β-amyloid production in cultures. *Neurosci Lett* 188: 70–74
44 Gitter BD, Cox LM, Rydel RE, May PC (1995) Amyloid beta peptide potentiates cytokine secretion by interleukin-1 beta-activated human astrocytoma cells. *Proc Natl Acad Sci USA* 92: 10 738–10 741
45 Fiebich BL, Lieb K, Berger M, Bauer J (1995) Stimulation of the sphingomyelin pathway induces interleukin-6 gene expression in human astrocytoma cells. *J Neuroimmunol* 63: 207–211
46 Lieb K, Kaltschmidt C, Kaltschmidt B, Baeuerle PA, Berger M, Bauer J, Fiebich BL (1996) Interleukin-1 beta uses common and distinct signaling pathways for induction of the interleukin-6 and tumor necrosis factor alpha genes in the human astrocytoma cell line U373. *J Neurochem* 66: 1496–1503
47 Fiebich BL, Hüll M, Lieb K, Gyufko K, Berger M, Bauer (1997) Prostaglandin E2 induces interleukin-6 synthesis in human astrocytoma cells. *J Neurochem* 68: 704–709
48 Wood JA, Wood PL, Ryan R, Graff-Radford NR, Pilapil C, Robitaille Y, Quirion R (1993) Cytokine indices in Alzheimer's temporal cortex: no changes in mature IL-1 beta or IL-1 RA but increases in the associated acute phase proteins IL-6, alpha-2-macroglobulin and C-reactive protein. *Brain Res* 629: 245–252
49 Abraham CR, Selkoe DJ, Potter H (1988) Immunochemical identification of the serine protease inhibitor alpha 1-antichymotrypsin in the brain amyloid deposits of Alzheimer's disease. *Cell* 52: 487–501
50 Strauss S, Bauer J, Ganter U, Jonas U, Berger M, Volk B (1992) Detection of interleukin 6 and alpha-2-macroglobulin immunoreactivity in cortex and hippocampus of Alzheimer's disease patients. *Lab Invest* 66: 223–230
51 Rozemuller JM, Stam FC, Eikelenboom P (1990) Acute phase proteins are present in amorphous plaques in the cerebral but not in the cerebellar cortex of patients with Alzheimer's disease. *Neurosci Lett* 119: 75–78
52 Ma J, Yee A, Brewer HB, Das S, Potter H (1994) Amyloid-associated proteins α1-antichy-

motrypsin and apolipoprotein E promote assembly of Alzheimer β-protein into filaments. *Nature* 372: 92–94
53 Giulian D, Haverkamp LJ, Yu JH, Karshin W, Tom D, Li J, Kirkpatrick J, Kuo LM, Roher AE (1996) Specific domains of beta-amyloid from Alzheimer plaque elicit neuron killing in human microglia. *J Neurosci* 16: 6021–6037
54 Fagarasan MO, Aisen PS (1996) IL-1 and anti-inflammatory drugs modulate Aβ cytotoxicity in PC12 cells. *Brain Res* 723: 231–234
55 Gadient RA, UH Otten (1997) Interleukin-6 (IL-6) – a molecule with both beneficial and destructive potentials. *Prog Neurobiol* 52: 379–390
56 Campbell IL, Abraham CR, Masliah E, Kemper P, Inglis JD, Oldstone MBA, Mucke L (1993) Neurologic disease induced in transgenic mice by cerebral overexpression of interleukin 6. *Proc Natl Acad Sci USA* 90: 10 061–10 065
57 Steffenson SC, Campbell IL, Henriksen SJ (1994) Site-specific hippocampal pathophysiology due to cerebral overexpression of interleukin-6 in transgenic mice. *Brain Res* 652: 149–153
58 Heyser CJ, Masliah E, Samimi A, Campbell IL, Gold LH (1997) Progressive decline in avoidance learning paralleled by inflammatory neurodegeneration in transgenic mice expressing interleukin 6 in the brain. *Proc Natl Acad Sci USA* 94: 1500–1505
59 Chiang C-S, Stalder A, Samimi A, Campbell I (1994) Reactive gliosis as a consequence of interleukin-6 expression in the bra*In*: studies in transgenic mice. *Develop Neurosci* 16: 212–221
60 O'Banion MK, Chang JW, Coleman PD (1996) Decreased expression of prostaglandin G/H synthase-2 (PGHS-2) in Alzheimer's disease brain. *Adv Exp Med Biol* 407: 171–177
61 Oka A, Takashima S (1997) Induction of cyclo-oxygenase 2 in brains of patients with Down's syndrome and dementia of Alzheimer type: Specific localization in affected neurones and axons. *Neuroreport* 8: 1161–1164
62 Lukiw WJ, Bazan NG (1997) Cyclooxygenase 2 RNA message abundance, stability, and hypervariability in sporadic Alzheimer neocortex. *J Neurosci* 50: 937–945
63 Roe MT, Dawson DV, Hulette CM, Einstein G, Crain BJ (1996) Microglia are not exclusively associated with plaque-rich regions of the dentate gyrus in Alzheimer's disease. *J Neuropathol Exp Neurol* 55: 366–371
64 DiPatre PL, Gelman BB (1997) Microglial cell activation in aging and Alzheimer disease: Partial linkage with neurofibrillary tangle burden in the hippocampus. *J Neuropathol Exp Neurol* 56: 143–149
65 Voytko ML (1996) Cognitive functions of the basal forebrain cholinergic system in monkeys: Memory or attention? *Behav Brain Res* 75: 13–25
66 Vnek N, Kromer LF, Wiley RG, Rothblat LA (1996) The basal forebrain cholinergic system and object memory in the rat. *Brain Res* 710: 265–270
67 Geula C, Mesulam MM (1996) Systematic regional variations in the loss of cortical cholinergic fibers in Alzheimer's disease. *Cereb Cortex* 6: 165–177
68 Barger SW, Smith-Swintosky VL, Rydel RE, Mattson MP (1993) beta-Amyloid precursor protein mismetabolism and loss of calcium homeostasis in Alzheimer's disease. [Review]. *Ann N Y Acad Sci* 695: 158–64:158–164
69 Hartmann T, Bieger SC, Brühl B, Tienari PJ, Ida N, Allsop D, Roberts GW, Masters CL, Dotti CG, Unsicker K et al (1997) Distinct sites of intracellular production for Alzheimer's disease Aβ40/42 amyloid peptides. *Nat Med* 3: 1016–1020
70 Müller DM, Mendla K, Farber SA, Nitsch RM (1997) Muscarinic M1 receptor agonists increase the secretion of the amyloid precursor protein ectodomain. *Life Sci* 60: 985–991
71 Nitsch RM, Deng A, Wurtman RJ, Growdon JH (1997) Metabotropic glutamate receptor subtype mGluR1α stimulates the secretion of the amyloid β-protein precursor ectodomain. *J Neurochem* 69: 704–712
72 Nitsch RM, Deng M, Growdon JH, Wurtman RJ (1996) Serotonin 5-HT2a and 5-HT2c receptors stimulate amyloid precursor protein ectodomain secretion. *J Biol Chem* 271: 4188–4194
73 Nitsch RM, Kim C, Growdon JH (1998) Vasopressin and bradykinin regulate secretory processing of the amyloid protein precursor of Alzheimer's disease. *Neurochem Res* 23: 807–814
74 Lahiri DK, Farlow MR, Nurnberger JIJ, Greig NH (1997) Effects of cholinesterase inhibitors on the secretion of beta-amyloid precursor protein in cell cultures. *Ann N Y Acad Sci* 826: 416–421
75 DeLapp N, Wu S, Belagaje R, Johnstone E, Little S, Shannon H, Bymaster, Calligaro D, Mitch C, Whitesitt C et al(1998) Effects of the M1 agonist xanomeline on processing of human beta-amyloid precursor protein (FAD, Swedish mutant) transfected into Chinese hamster ovary-m1 cells.

Biochem Biophys Res Commun 244: 156–160
76 Bymaster FP, Whitesitt CA, Shannon HE, DeLapp N, Ward JS, Calligaro DO, Shipley LA, Buelke-Sam JL, Bodick NC, Farde L et al (1997) Xanomeline: A selective muscarinic agonist for the treatment of Alzheimer's disease. *Drug Dev Res* 40: 158–170
77 LeBlanc AC, Koutroumanis M, Goodyer CG (1998) Protein kinase C activation increases release of secreted amyloid precursor protein without decreasing Abeta production in human primary neuron cultures. *J Neurosci* 18: 2907–2913
78 Bandiera T, Lansen J, Post C, Varasi M (1997) Inhibitors of Aβ peptide aggregation as potential anti-Alzheimer agents. *Curr Med Chem* 4: 159–170
79 Abraham CR (1992) The role of the acute-phase protein alpha 1-antichymotrypsin in brain dysfunction and injury. *Res Immunol* 143: 631–636
80 Ma JY, Brewer HB Jr, Potter H (1996) Alzheimer Aβ neurotoxicity: Promotion by antichymotrypsin, ApoE4; Inhibition by Aβ-related peptides. *Neurobiol Aging* 17: 773–780
81 Machein U, Lieb K, Hüll M, Fiebich BL (1995) IL-1 beta and TNF alpha, but not IL-6, induce alpha 1-antichymotrypsin expression in the human astrocytoma cell line U373 MG. *Neuroreport* 6: 2283–2286
82 Das S, Potter H (1995) Expression of the Alzheimer amyloid-promoting factor antichymotrypsin is induced in human astrocytes by IL-1. *Neuron* 14: 447–456
83 Fiebich BL, Hofer TJ, Butcher RD, Lieb K, Hüll M, Schumann G, Berger M, Schulze-Osthoff K, Bauer J (1999) Tepoxalin is a potent inhibitor of interleukin-6 and a1-antichymotrypsin synthesis in human astrocytoma cells. *Neuropharmacology; in press*
84 Billingsley ML, Kincaid RL (1997) Regulated phosphorylation and dephosphorylation of tau prote*In*: effects on microtubule interaction, intracellular trafficking and neurodegeneration. *Biochem J* 323: 577–591
85 Hong M, Chen DC, Klein PS, Lee VM (1997) Lithium reduces tau phosphorylation by inhibition of glycogen synthase kinase-3. *J Biol Chem* 272: 25 326–25 332
86 Munoz-Montano JR, Moreno FJ, Avila J, Diaz-Nido J (1997) Lithium inhibits Alzheimer's disease-like tau protein phosphorylation in neurons. *FEBS Lett* 411: 183–188
87 Adamec E, Mercken M, Beermann ML, Didier M, Nixon RA (1997) Acute rise in the concentration of free cytoplasmic calcium leads to dephosphorylation of the microtubule-associated protein tau. *Brain Res* 757: 93–101
88 Gray CW, Patel AJ (1995) Neurodegeneration mediated by glutamate and β-amyloid peptide: A comparison and possible interaction. *Brain Res* 691: 169–179
89 Wischik CM, Edwards PC, Lai RY, Roth M, Harrington CR (1996) Selective inhibition of Alzheimer disease-like tau aggregation by phenothiazines. *Proc Natl Acad Sci USA* 93: 11 213–11 218
90 Combs CK, Johnson DE, Cannady SB, Lehman TM, Landreth GE (1999) Identification of microglial signal transduction pathways mediating a neurotoxic response to amyloidogenic fragments of beta-amyloid and prion proteins. *J Neurosci* 19: 928–939
91 McDonald DR, Bamberger ME, Combs CK, Landreth GE (1998) beta-Amyloid fibrils activate parallel mitogen-activated protein kinase pathways in microglia and THP1 monocytes. *J Neurosci* 18: 4451–4460
92 McDonald DR, Brunden KR, Landreth GE (1997) Amyloid fibrils activate tyrosine kinase-dependent signaling and superoxide production in microglia. *J Neurosci* 17: 2284–2294
93 Pyo H, Jou I, Jung S, Hong S, Joe EH (1998) Mitogen-activated protein kinases activated by lipopolysaccharide and beta-amyloid in cultured rat microglia. *Neuroreport* 9: 871–874
94 Nakai M, Hojo K, Taniguchi T, Terashima A, Kawamata T, Hashimoto T, Maeda K, Tanaka C (1998) PKC and tyrosine kinase involvement in amyloid beta (25-35)-induced chemotaxis of microglia. *Neuroreport* 9: 3467–3470
95 Netland EE, Newton JL, Majocha RE, Tate BA (1998) Indomethacin reverses the microglial response to amyloid beta-protein. *Neurobiol Aging* 19: 201–204
96 Rudolphi KA, Schubert P (1997) Modulation of neuronal and glial cell function by adenosine and neuroprotection in vascular dementia. *Behav Brain Res* 83: 123–128
97 Si QS, Nakamura Y, Schubert P, Rudolphi K, Kataoka K (1996) Adenosine and propentofylline inhibit the proliferation of cultured microglial cells. *Exp Neurol* 137: 345–349
98 Fiebich BL, Biber K, Gyufko K, Berger M, Bauer J, van Calker D (1996) Adenosine A2b receptors mediate an increase in interleukin (IL)-6 mRNA and IL-6 protein synthesis in human astroglioma cells. *J Neurochem* 66: 1426–1431

99 Fiebich BL, Biber K, Lieb K, van Calker D, Berger M, Bauer J, Gebicke-Haerter PJ (1996) Cyclooxygenase-2 expression in rat microglia is induced by adenosine A2a-receptors. *Glia* 18: 152–160
100 Kittner B, Rössner M, Rother M (1997) Clinical trials in dementia with propentofylline. *Ann.NY Acad.Sci.* 826: 307–316
101 Breitner JCS (1996) Inflammatory processes and antiinflammatory drugs in Alzheimer's disease: A current appraisal. *Neurobiol Aging* 17: 789–794
102 Mackenzie IR, Munoz DG (1998) Nonsteroidal anti-inflammatory drug use and Alzheimer-type pathology in aging. *Neurology* 50: 986–990
103 Rogers J, Kirby LC, Hempelman SR, Berry DL, McGeer PL, Kaszniak AW, Zalinski J, Cofield M, Mansukhani L, Willson P et al (1993) Clinical trial of indomethacin in Alzheimer's disease. *Neurology* 43: 1609–1611
104 Penning TD, Talley JJ, Bertenshaw SR, Carter JS, Collins PW, Docter S, Graneto MJ, Lee LF, Malecha JW, Miyashiro JM et al (1997) Synthesis and biological evaluation of the 1,5-diarylpyrazole class of cyclooxygenase-2 inhibitors: identification of 4-[5-(4-methylphenyl)-3-(trifluoromethyl)-1H-pyrazol-1-yl]benze nesulfonamide (SC-58635, celecoxib). *J Med Chem* 40: 1347–1365
105 Fiebich BL, Lieb K, Hüll M, Berger M, Bauer J (1996) Effects of NSAIDs on IL-1 beta-induced IL-6 mRNA and protein synthesis in human astrocytoma cells. *Neuroreport* 7: 1209–1213
106 Henderson VW (1997) The epidemiology of estrogen replacement therapy and Alzheimer's disease. *Neurology* 48:S27–S35
107 Paganini-Hill A, Henderson VW (1996) Estrogen replacement therapy and risk of Alzheimer disease. *Arch Intern Med* 156: 2213–2217
108 Manolagas SC, T Bellido, RL Jilka (1995) New insights into the cellular, biochemical, and molecular basis of postmenopausal and senile osteoporosis: roles of IL-6 and gp130. *Int J Immunopharmacol* 17: 109–116
109 Birge SJ (1997) The role of estrogen in the treatment of Alzheimer's disease. *Neurology* 48:S36–S41
110 Aisen PS, Marin D, Altstiel L, Goodwin C, Baruch B, Jacobson R, Ryan T, Davis KL (1996) A pilot study of prednisone in Alzheimer's disease. *Dementia* 7: 201–206
111 Aisen PS, Pasinetti GM (1998) Glucocorticoids in Alzheimer's disease. The story so far. *Drug Aging* 12: 1–6
112 McEwen BS (1996) Gonadal and adrenal steroids regulate neurochemical and structural plasticityof the hippocampus via cellular mechanisms involving NMDA receptors. *Cell Mol Neurobiol* 16: 103–116
113 Lupien SJ, Gaudreau S, Tchiteya BM, Maheu F, Sharma S, Nair NPV, Hauger RL, McEwen BS, Meaney MJ (1997) Stress-induced declarative memory impairment healthy elderly subjects: Relationship to cortisol reactivity. *J Clin Endocrinol Metab* 82: 2070–2075
114 De Leon MJ, McRae T, Rusinek H, Convit A, De Santi S, Tarshish C, Golomb J, Volkow N, Daisley K, Orentreich N et al (1997) Cortisol reduces hippocampal glucose metabolism in normal elderly, but not in Alzheimer's disease. *J Clin Endocrinol Metab* 82: 3251–3259
115 Goodman YD, Bruce AJ, Cheng B, Mattson MP (1996) Estrogens attenuate and corticosterone exacerbates excitotoxicity, oxidative injury, and amyloid β-peptide toxicity in hippocampal neurons. *J Neurochem* 66: 1836–1844

New Approaches to Drug Development
ed. by P. Jollès

Delivery systems and adjuvants for vaccination against HIV

Dominique Velin[1] and Jean-Pierre Kraehenbuhl[2, 3]

[1] *Centre d'Immunologie Pierre Fabre (CIPF), F-74164 St. Julien en Genevois, France*
[2] *Institute for Biochemistry,* [3] *Swiss Institute for Experimental Cancer Research, University of Lausanne, CH-1066 Epalinges, Switzerland*

Summary. Epidemiological studies have revealed that HIV-1 infections occur through contact with contaminated blood or during unprotected vaginal or anal intercourse. Hence, to protect against HIV infection, vaccines should ideally induce both mucosal and systemic immune responses. We present a brief review of the different delivery systems and adjuvants which can be used to elicit mucosal immune responses. Oral or nasal administration of recombinant attenuated *bacteria or viruses* can induce both mucosal and systemic immune responses against the carried antigen. The oral delivery of mucosal adjuvants (such as cholera toxin) in association with antigens has been shown to enhance mucosal and systemic immune responses against them. Recently developed vaccination strategies using naked DNA or other antigen delivery systems are also discussed.

Introduction

In the last decade many efforts have been made to develop effective vaccines against human immunodeficiency virus (HIV). Their design was based on knowledge gained from previously developed vaccines directed against other pathogens which are different from HIV in structure, pathogenesis and mode of transmission. Unfortunately, HIV is a uniquely difficult target for vaccine development, since immune correlates of protective immunity against HIV infection remain uncertain [1]. Moreover, this virus persistently replicates in the infected individual, leading to disease despite the generation of vigorous humoral and cellular responses. HIV mutates during infection, resulting in the generation of viruses that can escape immune recognition. It is now well documented that HIV infections may occur through contact with contaminated blood or through unprotected vaginal or anal intercourse. Hence, to be efficient against both systemic and mucosal viral infection, vaccines must induce long-lasting immune responses in both the mucosal and systemic compartments of the immune system [1, 2]. This paper will summarize our current knowledge of HIV mucosal infection, the presumed immune effectors capable of controlling the viral infection and the different adjuvants and delivery systems which can be used to trigger a protective mucosal immune response against HIV by vaccination.

HIV infections mainly occur through mucosal surfaces

Epidemiological studies have revealed that most adult HIV-1 infections worldwide occur through unprotected vaginal or anal intercourse [3]. The secretions that act as a vector for HIV transmission contain free HIV particles and numerous HIV-infected mononuclear cells [4]. Free or cell-associated HIV penetrates the mucosa by passing through either stratified (vagina, exocervix, anus) or simple (intestine, rectum, endocervix) epithelia. HIV can cross the mucosa by different mechanisms: through physical trauma, by transport through stratified epithelia via Langerhans cells [5], or by transcytosis through tight epithelial monolayers. The transcytosis of HIV generated from infected cells across polarized simple epithelial monolayers was demonstrated by Bomsel et al. [6]. Alternatively, HIV could cross simple epithelia through M-cells [7]. M-cells are specialized epithelial cells which deliver samples of foreign material by transepithelial transport from the lumen to the underlying organized mucosa-associated lymphoid tissue. HIV crossing the mucosal epithelium via this route would therefore be in close contact with its potential host cells: dendritic cells, macrophages and CD4 T-lymphocytes. It has been recently found that epithelial, stromal, dendritic and T cells of the upper and lower human female reproductive tract can be infected by HIV-1 [8].

The immune effectors in HIV infection

Our knowledge concerning the potential immune protective mechanisms against HIV infection comes from studies of the immune responses detected in patients with established HIV infection [9, 10] and in animal models of HIV infection.

CD8 T cells

CD8 cytotoxic T-lymphocytes (CTLs) appear to be important in containing the spread of HIV-1 in infected individuals. The replication of HIV-1 in CD4 T cells can be inhibited by autologous CD8 CTL mechanisms that probably include both lysis and release of chemokines and cytokines [11]. Moreover, in patients chronically infected with HIV-1, a high-frequency CTL response is correlated with the maintenance of low virus load and a stable clinical status [10]. A recent animal study done by Schmitz et al. demonstrates clearly the control of viremia in simian immunodeficiency (SIV) virus infection by CD8 lymphocytes [12]. These observations suggest that an effective HIV-1 vaccine should stimulate HIV-1-specific CTLs. This presents an important challenge in HIV-1 vaccine development, because most vaccines in common use for the prevention of other infectious diseases have not needed to induce effector T cells [1]. Furthermore, the vaccine should elicit strong CTL responses against multiple codominant viral epitopes, since it has already been observed that during primary infection, HIV-1 can generate CTL escape viruses [13].

CD4 T-helper cells

The most glaring defect in the immune repertoire in HIV-1 infection is the lack of virus-specific T-helper cell responses. Moreover, studies showed that HIV-1 is capable of inducing T-helper cell responses, and that the magnitude of T-helper cell responses to the HIV-1 Gag protein is associated with control viremia in untreated persons [14]. It therefore follows that HIV-1-specific helper cells are likely to be important in immunotherapeutic interventions and vaccine development. The recent work demonstrating that activated CD4 cells are able to condition dendritic cells to stimulate CD8 cells through CD40-CD40L interactions, raises the possibility that in the absence of sufficient helper activity, dendritic cells may not provide sufficient costimulatory signals to CD8 cells, thus disrupting their normal functions [15]. In addition to their potential roles in enhancing CTL precursor activity, the CD4 cells may contribute to the maintenance of effective antiviral immunity by increased production of antiviral cytokines, or augmentation of humoral immune response.

Neutralizing antibodies

Neutralizing antibodies can, in some circumstances, prevent HIV-1 infection and the induction of an effective antibody response has been an important goal of those developing HIV-1 vaccines [16]. Two recent studies provide evidence that neutralizing envelope-specific antibodies in monkeys can protect monkeys against chimeric virus ($SHIV_{DH12}$) carrying the envelope glycoproteins of a primary HIV-1 isolate (HIV-$1D_{H12}$) [17, 18]. Protection, however, required antibody titers that cannot be achieved by vaccination. The early enthusiasm slowly waned as it became increasingly clear that a principal neutralizing domain (the V3 loop of the surface glycoprotein gp120) present on HIV-1 laboratory strains, which was potentially a powerful immunogen, was a poor target on viruses recently isolated from patients [19, 20]. The design of an immunogen that can generate an antibody response able to neutralize a broad spectrum of primary isolates is therefore still a major goal. LaCasse et al. may have identified a solution to this problem that could eventually be exploited for vaccine development. They proposed that epitopes with superior immunogenicity and neutralizing activity might be exposed or created as HIV-1 begins to fuse with cell membranes [21].

Mucosal immune response

The mucosal immune response seems also to be one important aspect to consider in HIV protective immunity. Indeed, an epidemiological study recently showed that mucosal IgA directed against HIV can be detected in the urine and vaginal washes of HIV-seronegative partners of HIV-seropositive individuals (heterosexual couples). This result suggests that mucosal IgA might be involved in the absence of seroconversion of HIV-seronegatives who are heavily exposed to repeated HIV mucosal contamination [22]. Taken together, these observations suggest that the development of vaccines against HIV should at least induce long-lasting mucosal IgA responses.

Mucosal immune mechanisms of protection

The immune defense of the mucosa is composed by innate mechanisms (tight junctions, mucus, enzymes, defensins, inflammatory cells) and by adaptive humoral and cell-mediated immunity [23]. Cell-mediated immunity has been shown to prevent pathogen invasions of the mucosa [24]. The antibodies that are responsible for protecting mucosal surfaces are secretory IgA (sIgA) [23]. sIgA is particularly well suited for mucosal surfaces due to its high avidity and resistance to proteolytic cleavage. Although the primary function of sIgA has been regarded as the ability to neutralize pathogens at mucosal surfaces, more recently, *in vitro* [25] and *in vivo* [26] models have demonstrated that sIgA is capable of forming intracellular complexes with viruses and inhibiting viral replication, subsequently preventing primary infections or resolving chronic infections. Recently, Bomsel et al. demonstrated *in vitro* an intracellular neutralization of HIV transcytosis across tight epithelial barriers by anti-HIV envelope proteins sIgA or IgM [27]. These results suggest that induction of mucosal immunity to HIV envelope proteins may impair the transcytotic route of HIV mucosal transmission.

Taken together, it can be postulated that an HIV vaccine must induce specific sIgA, since sIgA is thought to interact specifically with mucosal pathogens in the secretions and may consequently decrease the viral load by immune exclusion. The vaccine should also induce a mucosal CD8 T-cell immune response to destroy the infected cells present in the simple or stratified epithelia either by lysis or by production of soluble suppressor factors (such as chemokines) [11].

Adjuvants and delivery systems used to trigger mucosal immune responses

To elicit an efficient long-term mucosal immune response, the antigen has to be delivered to a mucosal antigen-sampling site, such as the Peyer's patches of the small intestine or the mucosal-associated lymphoid tissue of the nasal cavity [23].

Mucosal immunization is desirable via oral and nasal routes. In contrast, the anal route will certainly not be well accepted in humans, and the female genital tract is generally unable to respond well to nonreplicating antigens following intravaginal immunization [28]. The two major disadvantages of oral vaccination protocols are: the state of immunological tolerance which normally occurs in response to most soluble antigens and the proteolytic degradation of antigen preparations in the digestive tract. The nasal route of antigenic preparation delivery is certainly one of the most promising ways to induce both mucosal and systemic immune responses. The mucosa of the nasal cavity is very sensitive and further studies are needed to investigate the possible induction of an allergic response by the delivery of vaccines via the nasal route.

In order to deliver a vaccine via the mucosal route, different mucosal adjuvants and delivery systems have been used and many of them have already been tested in human trials. Cholera toxin or heat-labile enterotoxin of *Escherichia coli*, which enhance immune responses to the antigen of interest, are the most powerful known mucosal adjuvants [29]. The use of live attenuated microrganisms as vectors to express foreign antigens has also been shown to be successful. The induction of a restricted infection by such attenuated organisms at the mucosal level enables the mucosal and systemic immune systems to be primed against both the vector and the antigen without, ideally, any harmful side-effects.

Attenuated bacteria trigger mucosal and systemic immune responses

Vaccination of BALB/c mice via the oral or nasal route with attenuated *Salmonella* expressing a foreign antigen, such as the nucleocapsid protein of Hepatitis B virus, has been shown to induce a strong mucosal and systemic immune response against both the carrier and the carried antigen [30]. Perhaps one of the most interesting properties of *Salmonella* vectors is their ability to stimulate antigen-specific cytotoxic $CD8^+$ T-cell responses [31]. Hence, immunization with live oral attenuated *Salmonella* could provide a safe and inexpensive means for delivery of HIV antigens to the mucosal and systemic compartments of the immune system.

Recombinant live *Mycobacterium bovis* BCG strains (rBCG) expressing different HIV or SIV antigens could be good candidates for the development of vaccines against AIDS. In BALB/c mice immunized via different mucosal routes (oral, nasal, and rectal) with a mixture of three rBCG strains expressing, respectively, the entire SIVmac251 Nef protein, and large fragments of the Env and Gag proteins, all routes of immunization studied induced IgA antibodies against mycobacterial SIV Env, and SIV Gag antigens in faeces and bronchial lavages as well as specific IgG in serum. Strong, specific cytotoxic responses of splenocytes against Nef, Env, and Gag were observed whatever the mucosal route of immunization. Therefore, mucosal vaccination with a cocktail of rBCG strains induces local, specific IgA, systemic IgG, and systemic CTLs against the three SIV antigens expressed [32].

Recombinant virus

Adenovirus

Adenovirus vectors are an attractive means for inducing long-lasting systemic and mucosal immune responses. Indeed, Gallichan and Rosenthal have shown that intranasal immunization with a recombinant adenovirus triggers both specific secretory and systemic antibodies in addition to systemic and mucosal CTL responses [33, 34]. Adenovirus has proven to be an excellent mucosal vaccine vector in humans. Unattenuated human adenovirus type 4 and 7 have been orally administered in enteric-coated capsules to millions of US military recruits and have proven effective in preventing adenovirus-induced acute respiratory disease without evidence of adverse effects [35, 36]. Furthermore, it has been recently reported that a combination AIDS vaccine approach consisting of priming with adenovirus-HIV-1gp160 recombinants followed by boosting with HIV-1 gp120 could protect chimpanzees against systemic high-dose HIV-1 challenge [37].

Pox virus

Recombinant organisms created by inserting AIDS virus genes into vaccinia (the live, attenuated vaccine virus that has eliminated smallpox infections worldwide) have elicited AIDS virus-specific cellular and humoral immunity in macaques. When immunization with such constructs have been followed by boosting with recombinant proteins, vaccinated monkeys has been protected against infection by some SIV isolates [38]. Intrarectal immunization with such recombinant vaccinia-expressing envelope proteins of HIV has been reported to initiate CTL responses in the mucosa [39]. However, since HIV-infected patients are immunosuppressed, there is a reluctance to use this vector system in large-scale human trials, since vaccinia could develop in these patients. Much interest has therefore turned to pox viruses with limited *in vivo* replicative capacity and, therefore, limited pathogenic potential in human such as Ankara vaccinia.

Recombinant Canarypox virus expressing different HIV genes is also under development; these vectors could also be interesting as a mucosal vaccine, since infection occurs without replication of the virus, generating immune responses against both the virus and the recombinant protein [40]. However, it has yet to be demonstrated that this delivery system is sufficiently immunogenic in humans [41].

Oral or nasal administration of cholera toxin with antigen induces mucosal and systemic immune responses

Cholera toxin, an enterotoxin produced by *Vibrio cholerae*, is a potent mucosal adjuvant for the induction of systemic and mucosal humoral and cytotoxic responses when coadministered with proteins [42, 43]. The toxin is composed of separate A and B subunits. The B subunit consists of five identical peptides and binds to the GM1 ganglioside expressed by epithelial cells [44]. The A subunit, once translocated into the cell cytoplasm, binds to NAD and catalyzes ADP ribosylation of the G protein, Gsα [45]. This GTP-binding protein activates adenylate cyclase, resulting in the elevation of intracellular cAMP levels, which in epithelial cells causes secretion of water and chloride ions into the small intestine [46]. Recently, a nontoxic mutant of cholera toxin has been generated and used as a mucosal adjuvant in mice. The mutant was made by substitution of serine with phenylalanine at position 61 of the A subunit, which resulted in the loss of ADP ribosyltransferase activity and toxicity [47].

Female rhesus macaques have been nasally immunized with p55gag (p55) of SIV and cholera toxin as a mucosal adjuvant. Nasal immunization induced Ag-specific IgA and IgG Abs in mucosal secretions (e.g., cervicovaginal secretions, rectal washes and saliva) and serum. Furthermore, high numbers of p55-specific IgA and IgG Ab-forming cells were induced in mucosal effector sites, i.e., the uterine cervix, intestinal *lamina propria*, and the nasal passage. p55-specific CD4 T cells in both systemic and mucosal compartments expressed interferon (IFN)-gamma and interleukin (IL)-2 (Th1-type)- as well as IL-5, IL-6, and IL-10 (Th2-type)-specific mRNA. Moreover, p55-specific CTL activity was demonstrated in lymphocytes from blood, tonsils and other lymphoid tissues. These results show that nasal immunization with SIV p55 using cholera toxin as an adjuvant elicits both Th1- and selective Th2-type cytokine responses associated with the induction of SIV-specific mucosal and serum Abs, and CTL activity [48].

Proteosomes

Nasal immunization of mice with oligomeric gp160, formulated with liposomes containing monophosphoryl lipid A, proteosomes, emulsomes or proteosomes with emulsomes, elicited strong gp160-specific IgG and IgA responses in serum as well as vaginal, lung, and intestinal washes and faecal pellets. These data demonstrate the feasibility of inducing both systemic and mucosal HIV-1-neutralizing antibodies by intranasal immunization with an oligomeric gp160 protein [49]. Although this nasal immunization is very efficient in a mouse model, its promise needs to be confirmed in human trials [50].

DNA vaccination

Promising results have also been obtained with DNA vaccination. Vaccination with an HIV-1 DNA vaccine encoding env, rev, gag/pol was shown to protect chimpanzees from systemic HIV-1 challenge [51]. DNA vaccines can generate HIV-1- and SIV-specific CTL, Helper T-cells and antibodies in mice and non-human primates. Since it has been described that DNA vaccination can also be efficient when administered via mucosal surfaces [52], it is tempting to speculate that the delivery of DNA vaccines intramuscularly and mucosally will be protective against systemic and mucosal HIV-1 challenge.

A DNA-based vaccine containing HIV-1 env and rev genes has been tested for safety and host immune responses in 15 asymptomatic HIV-infected patients. Vaccine administration induced no local or systemic reactions, and no laboratory abnormalities were detected. Specifically, no patients developed anti-DNA antibodies or muscle enzyme elevations. Increased levels of gp120 antibodies could detected in some patients. Some increases were noted in cytotoxic T-lymphocyte activity against gp160-bearing targets and in lymphocyte-proliferative activity. This demonstrates the safety and potential immunogenicity of an HIV-directed DNA-based vaccine in humans [53, 54].

Conclusions

From studies of the different modes of HIV transmission and the anti-HIV immune responses developed by infected patients, we can postulate that vaccines against HIV should induce both mucosal and systemic humoral and cellular immune responses.

Much progress has been made in the induction of mucosal and systemic immune responses by vaccination. The different recombinant vectors (live attenuated bacteria, adenovirus, DNA) and mucosal adjuvants (cholera toxin or heat-labile enterotoxin) have been better characterized and many of them are now in clinical trials. Before succeeding in developing HIV vaccines, huge clinical studies will have to be conducted, and much effort will have to be made in order to complete our knowledge of HIV mucosal infection and the mucosal immune system.

Acknowledgements
This work was supported by grants from the Swiss National Science Foundation SNSF (31-37612.93), the Swiss AIDS program (3139-37155.93) and the Swiss Research against Cancer Foundation (AKT 622) to Jean-Pierre Kraehenbuhl.

References

1 Letvin NL (1998) Progress in the development of an HIV-1 vaccine. *Science* 280: 1875–80
2 Velin D, Hopkins S, Kraehenbuhl JP (1998) Delivery systems and adjuvants for vaccination

against HIV. *Pathobiology* 66: 170–175
3 Chin J (1990) Global estimates of AIDS and HIV infections. *AIDS (London)* 4: S277–S278
4 Milman G, Sharma O (1994) Mechanisms of HIV/SIV mucosal transmission. *AIDS Res Hum Retrovirus* 10: 1305–1309
5 Spira AI, Marx PA, Patterson BK, Mahoney J, Koup RA, Wolinsky SM, Ho DD (1996) Cellular targets of infection and route of viral dissemination after an intravaginal inoculation of simian immunodeficiency virus into rhesus macaques. *J Exp Med* 183: 215–225
6 Bomsel M (1997) Transcytosis of infectious human immunodeficiency virus across a tight human epithelial cell line barrier. *Nat Med* 3: 42–47
7 Amerongen HM, Weltzin RA, Farnet CM, Michetti P, Haseltine WA, Neutra MR (1991) Transepithelial transport of HIV-1 by intestinal M-cells: a mechanism for transmission of AIDS. *J Acq Immun Defic Synd Hum R* 4: 760–765
8 Howell AL, Edkins RD, Rier SE, Yeaman GR, Stern JE, Fanger MW, Wira CR (1997) Human immunodeficiency virus type 1 infection of cells and tissues from the upper and lower human female reproductive tract. *J Virol* 71: 3498–3506
9 Haynes BF, Pantaleo G, Fauci AA (1996) Toward and understanding of the corrElates of protective immunity to HIV infection. *Science* 271: 324–328
10 Ogg GS, Jin X, Bonhoeffer S, Dunbar PR, Nowak MA, Monard S, Segal JP, Cao Y, Rowland-Jones SL, Cerundolo V et al (1998) Quantitation of HIV-1-specific cytotoxic T lymphocytes and plasma load of viral RNA. *Science* 279: 2103–2106
11 Walker CM, Moody DJ, Stites DP, Levy JA (1986) CD8+ lymphocytes can control HIV infection *in vitro* by suppressing virus replication. *Science* 234: 1563–1566
12 Schmitz JE, Kuroda MJ, Santra S, Sasseville VG, Simon MA, Lifton MA, Racz P, Tenner-Racz K, Dalesandro M, Scallon BJ et al (1999) Control of viremia in simian immunodeficiency virus infection by CD8+ lymphocytes. *Science* 283: 857–860
13 Borrow P, Lewicki H, Wei X, Horwitz MS, Peffer N, Meyers H, Nelson JA, Gairin JE, Hahn BH, Oldstone MB ET AL (1997) Antiviral pressure exerted by HIV-1-specific cytotoxic T lymphocytes (CTLs) during primary infection demonstrated by rapid selection of CTL escape virus. *Nat Med* 3: 205–211
14 Rosenberg ES, Billingsley JM, Caliendo AM, Boswell SL, Sax PE, Kalams SA, Walker BD (1997) Vigorous HIV-1-specific CD4+ T cell responses associated with control of viremia. *Science* 278: 1447–1450
15 Ridge JP, Di Rosa F, Matzinger P (1998) A conditioned dendritic cell can be a temporal bridge between a CD4+ T-helper and a T-killer cell. *Nature* 393: 474–478
16 Emini EA, Schleif WA, Nunberg JH, Conley AJ, Eda Y, Tokiyoshi S, Putney SD, Matsushita S, Cobb KE, Jett CM et al (1992) Prevention of HIV-1 infection in chimpanzees by gp120 V3 domain-specific monoclonal antibody *Nature* 355: 728–730
17 T, Brown C, Azadegan A, Haigwood N, Dimitrov D, Martin MA, Shibata R (1999) Human immunodeficiency virus type 1 neutralizing antibodies accelerate clearance of cell-free virions from blood plasma. *Nat Med* 5: 211–216
18 Shibata R, Igarashi T, Haigwood N, Buckler-White A, Ogert R, Ross W, Willey R, Cho MW, Martin MA (1999) Neutralizing antibody directed against the HIV-1 envelope glycoprotein can completely block HIV-1/SIV chimeric virus infections of macaque monkeys. *Nat Med* 5: 204–210
19 Bou-Habib DC, Roderiquez G, Oravecz T, Berman PW, Lusso P, Norcross MA (1994) Cryptic nature of envelope V3 region epitopes protects primary monocytotropic human immunodeficiency virus type 1 from antibody neutralization. *J Virol* 68: 6006–6013
20 Parren PW, Wang M, Trkola A, Binley JM, Purtscher M, Katinger H, Moore JP, Burton DR (1998) Antibody neutralization-resistant primary isolates of human immunodeficiency virus type 1. *J Virol* 72: 10 270–10 274
21 LaCasse RA, Follis KE, Trahey M, Scarborough JD, Littman DR, Nunberg JH (1999) Fusion-competent vaccines: broad neutralization of primary isolates of HIV. *Science* 283: 357–362
22 Mazzoli S, Trabattoni D, Lo Caputo S, Piconi S, Ble C, Meacci F, Ruzzante S, Salvi A, Semplici F, Longhi R et al (1997) HIV-specific mucosal and cellular immunity in HIV-seronegative partners of HIV-seropositive individuals. *Nat Med* 3: 1250–1257
23 Neutra MR, Pringault E, Kraehenbuhl JP (1996) Antigen sampling across epithelial barriers and induction of mucosal immune responses *Annu Rev Immunol* 14: 275–300
24 Munoz JL, McCarthy CA, Clark ME, Hall CB (1991) Respiratory syncytial virus infection in C57BL/6 mice: clearance of virus from the lungs with virus-specific cytotoxic T cells. *J Virol* 65:

4494–4497
25 Mazanec MB, Coudret CL, Fletcher DR (1995) Intracellular neutralization of influenza virus by immunoglobulin A anti-hemagglutinin monoclonal antibodies. *J Virol* 69: 1339–1343
26 Burns JW Siadat-Pajouh M, Krishnaney AA, Breenberg HB (1996) Protective effect of rotavirus VP6-specific IgA monoclonal antibodies that lack neutralizing activity. *Science* 272: 104–107
27 Bomsel M, Heyman M, Hocini H, Lagaye S, Belec L, Dupont C, Desgranges C (1998) Intracellular neutralization of HIV transcytosis across tight epithelial barriers by anti-HIV envelope protein dIgA or IgM. *Immunity* 9: 277–287
28 Parr EL, Parr MB (1992) Immunization for secretory immunity in the female reproductive tract. *Vaccine Res* 1: 221–225
29 Williams NA, Hirst TR, Nashar TO (1999) Immune modulation by the cholera-like enterotoxins: from adjuvant to therapeutic. *Immunol Today* 20: 95–101
30 Hopkins S, Kraehenbuhl JP, Schödel F, Potts A, Peterson D, De Grandi P, Nardelli-Haefliger D (1995) A recombinant *S.typhimurium* vaccine induces local immunity by four different routes of immunization. *Infect Immunity* 63: 3279–3286
31 Aggarwal AS, Jaffe R, Hone D, Gross M, Sadoff J (1990) Oral *Salmonella*: Malaria circumsporozoite recombinants induce specific CD8+ cytotoxic T cells. *J Exp Med* 172: 1083–1900
32 Lagranderie M, Winter N, Balazuc AM, Gicquel B, Gheorghiu M (1998) A cocktail of *Mycobacterium bovis* BCG recombinants expressing the SIV Nef, Env, and Gag antigens induces antibody and cytotoxic responses in mice vaccinated by different mucosal routes. *AIDS Res Hum Retrovirus* 14: 1625–1633
33 Gallichan WS, Rosenthal KL (1996) Long-lived cytotoxic T lymphocyte memory in mucosal tissues after mucosal but not systemic immunization. *J Exp Med* 184: 1879–1890
34 Gallichan WS, Rosenthal K (1995) Specific secretory immune responses in the female genital tract following intranasal immunization with recombinant adenovirus expressing glycoprotein B of *Herpes simplex* virus. *Vaccine* 13: 1589–1595
35 Top FH, Grossman RA, Barelloni PJ, Segal HE, Dudding BA, Russell PK, Buesher BL (1971) Immunization with live types 7 and 4 adenovirus vaccines. I. Safety, infectivity, antigenicity, and potency of adenovirus type 7 vaccine in humans. *J Infect Dis* 124: 148–154
36 Schwartz AR, Togo Y, Hornick RB (1974) Clinical evaluation of live, oral type 1, 2 and 5 adenovirus vaccines. *Amer Rev Respir Dis* 109: 233–238
37 Lubeck MD, Natuk R, Myagkikh M, Kalyan N, Aldrich K, Sinangil F, Alipanah S, Murthy SCS, Chanda PK, Nigida SM et al (1997) Long term protection of chimpanzees against high-dose HIV-1 challenge induced by immunization. *Nat Med* 3: 651–658
38 Hu SL, Abrams K, Barber GN, Moran P, Zarling JM, Langlois AJ, Kuller L, Morton WR, Benveniste RE (1992) Protection of macaques against SIV infection by subunit vaccines of SIV envelope glycoprotein gp160. *Science* 255: 456–459
39 Belyakov IM, Wyatt LS, Ahlers JD, Earl P, Pendleton CD, Kelsall BL, Strober W, Moss B, Berzofsky JA (1998) Induction of a mucosal cytotoxic T-lymphocyte response by intrarectal immunization with a replication-deficient recombinant vaccinia virus expressing human immunodeficiency virus 89.6 envelope protein. *J Virol* 72: 8264–8272
40 Clements-Mann ML, Weinhold K, Matthews TJ, Graham BS, Gorse GJ, Keefer MC, McElrath MJ, Hsieh RH, Mestecky J, Zolla-Pazner S et al NIAIDAIDSVaccine Evaluation Group (1998) Immune responses to human immunodeficiency virus (HIV) type 1 induced by canarypox expressing HIV-1MN gp120, HIV-1SF2 recombinant gp120, or both vaccines in seronegative adults. *J Infect Dis* 177: 1230–1246
41 Welter J, Taylor J, Tartaglia J, Paoletti E, Stephensen CB (1999) Mucosal vaccination with recombinant poxvirus vaccines protects ferrets against symptomatic CDV infection. *Vaccine* 17: 308–318
42 Elson CO, Dertzbaugh MT (1994) Mucosal adjuvants. *In*: PL Ogra, ME Lamm, JR McGhee, J Mestecky, W Strober, J Bienenstock (eds): *Handbook of mucosal immunology*. Academic Press, San Diego, 391
43 Bowen JC, Nair SK, Reddy R, Rouse BT (1994) Cholera toxin acts as a potent adjuvant for the induction of cytotoxic T-lymphocyte responses with non-replicating antigens. *Immunology* 81: 338–342
44 Van Heyningen S (1974) Cholera Tox*In*: interaction of subunits with ganglioside aGM1. *Science* 183: 656–657
45 Gill DM, King CA (1975) The mechanism of the action of cholera toxin in pigeon erythrocyte

lysates. *J Biol Chem* 250: 6424–6432
46 Field M, Rao MC, Chang EB (1989) Intestinal electrolyte transport and diarrheal disease. *N Engl J Med* 321: 800–806
47 Yamamoto S, Kiyono H, Yamamoto M, Imaoka K, Yamamoto M, Fujihashi K, Van Ginkel FW, Noda M, Takeda Y, McGhee JR (1997) A nontoxic mutant of cholera toxin elicits Th2-type responses for enhanced mucosal immunity. *Proc Natl Acad Sci USA* 94: 5267–5272
48 Imaoka K, Miller CJ, Kubota M, McChesney MB, Lohman B, Yamamoto M, Fujihashi K, Someya K, Honda M, McGhee JR et al (1998) Nasal immunization of nonhuman primates with simian immunodeficiency virus p55gag and cholera toxin adjuvant induces Th1/Th2 help for virus-specific immune responses in reproductive tissues. *J Immunol* 161: 5952–5958
49 VanCott TC, Kaminski RW, Mascola JR, Kalyanaraman VS, Wassef NM, Alving CR, Ulrich JT, Lowell GH, Birx DL (1998) HIV-1 neutralizing antibodies in the genital and respiratory tracts of mice intranasally immunized with oligomeric gp160. *J Immunol* 160: 2000–2012
50 Oftung F, Naess LM, Wetzler LM, Korsvold GE, Aase A, Hoiby EA, Dalseg R, Holst J, Michaelsen TE, Haneberg B (1999) Antigen-specific T-cell responses in humans after intranasal immunization with a meningococcal serogroup B outer membrane vesicle vaccine. *Infect Immunity* 67: 921–927
51 Boyer JDO, Ugen KE, Wang B, Agadjanyan M, Gilbert L, Bagarazzi ML, Chattergoon M, Frost P, Javadian A, Williams WV et al (1997) Protection of chimpanzees from high-dose heterologous HIV-1 challenge by DNA vaccination. *Nat Med* 3: 526–532
52 Fynan EF, Webster RG, Fuller DH, Haynes JR, Santoro JC, Robison HL (1993) DNA vaccines: Protective immunization by parenteral, mucosal, and gene-gun inoculations. *Proc Natl Acad Sci USA* 90: 11 478–11 482
53 MacGregor RR, Boyer JD, Ugen KE, Lacy KE, Gluckman SJ, Bagarazzi ML, Chattergoon MA, Baine Y, Higgins TJ, Ciccarelli RB et al (1998) First human trial of a DNA-based vaccine for treatment of human immunodeficiency virus type 1 infection: safety and host response. *J Infect Dis* 178: 92–100
54 Boyer JD, Chattergoon MA, Ugen KE, Shah A, Bennett M, Cohen A, Nyland S, Lacy KE, Bagarazzi ML, Higgins TJ et al (1999) Enhancement of cellular immune response in HIV-1 seropositive individuals: A DNA-based trial. *Clin Immunol* 90: 100–107

Subject index